Heinrich Obersteiner

The Anatomy of the Central Nervous Organs in Health and in Disease

Heinrich Obersteiner

The Anatomy of the Central Nervous Organs in Health and in Disease

ISBN/EAN: 9783744717540

Printed in Europe, USA, Canada, Australia, Japan

Cover: Foto ©berggeist007 / pixelio.de

More available books at **www.hansebooks.com**

THE ANATOMY

OF THE

CENTRAL NERVOUS ORGANS

In Health and in Disease.

BY

DR. HEINRICH OBERSTEINER,
Professor (Ext.) at the University of Vienna.

TRANSLATED, WITH ANNOTATIONS AND ADDITIONS,

BY

ALEX HILL, M.A., M.D., M.R.C.S.,
Master of Downing College, Cambridge; Examiner in Anatomy to the Universities
of Cambridge and Glasgow.

With 198 Illustrations.

LONDON:
CHARLES GRIFFIN & COMPANY,
EXETER STREET, STRAND.
1890.

[All Rights Reserved.]

PREFACE TO THE ENGLISH EDITION.

No apology is necessary for placing before the English student of neurology Professor Obersteiner's exact and impartial account of the anatomy of the central nervous system. The labour of selecting from the mass of literature, with which the subject is every year enriched, the facts of greatest importance, and the theories which harmonise most with one another, must have been immense. It would only be right that the students of all countries should be allowed to participate in the result.

In giving an English dress to Professor Obersteiner's text, we have attempted to transpose its forms of expression into the English mode. The translator has always held that to transform a book from one language into another needs the collaboration of members of both nationalities, and he wishes to express his indebtedness to Fraülein Kloss for having read through the whole of the German text with him. He has also been in constant communication with Professor Obersteiner, who has not only explained obscure passages, but has also added much new matter and made numerous alterations, which give to the English version the value of a new edition.

All additions to the text made by the translator are included in square brackets [ ]. He is also responsible for all footnotes, for the introduction of figs. 2, 3, 4, 5, 5a, 6, 8, 19a, 34, 57, 149, 150, 151, 156, 157, 192, 193, 194, 195, 196, 197, 198, and for the Appendix. In going through Professor Obersteiner's work, care has been taken to check the references in the text, and in the footnotes to the figures, where they have been re-arranged alphabetically. Some trouble has also been spent upon the Index, into which the terms used in the German edition have been introduced, in order that the reader, who is acquainted with such terms in their German dress—and some of them have not hitherto made their appearance in English—may have the opportunity of looking up the structures which they designate in this text-book. When doing this it seemed worth while to give the commoner synonyms and French equivalents. Not that the glossary has any pretentions to lexicographical completeness—it is merely the translator's working vocabulary.

No attempt has been made to keep step with the German edition in typography. Capital letters are used to call attention to the sections, while Italics are restricted to the names of observers quoted as authorities; personal names which are used as the appellations of structures or methods are printed in ordinary type.

The spelling of the word "neurogleia" will probably attract the attention of the reader. Neurogloea would undoubtedly be more correct, but would affect the pronunciation. In German the spelling "neuroglia" is perhaps unexceptionable, but it makes a terrible word when pronounced in the English fashion. Not only the spelling of the term but also its application is, however, open to discussion. It appears to the translator to be a useful term when applied to the connective-tissue of the central nervous system, which differs from other forms of connective-tissue in its origin from epiblast; whereas, when restricted to the "matrix" it gives an undesirable definiteness to what is, after all, a hypothetical substance.

DOWNING COLLEGE LODGE,
March, 1890.

PREFACE TO THE GERMAN EDITION.

SOME decades ago our knowledge of the intimate structure of the central nervous system was still very insufficient—so insufficient, indeed, that pathology was able to make little use of it. Hence we can understand how, of the little that was known, the practitioners of the time, with very few exceptions, made use of the most striking facts only, and had to be content with an extreme poverty of data.

Since then, however, a succession of distinguished observers, supported by the improvements made in method, have, with surprising rapidity, successively thrown more light into the chaos of manifold nerve-paths and their nodal points; and therefore it had to be acknowledged in practical medicine that the brain- and spinal cord-anatomy (until now so contemptuously set aside)—despite their difficulty—are worthy of the most exhaustive consideration. Nay more, regions which seem to stand far enough away from nerve-pathology—ophthalmology, osteology, and even dermatology—have come to feel the need of a fundamental orientation of the central nervous organs.

To meet this want we possess now, especially in German, a number of most excellent anatomical text-books. But as no part of anatomy (least of all, perhaps, the anatomy of the nervous system) can be learnt from books, students and physicians seek out the laboratories where opportunity is offered them of making themselves familiar with the structure of the brain and spinal cord. Certainly, the establishment of ideally-equipped laboratories for the study of brain anatomy, such as *His* wanted at the meeting of the Berlin Association of Naturalists of 1886, will long remain *pium desiderium*. At present, teachers and students must be content with the incomplete commencements of such institutions as already exist in some of the larger universities.

Experience has now taught me what are the justifiable claims which a beginner, who does not yet wish to become a specialist in the subject, may make upon a text-book. Especially must I assert that while, on the one hand, it is superfluous to go into incompletely established details (a course which is likely, indeed, to produce a depressing and confusing effect), yet, on the other hand, some information with regard to pathological processes should most certainly be given.

In the following pages I have tried to provide the student with a trustworthy and reliable guide, with which, in the absence of any other teacher, he may undertake the troublesome journey through the several regions of the central nervous system. Hence, I have continually introduced directions for making preparations: the numerous illustrations, although they are true to nature (with the exception of those which are purely diagrammatic), are only meant to facilitate the study of original preparations—not to replace them.

Any one who has the opportunity of visiting a laboratory with a good collection of ready-made preparations can with advantage use these, and so save himself much expenditure of time and patience in making a set of sections for himself. When, however, circumstances allow, working with the knife not only gives the dexterity necessary for undertaking independent investigations, but anatomical relations imprint themselves much more firmly upon the memory when one makes the sections for oneself, and, in especial, one obtains a clearer view of the situation of the several elements relatively to one another.

Good drawings and cleverly executed models facilitate the comprehension of difficult anatomical relations. With regard to models, however, it must be said that as yet we do not possess any that are completely satisfactory. Of the very artistic, but also very expensive, model of Aeby, *His* says, most truly, that although when we have it before us it seems very clear and transparent, it does not stand the test so soon as the eyes are removed from it.

The work under discussion, therefore, differs in many respects from existing text-books of brain-anatomy.

First, as to the manner in which the material is presented, the strictly didactic standpoint is maintained; whether the student makes preparations for himself or not, he can follow the route prescribed for him in the book. The more detailed histological relations are treated separately. The attempt has been made, while not overlooking any of the more important facts concerning the central nervous system, to avoid such minute details as should be left for special research.

The introduction of pathologico-anatomical observations, especially of the pathological changes in the elements, will prepare the road for the comprehension of the processes of disease in the central nervous system without its being in the least intended to work out an exhaustive pathological anatomy of these organs.

That a special value has been attributed to numerous and good illustrations has been already mentioned. In the choice of illustrations, which have been throughout executed in the xylographic establishment of *V. Eder* of Vienna in the most satisfactory way from original drawings, it is to be understood that a certain restraint had to be imposed to prevent the price of the book from becoming excessive. On this account, especially for drawings 118 to 136, the question had to be discussed whether the preparations chosen should be stained with carmine or according to Weigert's method. When I chose the former, I did so on the ground that I wished the illustrations to be true reproductions of the original preparations. Successful Weigert's-preparations from the adult are hardly to be made sufficiently instructive with low magnification; whereas preparations from the embryo were to be avoided on account of the difficulty which the student would find in getting the material.

I suppose I need not point out that the presentation of the material rests throughout upon autoptic observations; when facts are stated on the ground of the observations of other authors, this is in every case noted.

The usefulness of this book is further increased by the addition of an index.

HEINRICH OBERSTEINER.

Vienna, *October*, 1887.

CONTENTS.

	PAGE
INTRODUCTION,	1
METHODS OF EXAMINATION,	3
Defibering,	4
Section-series,	4
Celloidin,	8
Staining of nuclei,	11
,, of medullary sheath by *Weigert's* and other methods,	12
,, of the axis-cylinder,	16
Impregnation by *Golgi's* and other methods,	16
Embryological methods,	18
Study of degenerations,	19
Comparative method,	24
Physiological method,	24
MORPHOLOGY OF THE CENTRAL NERVOUS SYSTEM,	26
Histogeny,	28
Morphogeny,	34
Gross anatomy of the spinal cord,	38
,, ,, medulla oblongata,	42
,, ,, cerebellum,	47
,, its medullary centres,	55
Floor of the fourth ventricle,	55
Roof ,, ,,	59
,, ,, mid-brain,	60
,, ,, 'tween-brain,	63
Optic thalamus,	63
Floor of the third ventricle,	64
,, ,, cerebrum,	65
Nucleus caudatus,	65
Nucleus lenticularis,	67
Nucleus amygdaleus,	68
White matter,	68
Corpus callosum,	70
Fornix,	73
Anterior commissure,	74
Base of the fore-brain,	75
Ventricles,	77
Margin of the cortex,	81
Surface of the hemispheres,	84
Lobation ,,	89
Mesial surface ,,	98
Anomalies of convolution,	102
Physiological meaning of the convolutions,	104

CONTENTS.

	PAGE
HISTOLOGICAL ELEMENTS OF THE NERVOUS SYSTEM,	107
Nervous elements—	
Nerve-fibres,	107
,, pathological changes,	118
Nerve-cells,	121
,, pathological changes,	131
Non-nervous elements—	
Blood-vessels,	135
,, pathological changes,	140
Epithelium,	148
Connective-tissue,	149
,, pathological changes,	152
Neuroglia,	153
Fat-granule-cells, amyloid bodies, &c.,	155
ARRANGEMENT OF ELEMENTS IN THE NERVOUS SYSTEM,	158
Schemes of the nervous system—	
Luys,	168
Meynert,	168
Hill,	169
Aeby,	169
Flechsig,	169
TOPOGRAPHY OF THE SPINAL CORD,	170
Cross-section of its several areas,	177
Minute anatomy,	178
Course of fibres,	187
Blood-vessels,	200
Pathological anatomy,	203
TOPOGRAPHY OF THE BRAIN,	209
Examination by cross-section,	210
Medulla,	212
Pons,	225
Mid-brain,	235
'Tween-brain and fore-brain,	244
COURSE AND CONNECTIONS OF NERVE-FIBRES,	249
Tracts of the spinal cord,	249
Decussation of the pyramidal tracts,	250
Constituents of the crusta,	253
,, ,, internal capsule,	254
Fillet,	257
Cerebellar connections,	260
Gowers' tract,	263
Cranial nerves,	265
Olfactory,	265
Anterior commissure,	274
Cortical centres of olfaction,	275

	PAGE
Optic nerve,	279
The chiasm,	279
Cortical centres of vision,	284
Posterior commissure,	285
Corpora quadrigemina,	284
Oculomotor,	287
Trochlear,	290
Abducent,	292
Trigeminal,	293
Cortical connections,	290
Facial,	297
Auditory,	299
Corpus trapezoides,	306
Superior olive,	306
Cortical connections,	307
Glossopharyngeal,	308
Serial homology of cranial nerves,	309
Vagus,	312
Spinal accessory,	312
Hypoglossal,	314
Connections of the cerebellum,	315
Central nuclei,	315
Medullary substance,	317
Connections with other parts,	321
Cortex cerebelli,	323
Blood-vessels,	333
Pathological anatomy,	333
Connections of the cerebrum,	335
Ganglia at the base,	336
Optic thalamus,	336
Nuclei caudatus et lenticularis,	340
Corpus subthalamicum,	343
Substantia nigra,	344
Medullary substance,	344
Corona radiata,	344
Fornix,	345
Corpus mammillare,	345
Corpus callosum,	347
Anterior commissure,	348
Intrahemispheral fibres,	348
Cortex cerebri,	350
Local peculiarities of structure,	360
Cornu Ammonis,	362
Conarium,	369
Hypophysis,	370
Blood-vessels of the great brain,	371
Pathological anatomy,	372

CONTENTS.

	PAGE
MENINGES,	377
Dura mater,	378
Pathological changes,	381
Arachnoid,	383
Pathological changes,	385
Pia mater,	386
Pathological changes,	387
Telæ choroideæ,	387
THE GREAT VESSELS OF THE BRAIN,	388
Their diseases,	393
APPENDIX: ROTATION OF THE GREAT BRAIN,	395
GLOSSARIAL INDEX,	405

LIST OF ILLUSTRATIONS.

FIG.		PAGE
1.	Scheme illustrating the nutrition of a nerve-fibre (after *Schwalbe*),	19
2, 3.	Transverse sections, showing the formation of the sensory root-ganglia (after *Beard*),	28
4.	The epiblast involuted to form the central nervous system while still a single layer (after *His*),	29
5.	A group of spongioblasts (after *His*),	30
5a.	More advanced stage, showing neuroblasts with axis-cylinder processes,	30
6.	Transverse section of the spinal cord of a trout-embryo (after *His*),	31
7.	The cerebral vesicles,	35
8.	Diagram showing the connection of the nucleus caudatus with the cortex cerebri (after *Wernicke*),	38
9.	Caudal end of spinal cord from ventral surface,	39
10.	Cervical enlargement of the spinal cord from the dorsal side,	39
11.	The base of the brain as far as the optic tracts (dorsal view),	40
12.	Same as fig. 11, viewed laterally,	42
13.	Hind-brain, mid-brain, and 'tween-brain, from the dorsal surface,	44, 45
14.	Cerebellum, dorsal view,	47
15.	Cerebellum, from the ventral side,	48
16.	Sagittal section through the brain in the median line (right half),	50, 51
17.	Frontal section through the medulla oblongata and cerebellum of an ape,	54
18.	Hind-brain, mid-brain, and 'tween-brain, from the dorsal side,	56, 57
19.	Transverse section through the anterior corpora quadrigemina,	62
19a.	Diagram to show the cortical relations of the nucleus caudatus (after *Wernicke*),	66
20.	Section through the 'tween-brain and the neighbouring part of the fore-brain, half a centimeter beneath the upper surface of the optic thalamus and nucleus caudatus,	67
21.	Horizontal section of same, one centimeter deeper than in fig. 20,	69
22.	Frontal section through the human cerebral hemisphere,	71
23.	Sagittal section through the brain in the median line,	72, 73
24.	Part of a median section through the great brain,	75
25.	Left hemisphere of the brain in front of the optic chiasm,	76
26.	Diagram of the cerebral ventricles and the plexus choroideus,	77
27.	Frontal section through the right cerebral hemisphere, behind the splenium corporis callosi,	80
28.	Frontal section through the right hemisphere behind the uncus,	82
29.	Portion of a median section of the cerebrum,	83
30.	Left hemisphere from the side,	86
31.	Left hemisphere of a human embryo at the fifth month,	87
32.	Left hemisphere from above,	88
33.	Left hemisphere, mesial surface,	90
34.	Diagram showing the lobation of the cerebrum,	91

LIST OF ILLUSTRATIONS.

FIG.		PAGE
35.	Left hemisphere from the base,	94
36.	Left hemisphere, mesial surface,	100
37.	Cross-section of a small portion of the anterior columns of the spinal cord,	108
38.	Medullated peripheral fibre,	109
39.	Peripheral nerve-fibres of the frog,	110
40.	Axis-cylinder of a fibre from the white substance of the spinal cord,	110
41.	A fresh medullated nerve-fibre from the sciatic of the frog,	110
42.	Nerve-fibre from the sciatic of the frog,	113
43.	A short piece of a thin peripheral nerve, showing nodes of Ranvier,	113
44.	An isolated medullated fibre,	113
45.	Remak's fibres from the sympathetic of the neck of a rabbit,	115
46.	Central medullated nerve-fibre from the brain,	115
47.	Very fine varicose axis-cylinders from the bulbus olfactorius of the dog,	115
48.	Peripheral nerve-fibre from a new-born puppy, partially surrounded with myelin,	117
49.	Two fibres from the anterior roots springing from a softened spinal cord,	119
50.	Several forms of hypertrophic varicosity of axis-cylinder from softened foci in the spinal cord,	120
51.	A cell from the anterior horn of the spinal cord of the pike,	122
52.	A cell from the anterior horn of the human spinal cord,	122
53.	A pigmented cell from the substantia ferruginea (human brain),	122
54.	Two shrunken cells from a human spinal ganglion,	124
55.	Pyramidal cell from the cortex of human cerebrum,	124
56.	Granules from the cortex of the cerebellum,	127
57.	Diagram designed to show the homology of the granules of the olfactory bulb and retina and the cells of the spinal ganglia,	128
58.	Simple atrophy of a nerve-cell from the oculomotor nucleus (human),	131
59.	Commencing atrophy of a cell from the anterior horn of the spinal cord. Degeneration of the nucleus	131
60.	Fatty-pigmentary degeneration of a pyramidal cell of the cortex cerebri,	131
61.	Granular deterioration of a cell of the anterior horn in myelitis,	132
62.	Cell of anterior horn with ten vacuoles in myelitis,	132
63.	Colloid degeneration of a cell of the anterior horn in myelitis,	132
64.	Calcified nerve-cells from the cortex cerebri beneath an apoplexy,	134
65.	A cortical cell divided into a number of pieces,	134
66.	A middle-sized artery of the brain so torn as to expose each of its coats,	136
67.	A small artery from the brain, with clumps of pigment,	136
68.	A small vein from the brain-substance,	136
69.	An artery from the cortex cerebri,	138
70.	Section from the cornu Ammonis, showing perivascular and pericellular lymph spaces,	138
71.	Isolated capillaries from the cerebral cortex,	139
72.	Cells with hæmatoidin-crystals from the walls of an old apoplectic clot,	141
73.	A moderate-sized artery from the corpora striata, with numerous pigment cells,	141
74.	Capillary vessel from a case of melanæmia,	141
75.	Fatty degeneration of the muscular coat of a cerebral artery,	142
76.	Calcification of the muscular coat of a vessel of the brain,	142
77.	Calcification of an artery of the brain affecting the adventitia as well as its other coats,	142

LIST OF ILLUSTRATIONS.

xvii

FIG.		PAGE
78.	A vein from the brain showing fusiform hypertrophy of its lateral venules resulting in their obliteration,	143
79.	Pseudo-hypertrophy of the muscular coat of an artery of the brain,	143
80.	Atheromatous degeneration of the tunica intima of an artery of the brain,	143
81.	Beaded enlargement of a large artery of the brain,	144
82.	Miliary aneurysm of very small vessels,	144
83.	Ampullar dilatation of the adventitial lymph-space surrounding an artery of the brain,	145
84.	Packing of an adventitial lymph space with leucocytes,	145
85.	Ventricular epithelium of the frog,	148
86.	Isolated connective-tissue cell from the human spinal cord,	150
87.	Isolated connective-tissue cell from the ependyma of the lateral ventricle,	150
88.	Section of the white matter of the brain,	151
89.	Longitudinal section of the spinal cord,	151
90.	Connective-tissue cells with numerous short processes from a case of sclerosis,	154
91.	Two fat granule-cells from the spinal cord,	156
92, 93.	Scheme of a simple primitive nervous system,	162
94.	Scheme of a commissure and of a decussation,	163
95.	Diagram of motor nerve-roots,	164
96.	Diagram of sensory nerve-roots,	164
97.	Transverse section through the human spinal cord: at the level of the third cervical nerves,	172
98.	Transverse section through the human spinal cord: at the level of the sixth cervical nerves,	172
99.	Transverse section through the human spinal cord: at the level of the third dorsal nerves,	172
100.	Transverse section through the human spinal cord: at the level of the twelfth dorsal nerves,	173
101.	Transverse section through the human spinal cord: at the level of the fifth lumbar nerves,	173
102.	Transverse section through the human spinal cord: at the level of the third sacral nerves,	173
103.	Transverse section through the human spinal cord: through the inferior part of the conus medullaris at the origin of the nervus coccygeus,	173
104.	Cross-section of the anterior column of the spinal cord,	178
105.	Junction of the anterior root-bundle with the anterior horn, lumbar region,	181
106.	A nerve-cell from the anterior horn of the human spinal cord,	181
107.	A nerve-cell from Clarke's column as seen in longitudinal section,	185
108.	Diagram showing the course of fibres in the spinal cord,	188
109.	Diagram showing the subdivisions of the white columns of the cord,	194
110.	Descending degeneration of the spinal cord after a one-sided lesion of the brain,	197
111.	Ascending degeneration in the cervical swelling,	197
112.	Semidiagrammatic representation of the arteries in the interior of the spinal cord,	202
113.	So-called combined systemic disease of the spinal cord,	205
114.	Chronic transverse diffuse myelitis,	206

xviii LIST OF ILLUSTRATIONS.

FIG.		PAGE
115.	Syringo-myelia,	206
116.	Disseminated sclerosis in the cervical cord,	206
117.	Figure to show the level of the dorsal portions of the cross-sections represented in figs. 118-130, 132-136,	210
118.	Section of the medulla oblongata (at a, fig. 117),	212
119.	Section of the medulla oblongata (at b, fig. 117),	213
120.	Section of the medulla oblongata (at c, fig. 117),	214
121.	Section of the medulla oblongata (at d, fig. 117),	215
122.	Cross-section of the medulla oblongata (at e, fig. 117),	217
123.	Cross-section of the medulla oblongata (at f, fig. 117),	218
124.	Transverse section of the medulla oblongata (at g, fig. 117),	220
125.	Transverse section of the after-brain (at h, fig. 117),	222
126.	Cross-section of the after-brain (at i, fig. 117),	224
127.	Transverse section of the after-brain (at k, fig. 117),	227
128.	Transverse section of the after-brain (at l, fig. 117),	229
129.	Transverse section of the after-brain (at m, fig. 117),	230
130.	Transverse section of the after-brain (at n, fig. 117),	232
131.	Frontal section through the cerebellum and medulla oblongata of a monkey,	234
132.	Transverse section of the mid-brain (at o, fig. 117),	235
133.	Transverse section of the mid-brain (at p, fig. 117),	236
134.	Transverse section of the mid-brain (at q, fig. 117),	239
135.	Transverse section of the mid-brain (at r, fig. 117),	240
136.	Transverse section of the mid-brain (at s, fig. 117).	243
137.	Frontal section through brain of monkey,	244
138.	Frontal section through brain of monkey, passing through the middle of the optic thalamus,	246
139.	Frontal section through monkey's brain at level of front of thalamus,	247
140.	Scheme of pyramidal tracts,	250
141.	Diagram showing the constitution of the crus cerebri,	253
142.	Horizontal section through the internal capsule,	254
143.	Plan of the central connections of the posterior columns,	259
144.	Left hemisphere in front of the optic chiasm (base),	267
145.	Sagittal section of the bulbus olfactorius of the dog,	268
146.	Portion of the same,	268
147.	Transverse section of the human olfactory tract,	273
148.	Scheme of the central apparatus of smell,	274
149.	Outline of the brain of a dog,	278
150.	Outline of the brain of a cat,	278
151.	Outline of the brain of an otter,	278
152.	Scheme of the central apparatus of vision,	289
153.	Scheme of the central origin of the trigeminal nerve,	294
154.	Schematic projection of the medulla oblongata,	297
155.	Scheme of the central auditory apparatus,	301
156, 157.	Diagrammatic sections of the spinal cord and medulla showing the relative positions of the roots of a spinal and segmental cranial nerve,	311
158.	Diagram showing the disposition of the nervus accessorius Willisii in cross-section,	313
159.	Same, in longitudinal section,	313

LIST OF ILLUSTRATIONS.

FIG.		PAGE
160.	Sagittal section through the cerebellum some millimeters to one side of the middle line,	321
161.	Cross-section through a convolution of the cerebellum,	323
162.	Cross-section through a lobule of the cerebellum,	324
163.	Vertical section of cortex; from the lateral surface of a cerebellar convolution,	327
164.	A Purkinje's cell, as seen in a vertical section at right angles to the long axis of a convolution,	327
165.	A Purkinje's cell, as seen in a vertical section parallel with the long axis of a convolution,	328
166.	Section through a small heterotopia in the medullary substance of the cerebellum,	332
167.	Vertical section of the cortex cerebelli from a case of encephalitis,	332
168.	Diagrammatic frontal section through the great brain,	336
169.	Diagram of the associating tracts of the cortex cerebri,	348
170.	Vertical section through the human cortex cerebri where it covers the posterior part of the middle frontal convolution,	351
171.	Cortex of the lobulus paracentralis,	351
172.	Cortex of the cuneus in the fissura calcarina,	351
173.	Cortex of the gyrus cinguli,	351
174.	Cortex of the subiculum cornu Ammonis,	351
175.	Pyramidal cell from the cortex cerebri,	353
176.	Cortex cerebri, frontal lobe,	357
177.	Transverse section through the cornu Ammonis,	363
178.	Cortex of the cornu Ammonis and a part of the fascia dentata,	366
179.	Injected cortex cerebri of the dog,	371
180.	Encephalitic cysts of the cortex cerebri, with secondary degeneration of the white substances,	374
181.	Diagram of the membranes of the brain,	378
182.	Epithelium on the inner surface of the dura mater of the guinea-pig,	379
183.	Dura mater of a new-born puppy,	379
184.	A corpus arenaceum from the dura mater,	379
185.	Pseudo-membrane of the dura mater,	383
186.	Neo-membrane of the dura mater,	383
187.	Trabeculæ of the arachnoid,	384
188.	Pia mater in meningitis tuberculosa,	386
189.	Epithelium of the choroid plexus,	387
190.	Circle of Willis,	389
191.	Anomalies of the circle of Willis,	389
192.	Fornix, pyriform lobe, and olfactory bulb from the brain of the ox,	396
193.	Diagram showing the tracts of fibres which connect the central grey tube with the cortex,	398
194.	Diagrammatic view of the under side of the brain, after *Schwalbe*,	400
195.	Early fœtal brain,	402
196.	Brain of fœtal sheep with the rhinal fissure developed,	402
197.	Diagrammatic view of the under surface of the brain of a young rabbit,	402
198.	Brain of rabbit,	403

INTRODUCTION.

An investigation of the complicated characters which distinguish the fine structure of the brain and spinal cord is impossible without a previous acquaintance with the more obvious features of their external configuration. Details of fine structure are often difficult to grasp, and their comprehension is facilitated by filling them into a mental outline of the organ to which they belong. Therefore, as soon as the **first section** (devoted to methods of study most in vogue) is disposed of, we shall give an account of the more obvious microscopical features of the cerebro-spinal axis, especially its external mouldings, and of such details presented by cross-sections through the brain at various levels as can be recognised without further preparation (**second section**—morphology).

Before commencing the microscopical investigation of the central nervous system by means of transparent sections, acquaintance must be made with the characters of the histological elements of which it is made up. In the **third section** an account is given of the more important nervous and non-nervous constituents of the system in health, and also of the changes to which they are subject in disease.

Next, the spinal cord is described as being relatively the simplest part of the central nervous organs (**fourth section**).

After this, we suppose (**fifth section**) that a number of cross-sections, not constituting an unbroken series, but useful for microscopical investigation, are made through the spinal cord and brain. During the preparation of such sections and their necessary examination with an ordinary magnifying-glass, one becomes acquainted with many facts concerning their organisation, especially as one can trace from section to section the changes in topographical distribution which their constituents undergo. The same routine must be followed by any one who investigates a series of sections which he has not cut for himself. He ought first to make a general survey of his preparations with a magnifying-glass, and try to obtain a stereoscopic picture of their more important features.

This done, we shall attempt, on the ground work of knowledge thus obtained, to follow individual bundles, trace their divisions and connections, and determine their end-points. This is the object of the **sixth section,** which treats first of the fibre-routes in the spinal cord, and then of the cranial nerves. Their finer relations are rendered intelligible by a study of the structure of the cerebellum and cerebrum.

The concluding section (the **seventh**) is devoted to the cavities of the central nervous system, which stand in such close anatomical and physiological relation to the brain and cord as to deserve a special description.

SECTION I.—METHODS.

The anatomical study of the central nervous system is fraught with difficulties such as are never met with in investigating other organs. The account of the structure of the brain and spinal cord about to be given, which goes somewhat further than the mere outlines of the subject, is based upon the most recent results of research in this field.

The causes of the difficulty are not far to seek. It might be anticipated that the structure of the organ to which the most various and complicated, and at the same time highest and noblest, functions are allotted, would obviously correspond in complexity to its work. It is also to be understood that this organ of relatively small size but complicated structure, made up as it is of minute nerve tracts and other parts, and composed of a delicate, soft, and destructible tissue, will hardly admit of investigation by the ordinary anatomical methods.

Such reflections alone suffice to account for the fact, that only since the introduction of special methods has this "book sealed with seven seals" been opened and its characters so difficult to read been forced to yield their meaning.

The methods in use up to the present, most of which are required for the investigation of the varying situation and connections of the elements rather than their structure, although very different in principle, yet support and complete one another.

Excluding simple anatomical inspection, we can arrange our methods in five groups, as follows:—

1. The teasing out of the fibres of a properly-prepared central nervous system.

2. The preparation of an uninterrupted series of sections through the normal organ.

3. The study of organs, the several parts of which either develop at different periods, or else have sustained retrogressive metamorphosis.

4. The comparison of homologous parts of the central nervous system in different animals.

5. The experimental observation of action, from which structure

may be inferred, or the study of localised disease of the central nervous system associated with functional anomalies.

Other methods of more limited application, but none the less valuable on this account, will be detailed in their proper places.

1. DEFIBERING.

Since the central nervous system when fresh presents a consistence which precludes the separation of fibres, it must, if this method is to be used, be subject to a preparation which, while hardening the bundles of fibres, softens the connective tissue binding them together. Such a result has not hitherto been satisfactorily obtained.

Simple hardening in alcohol to which saltpetre or hydrochloric acid (used by *Ruysch* and *Vicq d'Azyr*) or potash (*Reil*) is added has been known for a long time. Hardening in chromates with subsequent hardening in alcohol is better. External configuration is best studied after hardening in bichromate of potassium and subsequently in alcohol. J. *Stilling* places pieces of brain in Müller's fluid, dehydrates in spirit, and then leaves them in absolute alcohol until they attain a firm consistence. After this they are macerated in artificial wood-vinegar (200 grms. acetic acid, 800 grms. water, 20 drops of creosote). In this they remain, as a rule, for several weeks (it is impossible, however, to fix the time, for it can only be determined by experience); if the preparations become too soft they are placed for several days in crude wood-vinegar. We can with the help of forceps separate certain tracts of fibres in such pieces of brain and preserve them in Canada balsam, after treatment, in a watch-glass, with oil of cloves.

In all well-hardened spirit and chromic preparations, every artificial break in the white matter, and for the most part also in the grey masses of the brain, shows the course of fibres.

It must, however, be understood that all methods of defibering, especially when the preparation contains fibres crossing one another in different directions, are apt to yield misleading results.

2. THE PREPARATION OF SECTION-SERIES.

To *Stilling* belongs the merit of having introduced this most useful method into brain-anatomy.

If we imagine a piece of brain cut into such a series of sections as would, if put together again, completely reproduce the original structure, we shall see that it ought to be possible, were it not for special difficulties which present themselves, to follow any transversely cut band of fibres through the whole length of the series.

Although this ideal is not always attainable, it is only since the introduction of this method of making continuous series of sections that notable progress in the anatomy of the brain has been made. Further histological methods can be applied to any of the sections.

The reconstruction of an organ from the observation of a series of transparent sections presents no little difficulty, and the conception of structure which we gain by this method needs to be checked by seeing the object itself as exposed by dissection.

When such a series is to be made, the central nervous system must first be **hardened**.

Attempts to freeze the tissue and cut it when fresh have not been successful, for the natural brain-substance suffers too much in the process. The freezing method is, however, useful for tumours. Amongst hardening fluids, solutions of chromic salts stand first, and are to be preferred to simple chromic acid. Bichromate of potassium is most used. Fresh pieces of the central nervous system are placed in a 1 per cent. solution of this salt. The fluid is repeatedly changed for the first few days, and is rendered gradually stronger, until it is brought up to 2 or 3 per cent., at which strength the pieces are left until they are sufficiently hard. The time needed averages six to eight weeks, but depends on various circumstances, on the temperature of the room for example (it requires less time in summer than in winter); small pieces, too, take a shorter time than large ones. In an incubator, in which the temperature is maintained at from 35° to 40° C., it is possible to harden the tissue sufficiently for cutting in from eight days to a fortnight. The time required depends upon the particular part of the central nervous system under treatment. The hardening of spinal cord in chrome-salts requires especial care.

After the preparations are ready for cutting, they can still remain for some months in the chromic solution without injury; if they are to be preserved still longer they must be transferred to a weak solution (0·1 per cent.) of the salt, in which they can be kept for years. The formation of mould is no sign that the preparations are spoilt. The addition of a little carbolic acid does not prevent the formation of fungi, but checks their growth. The hardening is hastened by adding to the bichromate of potassium a little free chromic acid (20 or 30 drops of a 1 per cent. solution of chromic acid to 500 grms. of the bichromate solution).

Even though the pieces are not thoroughly hardened in chrome-salts, they can still bear the subsequent hardening in alcohol. This is accomplished usually by washing them out first for several days in water [it is not, as a rule, however, advisable to place them directly in water, but to transfer them from the solution of chromates to 25 or 30 per cent. spirit, which is changed every day until the liquor drawn off is almost colourless], and then for a like time in 50 per cent. alcohol, after which they are transferred to strong (95 per cent.) alcohol. To prevent precipitation it is recommended that the vessels should be kept in the dark (*H. Virchow*). A long maceration in alcohol renders such preparations easier to cut, but destroys certain details of structure. Owing to a partial solution of the myelin by the alcohol, various spots, spaces, and so forth are artificially produced. If we desire to stain the myelin-sheath (in osmic acid or by certain other methods) the dehydration by alcohol must be omitted. The preparations in this case are only washed in water. The use of alcohol from the commencement of hardening is to be avoided, except in certain cases in which one wishes to study details in the structure of nerve-cells (*Nissl, Trebinski*).

Müller's fluid consists of 10 parts bichromate of potassium, 5 parts sulphate of soda, and 500 parts water.

Erlitzky's fluid is made by mixing 5 parts bichromate of potassium, 1 part sulphate of copper, 200 parts water. It hardens more quickly than *Müller's* fluid or bichromate of ammonia (which is apt to make the preparation too hard), but sometimes produces dark precipitates in the preparation, which have already led to mistakes.

With some practice one can tell by touching, or gently pressing, a preparation, whether it has reached the proper consistence for cutting; the safest thing to do is to try it with the razor. It may comfort inexperienced persons to know that it sometimes happens that the preparation proves to be unfit to cut, although one cannot account for this mishap.

When it is desired to prepare very small pieces of brain or cord for cutting in a condition in which the most minute details of structure, as for example the nuclear figures, may be visible, it is necessary to take the pieces quite fresh from the living or recently-killed animal, and treat them with one of the so-called **fixing media**. *Fol's* modification of *Flemming's fluid* is the best of all the fixing media yet proposed; it consists of—

1 per cent. solution of	perosmic acid,	.	2	parts by vol.
1 ,, ,,	chromic acid,	.	25	,,
2 ,, ,,	acetic acid,	.	8	,,
Water,	. . . .	.	68	,,

It does not do to be economical with this fluid. It should be changed as soon as it appears cloudy. After some hours (even up to twenty-four or more) the preparation is carefully washed until all traces of the hardening fluid are removed, and then preserved in 80 per cent. alcohol.

The preparation of sections, which need often to be of very large size, used to require a skilful steady hand; but the difficulties are now very much reduced by the introduction of the **microtome**. Out of all those which have been introduced in recent years, only such microtomes as answer our purpose need be mentioned. In its simplest form the microtome consists of a hollow metal cylinder closed below with a movable floor, which is pushed up and down without rotation on its axis by a fine micrometer-screw. The preparation is fixed in the cylinder by means of an embedding mass. A perfectly flat broad glass or metal ring is fixed to the free edge of the tube. The preparation is raised by means of the screw through the required thickness of the section. The knife, which should be broad and light, with a biconcave or plano-concave surface, is kept wet with water or, better still, with alcohol as it is pushed over the smooth ring or plate.

Gudden's microtome is made on this principle. It must be placed so that its upper part is immersed in a vessel of water under the surface of which the sections are cut. It is intended for preparing large sections which swim away in the water as they are cut, and hence are preserved from traction. Some skill is required in using the knife in this microtome if faultless sections are to be obtained. It is made by *Katsch* of Munich.

The simplest embedding mass is made by melting wax and oil together and pouring them while hot into the tube of the microtome, usually a mixture of three parts wax and two parts oil suffices; but the proportions of the two substances must be regulated by the hardness of the tissue. Other substances also, such as stearin, paraffin, tallow, &c., can be employed in a similar way.

It is necessary to clear away the embedding mass from the margin of the preparation, so that the knife passes through little besides the tissue. The knife needs frequent stropping, and must be cleansed after each cut. When the section is large and perishable it is caught up out of the water on a piece of filter paper and at once covered by another piece of wet paper. The section remains in its envelope of paper throughout the procedure about to be described. Each envelope should be marked with a number.

Excellent results can also be obtained with other microtomes, particularly the so-called sledge-microtomes, in which the knife is carried on a sledge, while knife and preparation alike are kept moist with alcohol dropped from a wash-bottle. We especially recommend the sledge-microtome of *Reichert* of Vienna, with its automatic arrangement for lifting the sections. *Weigert's* modification of the "diving-microtome" of *Schanze* of Leipsic is useful for large preparations. It makes it possible to cut sections under alcohol.

[For general laboratory purposes the most convenient form of microtome is undoubtedly one which allows the tissue to be cut frozen. In some cases a piece of tissue can be frozen, cut, stained, and mounted in glycerin and water for hasty examination in the *post-mortem* room. If the tissue has been hardened in spirit it is necessary to throw a piece of suitable size and shape into water until all spirit is removed. This takes, as a rule, about an hour; but if the piece sinks (as does not usually occur with nervous tissue) this may be accepted as a sign that most of the spirit has diffused out into the water. It is then dipped in gum and placed on the freezing-plate of the microtome. If ether is used as a freezing agent a few minutes only are required to bring the tissue into a suitable condition of hardness, which means, for some tissues, the most complete freezing possible, for others, a condition of partial thaw. Nervous tissue, when frozen hard, is, as a rule, too brittle to cut. The surface may be partially thawed by touching it with a wet camel's-hair brush. No microtome is more suitable for cutting frozen nervous tissue than *Roy's;* the extremely oblique position of the razor and the circular movement enable one to avoid breaking the section transversely, as is very apt to happen when a knife is driven straight forwards through the extremely friable frozen nervous tissue. The sections are lifted from the razor into salt solution with a large wet camel's-hair brush.]

The tissues can be embedded for cutting with the sledge-microtome in pasteboard or metal boxes filled with wax and oil, or, if they are of no great depth, they may even be stuck on a piece of cork. If they are to be cut on cork they must be placed with the cork in a thick solution of gum, out of which both are lifted together into absolute alcohol in which they remain for twenty-four hours; or else they are placed in thick solution of celloidin, subsequently set by immersion in alcohol of 80 per cent. The alcohol commonly used in the laboratory is about 95 per cent.* It may be let down to the required strength by mixing with distilled water in the proportion of 9 to 1·5. Stronger alcohol, and especially absolute alcohol, dissolves celloidin.

When the tissues are unfit to cut in other ways, owing to the hardening being improperly carried out, or else owing to the tissue, softened by disease perhaps, being incapable of hardening, they may yet be made into excellent sections by adopting the **celloidin method.** Pieces of spinal cord about 1 centimeter thick, are placed for two or three days in common alcohol. They are then completely dehydrated in absolute alcohol, and subsequently placed in a thin solution of celloidin in equal parts of sulphuric ether and absolute

[* *Methylated spirit if free from resins, as shown by its giving a clear mixture with water, is equally useful.*]

alcohol. In this they remain for a variable time, according to thickness (usually three or four days). Next they are transferred to a syrupy solution of celloidin, in which they remain a few days. . After this a piece is lifted with adhering celloidin on to a cork, exposed under a glass until it is almost set, and then cork and preparation together are placed in 80 per cent. alcohol. This method of embedding in celloidin is indispensable for very friable preparations, or for preparations which are with difficulty held together, or which present cavities in their interior.

[The method of embedding in celloidin which we have adopted for the last six years is the same in principle with that described by *Barrett.** Its usefulness for all purposes depends upon the replacement by water of the alcohol in the celloidin-mass. Owing to the tissue-like permeability of celloidin, this is effected without perceptible alteration in form. The embedded preparation can afterwards be cut on a freezing-microtome. So simple is the procedure that it is well as a matter of routine to apply it invariably in investigations into the structure of the central nervous system. The tissue, either stained *en blocq* or left for staining after it is cut, is placed in absolute alcohol. This is replaced by a mixture of absolute alcohol *3 parts*, ether *1 part*. When this has soaked into the tissue a small piece of *Shering's* dry celloidin is placed in the vessel. The celloidin dissolves very slowly, and the gradually concentrating solution permeates the tissue far more thoroughly than even the weakest ready-made solution would do. Fresh pieces, or, to hasten the process, pieces of waste celloidin-jelly, are added daily until the solution flows with difficulty. It is then poured out with the tissue in its centre into a flat-bottomed glass dish. The dish is covered with a plate of smooth glass. By this arrangement a very slow evaporation is allowed, and the celloidin when set will be found to be of uniform consistency, and not prone to curl when cut. If it is important to save time, the celloidin is poured into a paper boat which is immersed in chloroform, which sets the celloidin in a few hours without measurable alteration in bulk (*Caldwell*). When set somewhat firmly, the tissue with a convenient bed is cut out with a knife, and the block thrown into water for an hour (or if saturated with chloroform, into spirit and then into water). When all the alcohol is replaced by water, the block can be frozen and cut with a facility quite unattainable in spirit-set celloidin. There is no pleasanter material to cut than frozen celloidin. In some cases it is desirable to embed the tissue in celloidin, even before cutting into series of sections in paraffin. If this is desired, the chloroform-

---

* *Journal of Anatomy and Physiology*, vol. xix., p. 94, 1885.

saturated block, or even the alcohol block, can be placed in melted paraffin.]

The investigation of the constitution of the central nervous system by means of sections only reached its full development when *Gerlach* showed us how the use of **staining agents** reacting differently to the several tissue-elements allow us to make a differentiation in the preparation. Ammonia-carmine, the stain which happened to be employed first, has not only yielded the most abundant results, but is still one of the best reagents. The best carmine which can be bought, and it is not always possible to obtain a good specimen, is mixed in a beaker with ammonia into a soft pap, to which so much distilled water is added as will yield a dark black-red solution. The solution is filtered, and exposed to the air until the surplus ammonia has evaporated. The solution improves on keeping. The fluid can always be filtered back into the bottle after use, and may be employed for years. Alcohol preparations are very quickly coloured in this solution; a few minutes only being needed. Chromic preparations take a varying time, increasing with the time the preparation has been kept; it may vary from an hour to several days, and each case must be treated on its own merits. When it is desired to accomplish the staining quickly, the watch-glass containing the preparation should be placed on a wire net over a vessel of boiling water; three to five minutes usually suffices under these circumstances. In the incubator the time depends upon the temperature. For slight magnification the sections are cut thick and slightly stained; for use with high powers, deep staining is necessary.

The stained sections are washed in distilled water (the addition of a few drops of acetic acid makes the nuclei more conspicuous), they are then placed for a quarter of an hour in common, and for a similar time in absolute, alcohol. From absolute alcohol they are transferred to clove-oil, in which they remain until transparent. Celloidin preparations can bear neither absolute nor clove-oil; they must be cleared in oil of thyme (origanum) [or oil of bergamot] or creosote; clove-oil dissolves celloidin. Creosote is expensive, but has the advantage that the dehydration with alcohol need not be so complete. Cedar-oil is not good for celloidin preparations as it makes them opaque, but otherwise it is a good clearing medium. Turpentine may be used by itself. Finally the section is spread out on a slide, the surplus oil is sucked up with blotting-paper, dammar varnish [or Canada balsam] is dropped on to it, and it is covered with a cover-slip.

Sections enclosed between two pieces of paper are placed with their envelope in the clove-oil. From this they are lifted with forceps on to the slide, and it is now easily possible to detach the upper paper.

The other piece of paper is then seized with the forceps and turned over, so that the section comes to lie upon the glass. The second paper is then easily removed, provided the surplus clove-oil has been dried off with a number of layers of blotting-paper.

Ammonia-carmine stains the axis-cylinders and all cells non-nervous as well as nervous [but especially their nuclei].

Hoyer's dry ammonia-carmine is useful for some purposes, since the common preparation is apt to become suddenly useless owing either to the formation of a bright-red precipitate, or else to the growth of fungus in the fluid. A freshly-prepared solution of the powder affords a useful substitute for the liquid ammonia-carmine. Picro-carmine is often recommended instead of ammonia-carmine. A good picro-carmine for colouring the central nervous system is prepared by *Löwenthal* in the following way :—In 100 grms. water, 0·05 grm. of caustic soda is dissolved, to this is added 0·4 grm. carmine; the mixture is then boiled for ten or fifteen minutes, and diluted to 200 cc. To this fluid is added just sufficient of a 1 per cent. watery solution of picric acid to redissolve the precipitate first formed. It then stands for two or three hours when it is filtered several times through the same filter-paper. After some weeks or months the solution is apt to become dim.

Beside the carmine-staining we use also—1, nuclear stains; 2, stains for medullary sheaths; 3, stains for the axis-cylinders alone.

A. NUCLEAR STAINS.

Alum-hæmatoxylin is the most used. As much hæmatoxylin as will lie on a three-penny bit, and the same quantity of alum are placed in a test-tube two-thirds full of distilled water, and boiled until they dissolve; the resulting claret-coloured fluid is then filtered. The solution will not stain until after it has been kept for some days. It is advisable to filter off the precipitates as they form. Colouring occurs very quickly, and it is often necessary to use the solution in a considerably diluted condition. The section, after it is washed, should only appear grey-blue, but all nuclei and amyloid bodies (when they occur) will be found strongly stained. All the rest of the section ought to be left almost or quite uncoloured. Various other preparations of hæmatoxylin act in a similar manner. If the section is overstained it can in some cases be set right by weak salt solution. Nuclear staining gives beautiful results in sections already stained in carmine or picro-carmine. The further treatment is the same as for carmine preparations; alcohol, clove-oil (thyme-oil), dammar varnish.

Csokor's carmine stains nuclei well. 50 grms. powdered cochineal and 5 grms. alum are dissolved in 500 grms. of water, and the solution reduced to two-thirds its bulk by boiling. A few drops of carbolic acid are added to prevent the formation of moulds.

Numerous other nuclear stains are useful in their place—*e.g.*, a watery solution of Bismarck-brown (1 to 300), *Grenacher's* carmine solution, nigrosin, &c., &c.

B. MEDULLARY-SHEATH STAINS.

1. *Exner's Perosmic Acid Method.*—Quite small pieces of nerve-tissue (at the outside not more than a centimetre thick) are placed in a sufficient quantity of 1 per cent. perosmic acid solution. In two days the solution is changed. In five to ten days the pieces are darkly coloured, but they may remain for a longer time in the solution. [Good results are often obtained in twelve to twenty-four hours; the preparation should, while in the osmic acid, be kept in the dark.] The preparation is then washed [a quarter to half an hour in running water is by no means too long to remove all traces of dissolved osmic acid and prevent subsequent precipitation on the addition of alcohol], placed for a few seconds in alcohol, embedded and cut. The sections, which must be very thin, are cleared in glycerin, lifted with adhering glycerin into a slide, on which a drop of strong ammonia-water has previously been placed, and covered with a cover-glass after exposure to the air for a few minutes.

Even the finest medullated fibres are stained black. The fault of this excellent method is that the preparations quickly degenerate and are often useless in a few days. [Permanent preparations, although with some faults incidental to the solution of the fat, are obtained by mounting the section after the usual dehydration, in Canada balsam.] The method can only be applied to very small pieces of tissue.

2. *Palladium and Gold.*—Alcohol is to be avoided in preparing the sections which are placed for five minutes in a watery solution of palladium chloride (about 1 in 2000). They are washed in distilled water, placed in a very weak solution of gold chloride (1 in 5000) made acid with hydrochloric acid, and exposed for twenty-four hours to a moderate light by which time the fibre-tracts assume a violet colour. After repeated washing in water they are treated with alcohol, clove-oil, and dammar varnish. The preparations are only to be used with a low power, but they give good general results, as only the coarser fibres are stained.

3. *Weigert's Hæmatoxylin-method.*—The tissue must be hardened in chromic salts; but it can be transferred through alcohol into celloidin,

although it is desirable that it should not be washed out in water. Fair results may be obtained with this method after washing out with water, and the tissue may even be cut under water on a Gudden's microtome, but the best results of which the method is capable are only possible when washing out in water has been avoided. The block of tissue is fastened on to cork with celloidin in the manner already described, and after being immersed for several hours in 80 per cent. alcohol it is placed in neutral solution of copper acetate (made by mixing equal parts of saturated copper acetate solution and water). In this it remains in an incubator at 35° to 40° C. for one or two days. The sections are then cut and placed in alcohol, from which they are lifted into a solution of hæmatoxylin, prepared by dissolving 1 part hæmatoxylin in 10 parts alcohol and 90 parts water. The solution is not fit to use for one or two weeks. Addition of 1 per cent. of a cold saturated solution of lithium carbonate ripens the fluid sooner. In this solution the sections remain a longer or shorter time (from two to twenty-four hours), according to the degree of colouration required—spinal-cord requires a shorter time, brain-cortex a longer time. The sections, now quite black, are washed in water, and then placed in a decolourising solution composed of borax 2 parts, ferridcyanide of potassium 2·5 parts, water 100 parts. Here the section remains until a differentiation between the nerve-fibres and grey matter is distinctly visible, the time necessary varying from a quarter of an hour to twenty-four hours. Owing to their blue-black colour the medullated fibres stand out sharply on a brown field. Often the decolourising fluid works too strongly, and it is advisable to thin it, even with fifty times its volume of water in the case of peripheral nerves (*Gelpke*). Since the sections which are cut after the tissue has been treated with copper-acetate are not amenable to staining with carmine it is frequently convenient to prepare a number of sections from the hardened tissue, and to submit only those to which it is desired to apply Weigert's method to the copper-acetate. Sections do not require to stay so long in the incubator as directed for the block of tissue. The sections should be rinsed in weak alcohol before they are transferred to the hæmatoxylin. In cases in which staining of the finest fibres is not necessary, it is possible to dispense with the copper solution, provided the colouring is conducted in the incubator. It is sometimes impossible when the sections are thick to effect a sufficient decolourising in the ferridcyanide solutions; after remaining, however, for twenty-four hours in alcohol the section can be again treated with ferridcyanide with better effect.

The preparation of an uninterrupted **series of sections** from tissue embedded in celloidin with subsequent staining in Weigert's hæmatoxylin is rendered possible by the following method recommended by this histologist. One or more glass-plates are carefully cleansed and covered with collodion as if for photography, next strips of tissue-paper are cut a little broader than the sections and a little longer than the glass-plate. The sections are taken off the razor with these strips in such a way that, by using a gentle traction, they are rotated on their axes, and arranged in the direction in which the series is proceeding. The strip travelling from right to left, each succeeding section is received on the right side of the one before, and the strips of sections when complete are kept in the same order. It is important to keep the section damp during the cutting, and until it is transferred to the glass-plate. This is accomplished by having near the microtome a shallow dish containing a number of layers of blotting-paper soaked in 80 per cent. alcohol, with a single sheet of tissue-paper on the top. The strips of paper bearing the sections are laid on this damp bed, the sections being on the upper surface. The collodion (solution of celloidin) on the glass-plates being now dry the strips of sections are inverted on to the plates, and being gently pressed the sections adhere to the celloidin. Usually two rows of sections can be laid side by side on the same plate. All superfluous alcohol being now removed without the sections being actually dried, a second film of collodion is quickly poured over them. As soon as this layer is dry on the surface the preparation can be marked with methyl-blue in such manner as to remind one of their orientation. The glass-plate is now either set aside in 80 per cent. alcohol, or *at once* (before it is really dry) transferred to the hæmatoxylin-solution. In this, especially if placed in an incubator, the strips of celloidin detach themselves easily from the glass. They must be carefully washed. The strips can be cut into convenient pieces, washed in alcohol of 90 to 95 per cent. (but not absolute), cleared in creosote or in a mixture of three parts xylol to one part of anhydrous carbolic acid and mounted in dammar varnish. Oil of thyme and oil of cloves are to be avoided.

[It is perhaps desirable to mount in xylol balsam or in dammar varnish without a cover-slip.]

Formerly *Weigert* introduced a method for staining medullated nerves in acid fuchsin; but this has now been superseded by the easy, and in every way excellent, hæmatoxylin-method, which is of inestimable value for studying degeneration of medullated nerves.

Since its introduction numerous modifications which cannot all be described here have been proposed. *Pal's* method deserves descriptions in detail, as it gives excellent results. Its value consists in the *complete* decolourisation of the tissue between the medullated nerves and the opportunity of subsequently staining it, which is not allowed by Weigert's method. Pieces of tissue are hardened in Müller's fluid. If this reagent has been completely washed out, or if the tissue has assumed a green colour, it must be put for a few hours into 0·5 per cent. chromic acid, or for a longer time in a 2 per cent. to 3 per cent. solution of bichromate of potassium before proceeding further. The sections are put for twenty-four to forty-eight hours in Weigert's hæmatoxylin-solution (part of the time, if need be, in an incubator at 35° to 45° C.); washed in water, to which, if the sections are not already stained deep-blue, some lithium carbonate solution is added; and then placed for 20 to 30 seconds or longer, until the section looks as if decolourised by Weigert's method, in a $\frac{1}{4}$ per cent. watery solution of permanganate of potash, after a previous washing in the following solution for a few seconds—

    1 part pure oxalic acid,
    1 part sulphide of potassium,
    200 parts distilled water.

Everything except the medullated nerves is completely decolourised by the permanganate of potash.

After careful washing, the sections can be further stained in various dyes (particularly picro-carmine). Dehydrate and clear in the usual way.

This method is excellent for a collodion-series.

The nerve-fibres are very sharply marked off by Pal's method; and other tissues can be defined with proper staining. Brown patches are unavoidable in some sections, but they do no harm. Other tissues besides the nerve-fibres are apt to be darkly stained by Weigert's method (or Pal's modification). The contents of blood-vessels especially stain; in some cases the colourisation affects the corpuscles, in other cases the plasma. Sometimes this staining only affects the vessels in a defined region, as, for instance, in the deepest layer of the cortex. At times coagulation-products, stained intensely dark, are seen in these vessels, and easily mistaken, when thread-like in form, for medullated nerves.

Calcified vessels and nerve-cells also stain. Within the nerve-cells the pigment assumes a darker colour. Especially after decolourisation with ferridcyanide we notice that all nerve-cells have not reacted in the same way to the colouring agents; an attempt has been made, as we shall show later on, to make use of this difference for the classification of nerve-cells, the difference being supposed to be associated with a difference in function.

C. AXIS-CYLINDER STAINS.

The axis-cylinders of nerves stain very distinctly with carmine; at the same time, however, owing to the staining of the connective tissue they are often unrecognisable, and it is of importance to find a method which, while staining the axis-cylinders and so rendering them conspicuous, leaves the connective tissue uncoloured. This is desirable, for instance, in the patches of disseminated sclerosis.

Freud's method of gold-staining answers the purpose, although occasionally the medullated sheath is stained. Sections from tissues hardened in chrome-salts are placed in 1 per cent. solution of gold chloride mixed with an equal volume of 95 per cent. alcohol. After four to six hours they are washed in distilled water and then placed in caustic soda (1 part to 5 or 6 of water). After two to five minutes, the sections are lifted out, drained, and put in 10 per cent. solution of potassic iodide; in five to ten minutes the sections, having acquired the proper colour, are washed in water and alcohol. To prevent swelling and crinkling, delicate sections, as soon as they leave the potassic iodide, need to be put on to a slide and dried with pieces of blotting-paper.

The results yielded by this method, which allows of the highest magnification, are excellent, although the method is a little troublesome. The nerve-fibres appear black, dark-blue or dark-red, according to the nature of the specimen.

OTHER METHODS OF COLOURING, IMPREGNATING WITH METALS, &c.

Of the principal methods of colouring by impregnation only a few of the chief need be mentioned.

The Sublimate Colouring of Golgi.—Little pieces of the central nervous system are, after thorough hardening in bichromate of potassium, placed in a 0·25 per cent. watery solution of corrosive sublimate. The fluid is renewed as often as it becomes coloured yellow, and the concentration of the solution may be raised to 0·5 or even 1 per cent. Small pieces are saturated in eight to ten days, but the longer they are left the more thoroughly are they permeated; and they may remain in the solution without harm for years. The pieces are now cut, and despite their very favourable consistence the sections need not be made very thin; they must be well washed, however, else after some weeks numerous acicular crystals of corrosive sublimates will be seen. Subsequent treatment as usual.

SAFFRANIN-STAINING.

With low and moderate magnification certain nerve and connective-tissue cells, but never all, as well as connective-tissue fibres appear intensely black. This colour is due to a fine crystalline precipitation, opaque to transmitted light, in the tissue spaces [or lymph spaces] around the tissue-elements. *(Ramon y Cajal in the*

No other method shows in such a conspicuous manner the continuity of the spaces around the cells and their branches. Its principal fault is its uncertainty, for in one preparation not more than a tenth part of the nerve-cells and probably fewer connective-tissue cells are stained, while another shows numerous connective-tissue cells but no nerve-cells. A similar method is also given by *Golgi* for impregnating with nitrate of silver, but it offers no special advantages. *Pal* has invented an improved sublimate-method. It consists in treating the sections with sodic sulphide (Na_2S), which gives more precise figures, black even with a high power. 10 grms. of caustic soda are dissolved in 1000 grms. of water; half of this is saturated with sulphuretted hydrodgen, mixed with the other half and kept in a well stoppered bottle. The sections are carefully lifted from the sublimate solution into this fluid, and remain there until the spots, at first white, become black. Subsequent treatment as usual.

Golgi's staining is especially successful when the brain has been hardened in the following manner:—A 2·5 per cent. solution of bichromate of potassium is injected into the carotids of an animal just killed, for the purpose of rinsing the brain; small pieces are then placed in Müller's fluid, which is frequently changed during eight to ten days; then for twenty-four hours in a mixture of 4 parts Müller's fluid to 1 part 1 per cent. solution of perosmic acid. The pieces may now be placed in the sublimate or silver solution. The preparations are supposed to keep better if not covered with a cover-slip [owing to the balsam, in the absence of the cover-slip, being allowed to dry].

Adamkiewicz' Staining in Saffranin.—The sections are placed in water weakly acidified with nitric acid. After a short time they are placed in the colouring solution (a deep burgundy-red watery solution of saffranin). Here they lie until they are overstained, when they are washed first in common, and then in absolute alcohol, which also is made acid with nitric acid. Lastly, they are placed in clove-oil when the red stain decreases; and mounted in Canada balsam. The nerve medulla is stained orange or red, the nuclei of the connective tissue violet. Degenerated parts come out very distinctly.

Thicker sections may be left **unstained** and mounted in glycerin, when a very clear general view is obtained, especially with the medulla oblongata; degenerated spots in the spinal cord are clearly distinguished from nerve-fibres. Such preparations mounted in glycerin are best ringed round with paraffin.

Lastly, it must be remarked that in many preparations in which nerve-fibres cross one another, by placing the plane mirror of the microscope in such a position that the **light** traverses the section **obliquely,** some sets of fibres are left dark, while others show up distinctly. This does not succeed with preparations made according to Weigert's method.

Flesch, also, has proposed the use of **coloured light;** where it is a matter of slight difference in colour in different parts of the section this artifice can be sometimes employed successfully.

3. THE INVESTIGATION OF THE CENTRAL NERVOUS SYSTEM IN EMBRYONIC AND PATHOLOGICAL CONDITIONS.

The methods arrange themselves in three groups :—

(*a.*) In the early periods of foetal life all nerve-fibres are destitute of medullary sheaths, so that to the unaided eye the central nervous system appears almost uniformly transparent, and of a red-grey colour. During further development all nerve-fibres are not surrounded simultaneously with medullary sheaths. White patches appear at different periods, owing to the successive acquisition by the nerves of their medullary sheaths, which occurs first in peripheral nerves and later in the central system; the ground-tissue remains grey. *Flechsig* was the first to show that the protection of the nerves with myelin does not take place at hap-hazard, but according to determined laws; hence important conclusions as to the structure and development of the system may be drawn from an observation of this process. By making use of this method of inspection it is possible to pick out and follow definite groups of fibres which later on are lost in the chaos of tracts. It is also possible by this method to distinguish in an apparently uniform nerve-tract constituents which, developing at different periods, must have separate functions. The conclusion, too, which may be safely accepted, that the fibres which first attain to a full development are the first to come into use is of the highest physiological importance. The particular stages at which they acquire their structural completeness must be noted.

Weigert's hæmatoxylin-method with its certainty and simplicity has greatly helped in the investigation of the time of myelination of nerves.

It is sometimes taken for granted that a nerve-fibre, no matter how long (reaching, it may be, from the brain to the lumbar cord), gets its myelin-sheath throughout its whole length at one time. This proposition is not proved, and it is well in using Flechsig's method to bear in

SECONDARY DEGENERATION.

mind the possibility of a slight difference in time in the acquisition of myelin by different parts of the nerve. It is also believed by many that a nerve acquires its myelin in the direction in which it subsequently conducts.

(*b.*) When a nerve is cut through, its peripheral end degenerates quickly. It is similarly the case that if a certain part of the white or grey substance of the cerebro-spinal axis is destroyed, by a tumour or hæmorrhage, for example, individual groups of fibres atrophy. The laws of this SECONDARY DEGENERATION—as this form of atrophy is called—are only partially known. We suppose, without being able to advance irrefutable proof, that every nerve-fibre is nourished by the cell with which it is connected—its trophic centre. If the trophic centre is destroyed, or if the nerve-fibre is severed from it, the nerve necessarily dies. When a nerve-route in the central system is cut through the part severed from its trophic centre degenerates. For most nerve-tracts, we cannot say for all, it is proved that degeneration progresses in the direction in which the impulses are in the habit of travelling along the fibre. *Rokitansky* first, in 1847, pointed out this secondary degeneration, and *Türck* so thoroughly worked at it that we have to thank him for a large part of the anatomical knowledge which we have acquired by this method, at any rate so far as the spinal cord is concerned. All injuries to the whole cross-section of the spinal cord result in the degeneration of certain fibre-tracts upwards from the seat of the disease, while others degenerate downwards. A third set of fibre-tracts remain apparently normal both above and below the injury. The first set of fibres have their nutrient centres below the injury, the second set above, while one ought to allow, with regard to the third set, that they are nourished from both sides. A more exact observation shows, however, that most of these fibres do not remain intact, but a small part of each of them, be it above or below the injury, does degenerate, and since we discover that the fibres of this portion of the cord have their trophic centres hard by the seat of injury, no matter where it occurs, we must conclude that the course of each individual fibre is a short one.

Fig. 1.—Scheme illustrating the nutrition of a nerve-fibre from two sides, after *Schwalbe*.

From this example we see in what manner secondary degeneration can yield information with regard to the course of fibres in the nervous system. We must, however, at the same time clearly be careful in making use of this class of evidence until the conditions of the degenerative process are better known. *Schwalbe* calls attention to possible sources of error. It is not inconceivable that a nerve-fibre which is connected at either end with a nerve-cell [a combination, the existence of which is only hypothetical] is influenced in its nutrition in such a way that the action of A in the direction B, diminishes the action of B in the direction of A. He believes that in the middle at i an indifferent point is to be found, at which the fibres can be cut through without resulting in secondary degeneration. If the fibre is cut through at a, the fibre dies from a to i, where the nutritive influence of B begins; and contrariwise when it is cut through at b. It is to be noticed, however, that such a diminishing nutrient influence is not proved; whereas we do know of cases in which fibres atrophy throughout the whole length of the central nervous system when their trophic centres are destroyed.

[Recent investigations into the histogeny of the nervous system have determined, almost with certainty, that the essential nerve-fibre, the axis-cylinder, is from origin to termination a process of a nerve-cell. The nerve-cells are united with one another by many branching processes. The central system consists of a plexus offering a variety of alternative routes to the afferent impulse. The nodal points of this meshwork are formed by cells which are for the most part small. Wherever a long efferent fibre starts from the plexus, the cell of which it is a process and from which it grows out, is found to present a size proportional to that of the fibre to which it gives origin, and over the nutrition of which it permanently presides. At its distal end, the fibre branches for the purpose of establishing a connection with the region towards which it grows out. No nutriment is received from this dendritic termination backwards along the fibre, but throughout its whole length the fibre depends for its vitality upon its connection with the cell from which it sprang. THE NERVE-FIBRE DIES WHEN CUT OFF FROM THE CELL OF WHICH IT IS A PROCESS. From this account of the growth of nerve-fibres, it will be understood that within the cerebro-spinal axis, motor (descending) fibres degenerate below the level of section; sensory (ascending) fibres degenerate above this level. The portions of the white columns, in the immediate vicinity of the grey matter, consist of fibres which connect together neighbouring regions of the plexus, and it is consequently very difficult to obtain evidence as to the direction in which they die; for the fibres which the lesion destroys are so short, that even in a section taken a little way above or below the lesion, the same situation in the axis is already occupied by other sets of fibres.

When an anterior spinal nerve-root is cut, the fibres below the section die, as shown by *Waller*. The distal portion of the root degenerates completely, as do also all the fibres which the anterior root yields to the mixed nerve. When the posterior root is cut proximally to its ganglion, all its fibres die between the section and the spinal cord; all the fibres between the section and the ganglion live. Until recently it was thought that if the posterior root is cut on the distal side of the ganglion, all the fibres between the section and the ganglion live, while all those which the section severs from the ganglion die. This has been shown by *Max Joseph* not to be quite correct; the greater number of the fibres of the posterior root depend for their nutrition upon the cells of the spinal ganglion, but a certain small proportion of them have their nutritive centres nearer to the periphery. So at least we may infer from the fact that some fibres die right through those portions of the root which are still attached to the ganglion on both its proximal and distal sides. The undegenerated fibres which should be found in the peripheral nerve have not been recognised as yet however. The difference between the fibres of the anterior and posterior roots, as regards their behaviour to section, has only been intelligible since *His*, *Froriep*, and *Beard* have shown us that while anterior roots grow out from the spinal cord, the spinal ganglia are formed from epiblastic thickenings outside and independent of the primitive neural plate, the cells of which give off processes which, growing inwards towards the cord and outwards towards the periphery, constitute the posterior roots and sensory nerves respectively.

If the degeneration-method is to be made use of, it is desirable that the alteration in appearance presented by the dying nerves, and the time of onset of the successive phases in these alterations should be understood.

In this investigation it is of the highest importance to bear in mind that, while the axis-cylinder is the process of a cell in the central plexus, the myelin-sheath by which it is surrounded is formed from many cells which have an independent, although epiblastic, origin.

Section of a nerve results in the death of its axis-cylinder. It seems, however, that as long as it lives the axis-cylinder exercises a restraining influence upon the nutrition of the myelin-cells by which it is surrounded; their tendency to form additional protoplasm, to grow and multiply is maintained at a minimum; their fatty metabolite, upon the existence of which depends their usefulness, is present in maximum quantity. When the axis-cylinder dies the myelin-cells enjoy a sudden exaltation of nutritive activity, their protoplasm accumulates, their fatty metabolite is absorbed; their nuclei increase in size, develop regular active chromatin skeins and initiate cell-division. This condition of sur-activity soon begins to wane, accumulation of protoplasm ceases, the cells shrink and assume a stable form. Finally, the degenerated nerve comes to resemble a cord of connective tissue.

A nerve-fibre has essentially the same structure whether it occurs within or without the axis. It consists of the real fibre, the cell-process or axis-cylinder, invested by myelin-cells, each of which is a hollow cylinder filled with phosphatic fat. While within the axis fibre and myelin-cells are supported by a neurogleia-sheath. When running through mesoblastic tissues, they are invested with a connective-tissue sheath, the sheath of Schwann.

A peripheral nerve completely loses its irritability (in a warm-blooded animal) within forty-eight hours after section. Even by this time a distinct change is visible to the naked eye. Owing to the already-commencing accumulation of protoplasm in the medullary sheath, at first about the nuclei, with coincident absorption of myelin, the fibre looks less solidly white, and glistening. In about twenty days (*Ranvier*) the myelin only remains in drops, which here and there distend the sheath of Schwann.

Within the fresh cord a degenerated area is recognisable by the fifth week after injury as a milky patch. In three or four months it becomes less white, then grey, transparent and gelatinous in appearance (*Sherrington*). Finally, it is indistinguishable until after hardening of the cord. In the cord, hardened in bichromate of potassium, the degenerated area is visible at an earlier stage than in the fresh cord (in the cervical and dorsal regions in nine days—*Sherrington*) as an area lighter in colour and yellower than the surrounding white matter. For about the first six months after injury the distinctness of the degenerated tract increases. After this time it begins to shrink. In sections stained with carmine or acid fuchsin degenerative changes can be recognised in about a week (in the posterior columns in three days, in the lateral pyramidal tract in five, in the direct cerebellar tract in seven—*Homén*). At first the axis-cylinder appears thicker than normal and granular, and stains less darkly with carmine and more darkly than normal with acid fuchsin. The myelin-sheath begins to stain, especially at its inner part, more distinctly with carmine and less distinctly with acid fuchsin or Weigert's hæmatoxylin. Absorption of fat and proliferation of the myelin and neurogleia cells then ensues, in the same manner as already described for peripheral nerves. In a carmine-stained section the degenerated area is recognisable for a long period with a low power as a dark-red patch, although its power of staining gradually diminishes.*]

* For further details see Langley, "Critical Digest," *Brain*, April, 1886.

It is self-evident that the same effects, which result from disease, will follow injury produced by the experimenter's knife; while in the latter case, it is possible to limit the injury to a single well-defined bundle.

This method of artificially-produced secondary degeneration was first used by *Waller* for the purpose of following fibre-tracts.

Information obtained by experiment upon animals can only be applied with qualifications to the human brain.

(c.) Essentially different to the methods we have just been describing is the plan introduced by *Gudden*, which also has yielded important results. There are points of similarity certainly between this method and the method of secondary degeneration; but in Gudden's method lesions are produced only on new-born animals (rabbits, puppies, and kittens). In such subjects the nervous system is still in a partly-embryonic condition, and we have, therefore, arrested development as well as secondary degeneration combined in the results of the injury.

The still-growing cell-groups stand in quite a different relation to the destroyed conducting paths to that in which they would stand to a fully-developed functional organ which had already obtained a certain stability of structure.

An advantage not to be overlooked in this method of Gudden is the facility with which operative interference can be undertaken. The animals are easily handled. The readier coagulation of blood causes the bleeding to stop soon, even when large vessels are cut. The wounds heal quickly without supuration; a few catgut sutures, which are absorbed, alone are necessary. The slight covering of hair to the new-born animal is even a help. The animal is given back to the parent's care after the operation. It may be allowed to live for six or eight weeks, the longer the better, as a rule, and then it is killed and its central nervous system examined. We may expect to deduce the most far reaching conclusions as to the structure of the nervous system from the results of this method.

The laws, according to which the results of these experimental injuries to new-born animals are obtained, are not sufficiently well established to enable us to use the method without precautions.

It cannot be certainly said how far the degeneration (an unsatisfactory term in this case) proceeds. From numerous examples it is determined that when a peripheral nerve is destroyed in a young animal, the groups of nerve-cells from which it takes its origin do not assume their proper form, but whether other routes springing from these cells on the opposite side suffer the same fate is an open question. Whether the degeneration is pathological or artificially set up, it is in all cases observed in the central system that it stops short at the nerve-cells in which the destroyed route terminates.

4. THE COMPARATIVE METHOD.

Since we can take for granted that the functional importance of an organ keeps pace with its anatomical prominence, we may expect many important disclosures from the comparative method.

First, we must study the system in the lower animals, in the hope that in them it will present a simpler organisation, and one, therefore, easier to understand them in man. Further, one may take into consideration the fact that certain functions, and the organs to which they belong, are not equally developed throughout the animal kingdom; the sense of smell is as deficient in man, for instance, as the sense of sight in the mole. The central organs associated with the senses above-named must show a corresponding feebleness. *Edinger* has successfully combined the comparative and historical methods by examining the embryos of lower vertebrates.

Animals which have strongly developed hind-legs (jumpers) can be compared with animals with large fore-legs (diggers), and also with others, such as the whale, in which the limbs are rudimentary. *Meynert* first attributed to this kind of observation its proper merit. There are many obvious differences in the relative importance of the several parts of the central nervous system in different animals, which we cannot yet bring into accord with their functional peculiarities.

5. PHYSIOLOGICAL METHOD.

All the methods by which injury is intentionally inflicted on animals might be included under this head; they have, however, been grouped with the injuries of disease. In the methods now under consideration we have to do with excitation or paralysis of muscles, resulting respectively from irritation and ablation of the brain or cord. For example, when a certain spot on the surface of the brain is stimulated, movement of a defined muscle-group results; or a particular region being excised, a sense organ is thrown out of gear. We may conclude that the part of the brain injured was in each case related to the organ over which it has lost its influence. No method, however, needs to be used with greater care, or leads more easily to mistakes.

The stimulation and the destruction of a part of the brain is apt to call into evidence, not the part especially implicated, but some neighbouring region sympathetically set in action; or, again, unless the stimulus is appropriate it is apt to be ineffective. This is not, however, the place to point out all the mistakes which have occurred; we must content ourselves with showing that the value of physiological experiments must only be estimated in careful association with anatomical data.

Perhaps the condition of electrotonus may be used for anatomical purposes.

The same results which we are able to produce in the central nervous system experimentally follow injury and disease; by the action of tumors, hæmorrhages, inflammation, &c., localised irritation and paralysis are induced, and the phenomena which result show the connection of the several parts of the system to one another. Even more caution is necessary in interpreting these results than in the case of injuries induced experimentally.

By the aid of the methods above-mentioned, we have been able recently to throw much light upon the structure, hitherto but little understood, of the highest organs of the body. New methods are still wanted, however, to enable us to unravel the tangle of conducting paths. It is due to *Türck, Gerlach, Stilling, Meynert, Flechsig, Gudden, Weigert* to point out the advances in the subject which the introduction of the methods which bear their names have produced.

So important is the differentiation which staining affords that one would expect useful results from staining the tissues of the living animal. *Ehrlich* has taken the first step by injecting methyl-blue into the circulation. It stains the endings of centripetal nerves and (for some minutes only) the ends of some centrifugal fibres. After-treatment with iodide of potassium (*Pal*) or biniodide of potassium (*Smirnow*) renders the preparations more durable Any method which would so fix the staining *intra vitam* as to make it available for study after death would be especially valuable. The fixing methods already mentioned make some progress towards this goal.

None of the methods at present in use will stand a strict criticism, and the same want may be anticipated in the case of methods devised in future. If it is recognised that no single method is sufficient in itself, but that all must be used in conjunction, a thorough exploration of the nerve paths may be finally counted on.

SECTION II.—MORPHOLOGY OF THE CENTRAL NERVOUS SYSTEM.

IN fresh preparations the difference in colour between the grey and white substances is sufficiently marked for anatomical purposes.

Previous hardening in alcohol gives to the soft nerve-mass a consistence which facilitates its handling, and, at the same time, brings out certain minute details in configuration. *Lenhossek* recommends that the specimens required for demonstrations should be furnished with a coating of celloidin. When this has been done they may be left exposed to the air for two hours or so without injury. They must be put away in spirit. The great disadvantage of alcohol-preparations is the loss of the natural colour. Artificial staining may be used to remedy this defect. No method is quite satisfactory, but if a section of an alcohol-hardened-brain is put in caustic potash the grey substance becomes much darker again. Soaking in such aniline colours as fuchsin or methyl-violet and subsequent washing does not afford durable preparations, but the staining is distinct at first.

[The following method of bringing out the grey and white matter in strong contrast is sometimes useful for class demonstrations:—Slices of spirit-brains, complete coronal sections for example, are placed in a watery solution of tannin or gallic acid; they are then washed for about an hour in running water, after which they are immersed in a solution of ferrous sulphate for a few minutes, and again washed. The large quantities of proteids in the grey matter fix the gallic acid, which subsequently combines with the iron. Very striking differentiation is thus obtained, but the black staining is seldom restricted to the grey matter; it is apt to sink into such bundles of fibres as are transversely cut.]

Brains hardened in bichromate of potassium, and subsequently, after the chrome-salt is washed out in water, placed in alcohol, are useful for macroscopic work. The chrome-salts effect a staining which varies with the direction and thickness of the nerve-fibres. During the process of hardening the brain, all pressure or traction upon the tissue must be prevented by careful support with cotton-wool and by the use of vessels of suitable form.

The differences in colour are made most visible if the preparations, after a month's hardening in Müller's fluid, are placed in alcohol, to which 1 per cent. of hydrochloric acid is added. Pieces so treated preserve for a long time in glycerin their characteristic staining (*Ageno* and *Beisso*). The various discomforts attending the use of wet preparations, such as the smell, the fumes of alcohol, the wetting of the fingers in handling them, have led to the introduction of dry methods, which are useful at least for the study of external form.

The following amongst these methods may be mentioned :—

The brain is hardened in alcohol or bichromate of potassium and alcohol, or it is put fresh into an almost concentrated solution of chloride of zinc, in which it remains until it sinks; after which it is placed in alcohol, changed several times, and left usually for about a fortnight. From alcohol, the brain is put in glycerin, where it remains until saturated; this takes from a fortnight to a month according to the size of the preparation. The brain is then lifted out of the glycerin and placed where it can drain and dry in the air (*Giacomini*). When dry it can be varnished. Different regions on its surface can then be painted.

Whole human brains are best hardened in chloride of zinc, as the centre is apt to become rotten if they are placed whole in bichromate of potassium. In the case of such thick pieces of tissue as the human brain-stem the central portions, in the pons especially, are apt to soften. The elastic feeling of the preparation shows when this is the case.

Schwalbe's method of preparing dry brains is to be recommended. They are hardened in zinc-chloride or spirit (if in zinc-chloride they must be thoroughly washed), dehydrated in strong alcohol (96-97 per cent.), placed in turpentine for about eight days, then in melted paraffin. The best paraffin for the purpose melts at from 45° to 50° C. They lie in the paraffin in an incubator for from five to eight days, by which time they are thoroughly soaked. The paraffin is allowed to drain off, and the preparation to cool in the position which best prevents distortion.

In the following account the terms "outer, inner, above, below, anterior, posterior" will only be employed in cases in which the use of the terms has become so universal as to be unavoidable without ambiguity; for instance, the "anterior" and "posterior nerve-roots."

The brain being looked upon as the centre we shall speak of proceeding towards it as proceeding brainwards or proximally, and away from it as travelling caudalwards or distally. The terms dorsal, ventral, lateral, and mesial (nearer to the middle line), and median (in the middle line) need no explanation.

DEVELOPMENT.

[The portion of the epiblast which is marked out as the seat of origin of the central nervous system constitutes the floor and sides of the medullary groove. It is not simply a uniform plate, but the central portion, out of which the spinal cord will be formed, is distinct from a row of thickenings which lies on either side of the large main fossa. It is these lateral thickenings of the epiblast which develop into the spinal ganglia (*His*).

The medullary folds grow up until meeting in the mid-dorsal line, they convert the medullary groove into a canal. Closure occurs in the neck-region first, and spreads rapidly forwards over the head and more

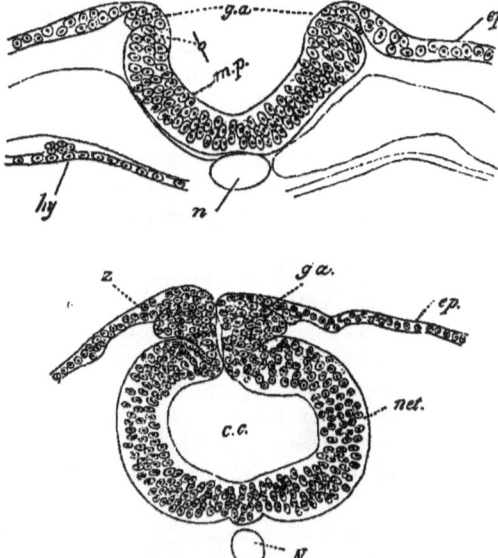

Figs. 2 and 3.—Transverse sections through a developing chick, showing the formation of the sensory root-ganglia (after *Beard*).—*ga*, Ganglia; *ep*, epiblast; *hy*, hypoblast; *m.p*, medullary plate; *n*, notochord; *net*, neuro-epithelial tube; *c.c*, central canal.

slowly backwards through the dorsal, lumbar, and sacral regions. The rudiments of the sensory ganglia (both cranial and spinal) are formed by delamination from the lateral thickenings just described (*Beard*).

When the medullary plates meet in the mid-dorsal line these thickenings are left outside the neural canal, or rather, to define their situation more accurately, they are just caught within the approaching lips of the medullary plate and rest upon the dorsal surface of the neural tube. Afterwards they sink down into their permanent positions on either side of the cerebro-spinal axis.

The whole of the central nervous system is formed from epiblast, and as the rest of this layer becomes the skin and sense organs, the portion of it which is set aside for the nervous system may be distinguished as neuro-epithelium. The layer of neuro-epithelium is at first only one cell thick, later on it becomes many layered, owing to proliferation of its cells; but it is particularly noticeable in sections which are stained, so as to bring out the chromatin figures of the nuclei, that these exhibit the changes which usher in cell division only in the cells which lie next to the central canal of the spinal cord, and in the case of the brain near its surface (cortex) as well as next to the ventricles.

In his recent researches, *His* has shown that even at the time when the neuro-epithelium constitutes but a single layer, the cells compos-

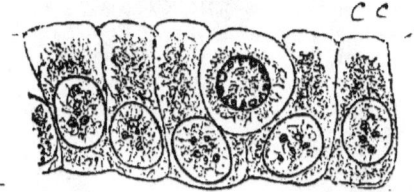

Fig. 4.—The epiblast involuted to form the central nervous system while still a single layer, rabbit (after *His*). A round germ-cell lies between the proximal ends of two supporting cells.

ing it are distinguishable into two classes. Although they all belong to the same layer, they exhibit from the first a distinction into the more important cells, "germ-cells," which develop into nerve-cells and the "spongioblasts" or supporting cells. The spongioblasts are elongated, palisade-like cells, the oval nuclei of which lie at some distance form the central canal. The germ-cells are round protoplasmic cells which lie amongst the inner non-nucleated segments of the spongioblast (figs. 4, 5, 5a).

As the spongioblasts become more elongated, their nuclei sheer over one another, but the supporting tissue of the system is formed by cells which reach throughout its whole thickness. The substance of these cells is not homogeneous, but consists of a formed part disposed in filaments and a soft transparent substance in which the filaments are embedded. The fibrillar elements are disposed so as to form a membrane, the "internal limiting" membrane which supports the epithelium of the central canal, and they also constitute the scaffolding for both grey and white matter. Throughout the grey matter the spongioblasts are disposed radially with tangential or oblique connecting bands. When they reach the outside of the grey matter, they break up into a close irregular plexus, the "border veil," as *His* has called it, an expression which may be Latinised as "velum confine," because

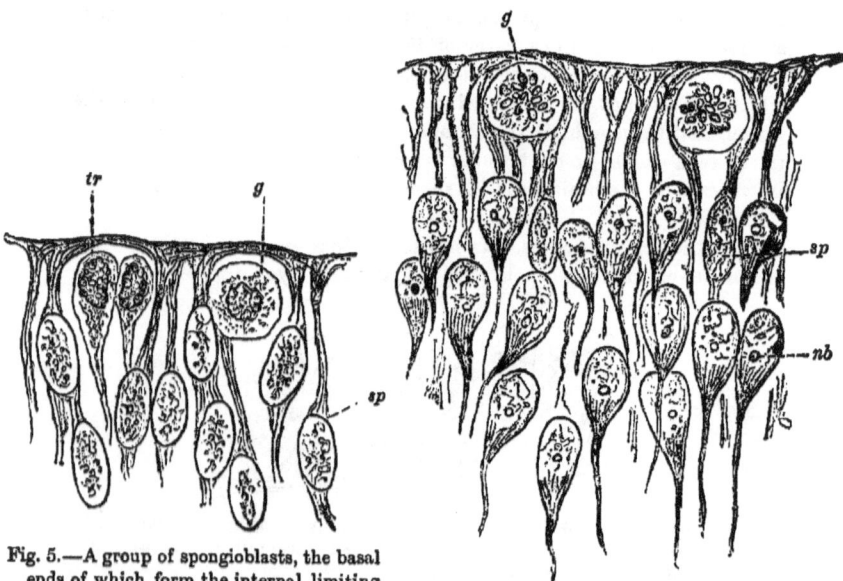

Fig. 5.—A group of spongioblasts, the basal ends of which form the internal limiting membrane (after *His*).—*g*, Germ-cell; *tr*, cells transitional between germ-cells and neuroblasts; *sp*, spongioblasts.

Fig. 5*a*.—Similar to 5, but somewhat more advanced. The neuroblasts (*nb*) are giving off axis-cylinder processes.

it prevents the migration farther outwards of the neuroblasts while it gives passage to thin processes. It is along this close outer network that the longitudinally-running fibres are directed, and it consequently becomes the scaffolding of the white matter also.

The nerve-cells, on the other hand, are formed by the division of the germ-cells, the daughter-cells of which, becoming pyriform, are termed "neuroblasts." For some considerable time the neuroblasts have one process only, the axis-cylinder process, which is directed outwards towards the anterior roots of the nerves. Subsequently the neuroblasts develop lateral dendritic processes.

In the earlier stages of development the cerebro-spinal axis is altogether occupied in forming motor-cells and fibres. Its neuroblasts take no part in the formation of sensory nerves. The fibres of the posterior root arise as outgrowths of the cells of the ganglia. These cells are at first bipolar; one of their processes extending outwards into the sensory nerve, the other inwards to the cord. Subsequently

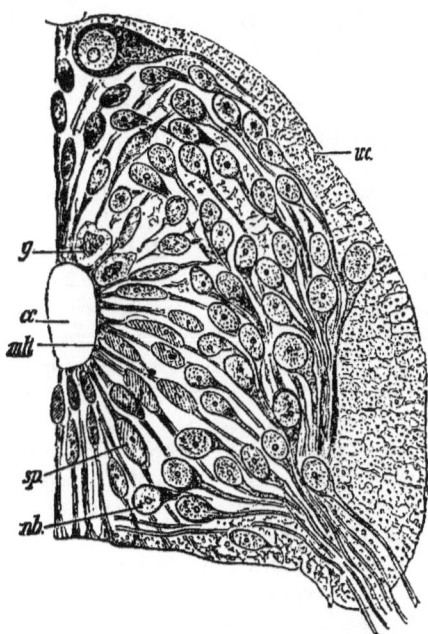

Fig. 6.—Transverse section of half the spinal cord of a trout-embryo (after *His*). —*c.c*, Central canal; *mli*, membrana limitans interna; *g*, germ-cell; *sp*, spongioblast; *nb*, neuroblast; *wc*, white columns, of which the supporting tissue is a close net-work formed from the ramifications of the outer segments of the spongioblasts.

the centripetal and centrifugal processes are placed in direct continuity by the cell body withdrawing itself away from them at right angles. Thus while the cells continue to provide for their nutrition through the vertical limb of the T, the passage of the afferent impulses through its body is rendered unnecessary.

The account of the early stages in the histogeny of the cells given by *His* still leaves some points unsettled, as, for instance, the origin of such nerve-cells as have no axis-cylinder processes, or of the neurogleia-cells of Deiters, unless these are all derived from spongioblasts. A similar obscurity enshrouds the origin of the myelin-cells that invest the axis-cylinder processes of the neuroblasts.

The origin of the neurogleia-cells has been much debated. By some they have been considered as immigrant-cells, white blood corpuscles, or cells of connective-tissue rank, but *Vignal* concludes from his observations that they, like the nerve-cells, are epithelial derivatives.

In the spinal cord, motor nerve-cells are recognisable as such, at the tenth week of intra-uterine life. They make their appearance in two principal groups corresponding to the anterior and lateral horns. As already remarked, the fibres of the posterior root are outgrowths of cells which lie in the sensory ganglia outside the cerebro-spinal axis.

Although dilated at its anterior end into the brain, the greymatter which borders the central canal is formed in the same manner throughout its whole length. Beneath the lining epithelium lies the layer of cells, which, as shown by the form and prominence of the chromatin-skeins in their nuclei, are undergoing active proliferation into nerve-cells. From the nerve-cells fibres extend outwards as motor nerves. It seems pretty certain that these fibres (axis-cylinders) extend without a break from the nerve-cells in the axis in which they take origin to the striated muscle which they innervate; the fibres destined for involuntary muscle, on the other hand, reach, in the first instance, to cells of sympathetic ganglia only, and are by means of these broken up into a number of finer filaments. Inside the cerebro-spinal axis, as well as throughout their peripheral course, the fibres are supported by myelin-cells, which, wrapping themselves round the axis-cylinder, develop in their substance a special phosphatic fat, and in this way constitute protecting and insulating tubes. Whether the myelin-cells arise from epiblastic or mesoblastic elements has not as yet been made out.

While within the central nervous system the axis-cylinders and their myelin-sheaths are supported by the neurogleia, and it is almost necessary to point out that the origin of the myelin-sheath of intraaxial fibres is far from clear, for, according to *Boveri*, it is not broken up into segments. *Schiefferdecker*, on the other hand, asserts that "nodes of Ranvier" are to be found within as well as without the cerebro-spinal axis.

The white columns appear later than the grey matter. Their origin has not as yet been satisfactorily made out. Probably the motor fibres grow downwards from cells of the cortex, while sensory fibres originate in cells of the cord and grow upwards.

The anterior end of the involuted tube of epiblast is dilated into the cerebral vesicles (fig. 6). From the anterior vesicle the cerebrum grows out as a secondary fore-brain. At first this new vesicle is single, but it soon divides into two, each communicating with the primary fore-brain by an aperture, the foramen of Monro. The formation of grey matter within the secondary vesicles is confined to the posterior and inferior parts of the wall of the ventricle. Here, however, it occurs extensively as the corpus striatum, which bulges into the ventricle, and is divided into two parts by a deep groove. The optic thalamus and corpus striatum are, therefore, widely separate in their origin, the former being a local development of the grey matter bordering the primary fore-brain, while the latter is formed in the wall of the secondary vesicle.

But little remains to be said with regard to the further development of the central grey matter—the formation of neuroblasts from the cells of its inner layer—their migration outwards—the radiation of their processes towards the periphery—the investment of the grey matter with a sheath of longitudinal fibres occur throughout its whole length. In the case of the brain, on the other hand, the seat of the chief formative activity is transferred to the surface of the vesicles where the neuroblasts, which are afterwards to become the cells of the cortex of the cerebrum, corpora quadrigemina and cerebellum, commence their existence.

To follow all the changes in the walls of the cerebral vesicles by which the formation of the brain is accomplished would require a special treatise on embryology, nor can the task be attempted at present owing to the incompleteness of our knowledge, but certain points which have the most important influence upon our conception of the fundamental structure of the central nervous system may be briefly referred to.

The cranial nerves are formed in the same way as the spinal nerves, their motor fibres being outgrowths from cells of the central grey matter, the sensory fibres originating in cells of the sensory ganglia, and growing inwards into the central grey matter and outwards to the surface. To this description the olfactory and optic nerves do not seem, at first sight, to correspond, for both these cranial nerves, or rather the "tracts," by means of which their connection with the brain is established, appear as hollow outgrowths from the brain; the optic vesicles being diverticula of the fore-brain; the olfactory vesicles being, apparently, diverticula of the hemispheres of the secondary fore-brain. In the case of the olfactory tracts, an earlier stage, in which they were connected with the dorsal wall of the primary fore-brain in the same manner as other sensory nerves, has been described by *Marshall*. The relation of the first two cranial nerves to the central system is, however, complicated by the fact that elements, which elsewhere lie in the sensory ganglia and cerebro-spinal axis, viz., bipolar cells, "gelatinous" substance, and multipolar cells, are situate in the olfactory bulb and retina in immediate juxtaposition with the epithelium of the sense organs. There appears to be this great phylogenetic difference between the nose and eye and other sense organs that, owing to their situation at the anterior end of the body and consequent advantages for obtaining information, they very early became highly specialised, and the portion of nerve plexus which lay beneath them was so intimately united to them that its subsequent withdrawal into the axial system, as in the case of other segmental sense organs, was impracticable.

The roofs of the anterior cerebral vesicle and the back part of the posterior cerebral vesicle (fig. 7, Zh and Nh) remain undeveloped. The ventricular epithelium is simply supported by pia mater, in which ramify the vessels of the choroid plexus. The roof of the anterior vesicle is a flat plate, triangular in shape, the "velum interpositum." The hemispheres of the cerebrum project backwards, and press up against the sides of the anterior vesicle. Where they touch the margins of the velum interpositum this membrane grows out sideways, pushing the wall of the secondary fore-brain in front of it, and so involuting it into its ventricle, where it rests as a free fold upon the corpus striatum. In this way a choroid plexus is carried into the lateral ventricles, the velum interpositum being grasped by the wall of the cerebral hemisphere from the back of the foramen of Monro to the uncus. By the subsequent downward growth of the hemisphere, as well as the increase in thickness of the crus cerebri, the posteroexternal angle of the velum interpositum is carried downwards and forwards round the crus. If the finger were placed beneath the wing of a butterfly to represent the crus cerebri and the back of the wing were then bent downwards round the finger, the form of the velum interpositum would be accurately represented. The margin of the curved slit, through which it pushes its way into the lateral ventricle, is thickened by longitudinal fibres. The bundle on the convexity of the slit is the fornix. The bundle in its concavity is the stria cornea.

The pineal gland arises as a hollow outgrowth from the back of the roof of the fore-brain; the pituitary body as a hollow downgrowth from its floor which applies itself to the back of a diverticulum from the buccal cavity, which comes up through the hole in the floor of the pituitary fossa of the sphenoid bone.

The corpus callosum is, in its full development, a secondary growth which breaks through the wall of the hemisphere, sweeping away the greater part of the arcuate convolution of Arnold. Its anterior

CEREBRAL VESICLES.

end is first formed and growth proceeds from before backwards. The remains of the arcuate convolution are seen in the striæ medullares *seu* obtectæ, or nerves of Lancisi, and in the subcallosal convolution. The portion of the wall of the hemisphere, which is intercepted between the corpus callosum and the fornix, remains undeveloped as the septum pellucidum or wall of the so-called fifth ventricle.*]

DIVISIONS OF THE CENTRAL NERVOUS SYSTEM.

From the earliest time the system has been described as the elongated **spinal cord** (medulla spinalis), and the more massive and globular **brain** (cerebrum in its wider sense or encephalon). Anatomically the brain and spinal cord are not sharply divided, and it has, therefore, been necessary to call the contents of the spinal canal the spinal cord, and to consider it as separated from the brain by a plane parallel to the anterior surface of the atlas.

The brain is again subdivided into great brain (cerebrum), small brain (cerebellum), and medulla oblongata. Usually the medulla oblongata is regarded as the segment between the proximal end of the spinal cord and the pons varolii, the latter being looked upon as belonging to the cerebellum. Some people (*e.g.*, *Merkel*) include the pons with the medulla oblongata. The part in front of the pons belongs to the great brain.

The division most widely accepted nowadays is based upon developmental study. In the embryo, the nervous system forms a tube closed anteriorly. This tube is swollen out into several smooth vesicles which divide it into, at first, three, and subsequently four portions, lying one behind the other.

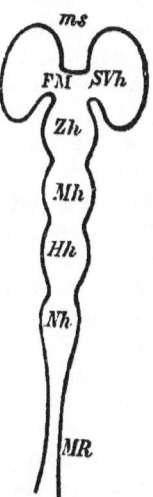

Fig. 7.—The cerebral vesicles.— SVh, Secondary fore-brain; Zh, 'tween-brain; Mh, mid-brain; Hh, hind-brain; Nh, after-brain; ms, longitudinal fissure; FM, foramen of Monro; MR, central canal.

The divisions are named from before backwards, the anterior, middle, and posterior cerebral vesicles. From the anterior vesicle grows out the secondary anterior vesicle. At first this is single, but soon it

---

* For a more detailed account of the development of the brain see *Quain's Anatomy*, Ninth Edition, vol. ii., pp. 818–850.

is divided by the downgrowth of the falx into the two cerebral hemispheres.

The several portions of the brain are developed from one or other of these five divisions.

1. Secondary anterior cerebral vesicle (Svh) forms the cerebrum with its cortex, corpus callosum, fornix and anterior commissure, nucleus lenticularis, and nucleus caudatus.

2. Primary cerebral vesicle (Zh), or 'tween-brain, includes the optic thalami, infundibulum, optic commissure, and corpora albicantia.

3. Middle cerebral vesicle (Mh) forms the mid-brain; corpora quadrigemina and peduncles of great brain.

4. The anterior of the two hinder vesicles (Hh) forms the hind-brain; cerebellum with its peduncles and the pons.

5. Posterior of the two hinder vesicles (Nh) forms the after-brain or medulla oblongata.

All the structures developed from the secondary anterior vesicles belong to the cortex or brain-mantle, while the structures to which the remaining four vesicles give rise constitute, with the exception of the cerebellum, the brain-stem (caudex). The nuclei lenticularis et caudatus are included, for the most part, in the description of the brain-stem, leaving only the cortex of the great brain for the mantle-formation; recently, however, it has been shown that the nucleus caudatus, as well as the lateral part [and perhaps the central part too] of the nucleus lenticularis, ought to be included amongst cortex-formations.

[The morphological relations of the cortex to the rest of the cerebro-spinal axis, is a matter of the greatest importance, involving, as it does, the question of the primary constitution of the nervous system. The brain is distinguished from the spinal cord by the possession of an envelope, not complete, but covering the greater part of the surface of its three vesicles, as the cortex or mantle-formation. Throughout the whole system the "axis" consists of a tube of grey matter bordering the central canal, invested by a sheath of longitudinally running white fibres. *Meynert* recognised the continuity of the grey matter bordering the central canal, and termed it "centrales Höhlengrau," without, however, giving to the term any strict morphological or anatomical limitations. From the translator's point of view, all the grey matter of the lower system, including therein the optic thalami, constitutes the central grey tube. This receives sensory, and gives origin to motor nerves, none of which appear to pass through its plexus without joining with it, but each nerve terminates in, or springs from, the portion of the grey matter belonging to the metamer which the nerve supplies. In its fore part the

CORPUS STRIATUM.

nervous system is dilated into the vesicles already described, and here a marked difference in anatomical arrangement is seen. The white matter which invests the central grey tube is in turn surrounded by a layer of grey matter of altogether different constitution. This second investment constitutes the cortex of the cerebellum, corpora quadrigemina, and cerebral hemispheres. In the lowest vertebrates the cerebellum is, as a rule, small; although, even amongst fishes, it may attain, as in the shark for instance, a considerable size and complex structure. The corpora bigemina are larger proportionally in lower vertebrates than they are in man. The cortex of the cerebrum is, however, a late development; only in mammals does it constitute an important layer. In reptiles and birds the mass of the cerebral hemisphere consists of what in mammals we know as corpus striatum. Does the corpus striatum belong to the central or the peripheral grey tube? It does not, like the optic thalamus, contain the primary centres of any sensory nerves. The nucleus caudatus is intimately connected with the cortex, for its head, as shown in the accompanying diagram after *Wernicke*, rests on the anterior perforated space, its tail is continuous with the cortex of the temporal lobe.

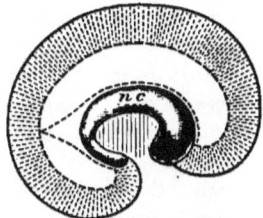

Fig. 8.—Diagram representing the connection of the nucleus caudatus with the cortex of the frontal and temporo-sphenoidal lobes.

Nucleus caudatus, nucleus amygdaleus, and claustrum are continuous at their temporal extremities. The nucleus lenticularis is sunk more deeply beneath the cortex, and completely invested in white matter. In structure and development, however, the lenticular and caudate nuclei strongly resemble one another, and all the evidence we at present possess is in favour of assigning them both to the same position in the system, as parts of the peripheral grey tube.

The optic thalamus, on the other hand, is to be regarded as the anterior extremity of the central grey tube.]

The entrance into the lateral vesicles from the fore-brain constitutes eventually the foramen of Monro; the cavities of the brain vesicles remain as the permanent **ventricles.**

The cavity of the secondary fore-brain becomes the lateral ventricle.
„ „ primary fore-brain   „   the third ventricle.
„ „ mid-brain   „   the aqueduct of Sylvius.
„ „ hind-brain   „   the fourth ventricle.
„ „ medullary canal   „   the central canal.

The several divisions of the central nervous system indicated in the account just given will be briefly passed in review in this section.

A. THE SPINAL CORD.

The human spinal cord (figs. 9 and 10) forms a cylindrical column of 40 to 50 cm. (18 inches) in length, reaching in the upright position from the first cervical to the first or second lumbar vertebra. In the child it reaches farther [the third lumbar at birth]. In the fœtus farther still. When the body is bent sharply forwards the lower end of the cord in the adult only reaches to the twelfth dorsal vertebra. *Heger* determined that when the body is strongly bent the spinal cord is stretched by 6·8 per cent. of its length.

The thickness of the cord varies but little in different individuals; treating its cross-section as a complete circle, one finds that its diameter above the cervical swelling varies from 8 to 11 mm.; in the dorsal cord, between the two swellings, from 6 to 9 mm.

In two situations it presents spindle-shaped swellings, due almost entirely to an increase in its transverse diameter. The first of these, the cervical swelling, has at the level of the fifth or sixth cervical vertebra a breadth of 15 mm., the lumbar enlargement lying at the level of the lower dorsal vertebræ attains a breadth of 11 to 12 mm. The antero-posterior diameter increases in these situations by 1 to 2 mm. only.

The lumbar swelling (intumescentia lumbalis) terminates directly in the conus medullaris. The latter forms the end of the cord. To it, however, is attached a thin thread of some 25 cm. in length, the filum terminale.

Flesch has shown that the cord is so constituted as to present—after the removal of the pressure of the vertebræ—the curves which are seen in it when *in situ*.

In the middle line of both dorsal and ventral surfaces it presents a fissure, the posterior and anterior longitudinal fissures. The former is distinct at the surface only, the latter shallow but broad. The dorsal roots originate in an almost uninterrupted line 2 to 3 mm. laterally to the posterior fissure. When they have been cut away, the place of their origin is still indicated by a furrow, the dorso-

SPINAL CORD. 39

lateral groove. It is shallow for the most part but somewhat deeply cut into the cord in the cervical region. The anterior roots arise in

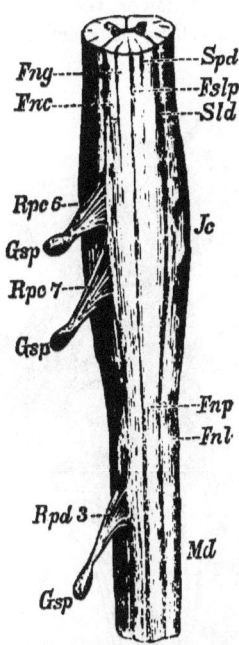

Fig. 9.—Caudal end of spinal cord from ventral surface (*natural size*).—*Jl*, Lumbar enlargement; *Cm*, conus medullaris; *Ft*, filum terminale. The anterior nerve-roots on the left side are removed, on the right side (*Ra*) they enter into the formation of the cauda equina (*Ce*). *Fsla*, fissura longitudinalis ant.; *Slv*, sulcus lateralis ventralis; *Fna*, funiculus anterior; *Fl*, funiculus lateralis.

Fig. 10.—Cervical enlargement of the spinal cord from the dorsal side (*natural size*). Besides the cervical enlargement (*Jc*) a portion of the dorsal cord is also visible (*Md*). All the posterior roots are cut away on the right side, and on the left side the sixth and seventh cervical (*Rpc* 6 and 7) and the third dorsal (*Rpd* 3) are left as far as the spinal ganglia (*Gsp*). [The division of the posterior roots into fasciculi, as shown in the picture, is hardly true to nature.]—*Fslp*, Fissura longitudinalis post.; *Spd*, sulcus paramedianus dorsalis; *Sld*, sulcus lateralis dorsalis; *Fnp*, funiculus posterior; *Fnl*, funiculus lateralis; *Fng*, funiculus gracilis; *Fnc*, funiculus cuneatus.

many separate bundles spread out transversely as well as longitudinally. The furrow left after their removal (the so-called antero-

40 SPINAL NERVE-ROOTS.

lateral groove) is very indistinct. The roots incline on leaving the cord caudally as well as laterally, the inclination downwards being sharper the lower their situation on the cord. By the time the lumbar swelling is reached the roots lie almost parallel with the

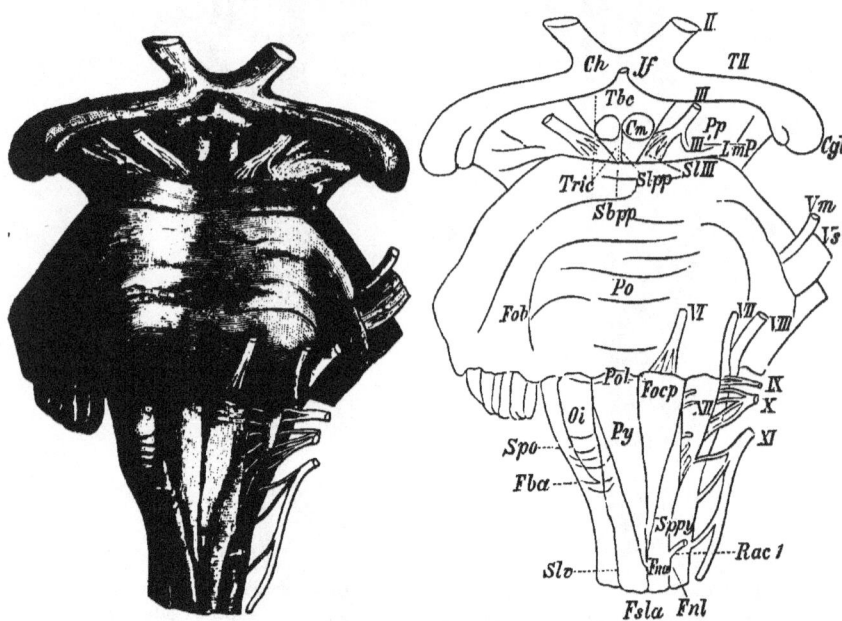

Fig. 11.—The base of the brain, as far as the optic tracts. The cerebellum is almost completely removed; the secondary fore-brain and all structures which lie in front of the optic tracts are cut away; on the left side the nerve-roots are retained; on the right side they are, with few exceptions, removed. *II*, Nervus opticus; *III*, nervus oculomotorius; *III'*, lateral accessory portion of oculomotorius; *V*, nervus trigeminus; *VI*, nervus abducens; *VII*, nervus facialis; *VIII*, nervus acusticus; *IX*, nervus glossopharyngeus; *X*, nervus vagus; *XI*, nervus accessorius Willisii; *XII*, nervus hypoglossus; *Cgl*, corpus geniculatum laterale; *Ch*, chiasma nervorum opticorum; *Cm*, corpus mammilare; *Fna*, funiculus anterior; *Fnl*, funiculus lateralis; *Fob*, fasciculus obliquus pontis; *Focp*, foramen cæcum posterius; *Fsla*, fissura longitudinalis anterior medullæ; *Jf*, infundibulum; *LmP*, tract connecting lemniscus and pedunculus cerebri; *Oi*, inferior olive; *Pp*, pes pedunculi cerebri; *Po*, pons; *Py*, pyramid; *Rac1*, anterior root of the first cervical nerve; *Sbpp*, substantia perforata posterior; *SlIII*, sulcus oculomotorii; *Slpp*, sulcus substantiæ perf. post.; *Slv*, sulcus lateralis ventralis; *Spo*, sulcus postolivaris; *Sppy*, sulcus parapyramidalis; *TII*, tractus nervi optici; *Tbc*, tuber cinereum; *Tric*, trigonum intercrurale; *Vm*, motor trigeminal root; *Vs*, sensory ditto.

cord; the conus medullaris and filum terminale lie in the middle of a considerable bundle of nerves, the whole constituting the cauda equina.

On account of the obliquity of the roots one can tell in any detached piece of cord which is its proximal and which its distal end. This, again, is of great use in helping us to distinguish the left side from the right in cases of unilateral lesion.

In the cervical cord a still more distinct furrow is to be seen about 1 mm. laterally to the posterior longitudinal fissure, becoming more distinct as we travel cerebralwards, the sulcus paramedianus dorsalis *seu* intermedius posterior, *Spd*.

The cord is divided by these furrows into several longitudinal columns distinct from one another on the surface.

1. Anterior column, *Fna*, lying between the anterior fissure and the line of exit of the anterior roots.

2. Lateral column, *Fnl*, on the outer side of the anterior column, between this and the postero-lateral sulcus.

3. Posterior column, *Fnp*, between the posterior longitudinal and dorso-lateral sulci. In those regions in which a paramedian sulcus is visible, the posterior column is again divided into two, Burdach's column, *Fnc* (funiculus cuneatus); and Goll's column (funiculus gracilis), *Fng*.

Usually 31 pairs of spinal nerves are reckoned, *viz.*, 8 cervical, 12 dorsal, 5 lumbar, 5 sacral, and 1 pair of coxygeal nerves. One or two microscopical coxygeal nerves may be usually found, however, in the filum terminale (*Rauber*).

The separate groups of anterior roots with the muscles which they innervate constitute not so much anatomical as physiological associations. *Ferrier* and *Yeo* have shown that stimulating a particular root in the ape's cord gives rise to a definite complex action corresponding to the habits of the animal. For example, stimulation of the first dorsal root, results in such a movement as in picking a fruit; the eighth cervical, the scalptor ani action; seventh cervical, a movement as if the body were drawn up on a branch clasped with the hands; sixth cervical, the hand is moved to the mouth.

B. THE BRAIN.

1. THE AFTER-BRAIN.

The cross-section of the system increases very rapidly in its transverse diameter in front of the first cervical nerve. The spinal cord forms into the **medulla oblongata**. This reaches to the back of

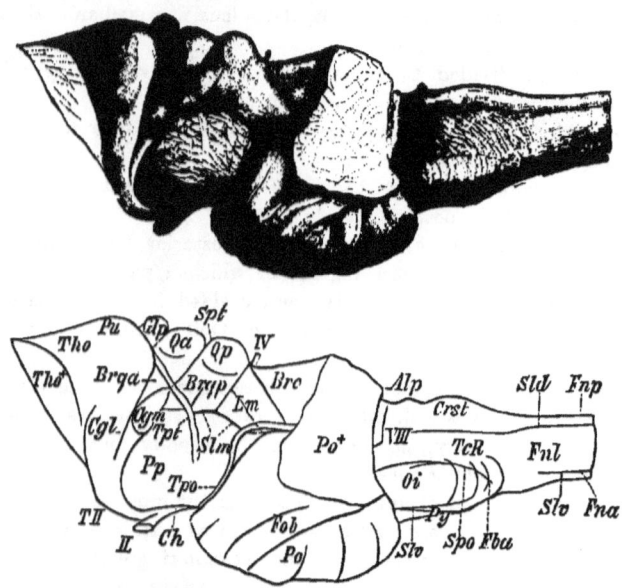

Fig. 12.—A preparation similar to that in fig. 11, seen from the left side. *Natural size.* Nerve roots are for the most part cut away.—*II*, Nervus opticus; *IV*, trochlearis; *VIII*, acusticus; *Alp*, ala pontis; *Brc*, brachium conjunctivum (superior cerebellar peduncle); *Brqa*, anterior brachium; *Brqp*, posterior brachium; *Cgm*, corpus geniculatum mediale; *Cgl*, corpus geniculatum laterale; *Ch*, chiasma nervorum opticorum; *Crst*, corpus restiforme; *Fba*, fibræ arcuatæ; *Fna*, funiculus anterior; *Fnl*, funiculus lateralis; *Fnp*, funiculus posterior; *Fob*, funiculus obliquus; *Glp*, glandula pinealis; *Lm*, lemniscus or fillet; *Oi*, inferior olive; *Po*, pons, cut across at *Po*+; *Pp*, pes pedunculi; *Pu*, pulvinar thalami; *Py*, pyramid; *Qa*, corpus quadrigeminum anterior; *Qp*, corpus quadrigeminum posterior; *Sld*, sulcus lateralis dorsalis; *Slm*, sulcus lateralis mesencephali; *Slv*, sulcus lateralis ventralis; *Spo*, sulcus postolivaris; *Spt*, sulcus corp. quad. transversus; *T II*, tractus opticus; *TcR*, tuberculum cinereum Rolandi; *Tpo*, tœnia pontis; *Tpt*, tractus peduncularis transversus; *Tho*, thalamus opticus, cut at *Tho*+.

the great cross-fibres of the pons. It attains a length of about 3 cm. On the surface of the medulla several details in moulding are to be noticed. We will describe the furrows first. For the most part the furrows are longitudinal, and continue upwards the sulci of the cervical cord.

The anterior fissure, *Fla* (fig. 11), extends on the ventral surface as far as the back of the pons; it is very shallow in the distal part, but deepens in front, and ends at last, where the pons fibres cross, in a blind hole, foramen cæcum posterius, *Focp*.

A rather shallow fissure forms an acute angle with the anterior fissure at the hinder end of the medulla, and extends forwards as far as the border of the pons, sulcus parapyramidalis, *Sppy*. The furrow corresponding to the anterior roots, hardly to be seen in the cord, is more distinct in some parts of the medulla oblongata, sulcus lateralis ventralis (*seu* internus olivæ), *Slv*. Here and there, however, it is obliterated by crossing fibres.

The following sulci belong to the dorsal surface (figs. 12 and 13):— 1, Sulcus lateralis dorsalis, *Sld;* 2, sulcus paramedianus dorsalis, *Spd;* 3, in the middle line, posterior longitudinal or dorsal fissure, *Fslp*. The first two incline laterally up the medulla, the sulcus lateralis can be followed to the pons, the sulcus paramedianus soon disappears. The fissura long. dors. ends suddenly where the central canal opens out into the fourth ventricle, and the posterior columns diverge to the two sides.

In the proximal part of the medulla appears a sharply marked fissure more than 1 cm. long, the sulcus postolivaris. It extends from the margin of the pons to the sulcus lat. vent. which it joins. In the rest of its extent, it lies between this fissure and the sulcus lat. dors.

The swellings which these furrows throw into relief are not equally prominent in all brains. The anterior columns of the cord are pushed aside by the pointed extremities of the pyramids (fig. 11), *Py*. Passing up beneath the pyramids they disappear from the surface. A very prominent swelling, the inferior olive (eminentia olivaris), 6 to 7 mm. broad by 12 to 14 mm. long lies between the sulci ventralis lateralis et postolivaris.

A bundle of fibres, fibræ arciformes, *Fba*, can invariably be seen arching round the hinder end of the olive and, to a certain extent, spread over it. It is not, as a rule, raised much above the surface. Especially in children's brains, a little rounded eminence, tuberculum cinereum Rolandi, *TcR*, is to be seen laterally to the olive near its distal end. The funiculi siliquæ are minute longitudinal columns occasionally described on the mesial or lateral side of the olive.

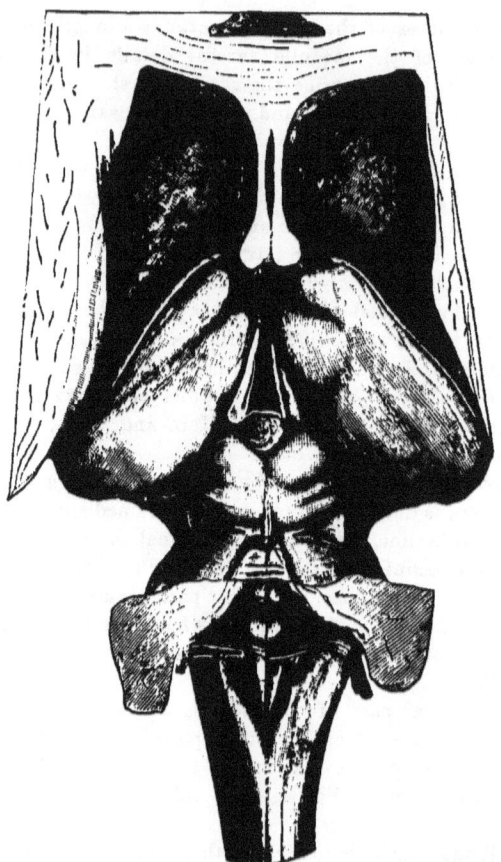

Fig. 13.—Hind-brain, mid-brain, and 'tween-brain from the dorsal surface (*natural size*). The greater part of the secondary fore-brain is removed by four sections, one horizontal, one frontal, and two sagittal. Most of the nerve-roots are cut away.—*IV*, N. trochlearis; *VII*, nervus facialis; *VIII*, nervus acusticus; *Ac*, ala cinerea; *Brc*, brachium cerebelli ad corp. quad. cut at *Brc+*; *Brqa*, brachium corp. quad. anterior; *Brqp*, brachium corp. quad. posterior; *Cgm*, corpus geniculatum mediale; *Cl*, clava; *Coa*, commissura anterior; *Com*, commissura mollis; *Crst*, corpus restiforme; *Cscr*, calamus scriptorius; *Et*, eminentia teres; *Fcl*, columnæ fornicis; *Fnc*, funiculus cuneatus; *Fng*, funiculus gracilis; *Fnl*, funiculus lateralis; *Foa*, fovea anterior; *Frv*, frenulum veli anterioris; *Fslp*, fissura longitud. posterior; *Gcc*, genu corporis callosi; *Gh*, ganglion habenulæ; *Glp*, glandula pinealis; *K*, conductor sonorus *vel* tractus auditorius; *Lc*, locus cœruleus; *Lg*, lingula; *Lm*, lemniscus or fillet; *M*, situation of foramen of Monro; *Nc*, nucleus caudatus; *Pdc*, pedunculus conarii; *Po*, pons at *Po+* cut across; *Pp*, pedunculus cerebri;

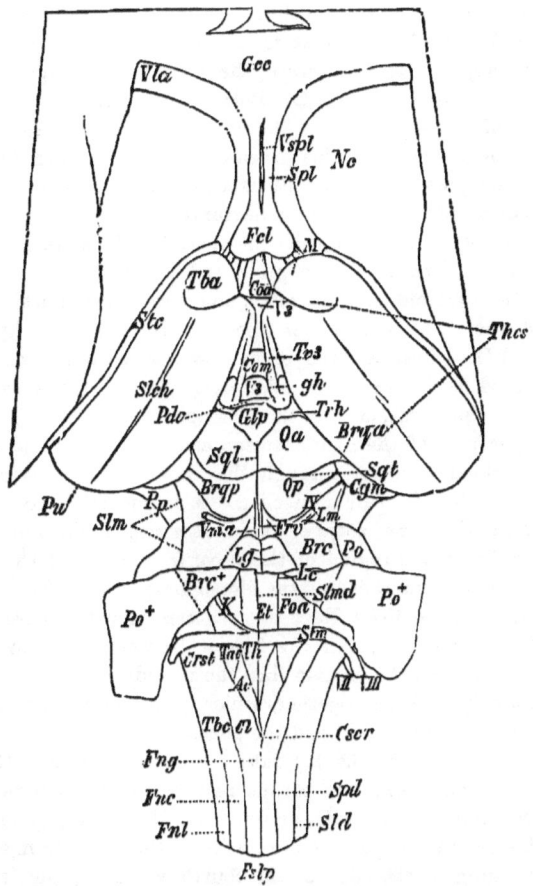

Pu, pulvinar; *Qa*, corp. quad. anteriora; *Qp*, corp. quad. posteriora; *Slch*, sulcus choroideus; *Sld*, sulcus lateralis dorsalis; *Slm*, sulcus longitudinalis mesencephali; *Slmd*, sulcus longitudinalis medianus ventriculi quarti; *Spd*, sulcus paramedianus dors.; *Spl*, septum pellucidum; *Sql*, sulcus corp. quadrig. longitudinalis; *Sqt*, sulcus corp. quadr. transversus; *Stc*, stria cornea; *Stm*, striæ medullares acusticæ; *Tac*, trigonum acustici; *Tba*, tuberculum anterius; *Tbc*, tuberculum cuneatum; *Th*, trigonum hypoglossi; *Thos*, thalamus opticus; *Trh*, trigonum habenulæ; *Tv3*, tænia ventriculi tertii; *Vla*, anterior horn of lateral ventricle; *Vma*, velum medullare anterius; *Vspl*, ventriculus septi pellucidi.

The part of the medulla which lies between the sulcus lat. dors. and the fourth ventricle, is named the restiform body (corpus restiforme, inferior peduncle of the cerebellum, brachium cerebelli ad medullam oblongatam), *Crst*. Looked at from the surface merely, the corpus restiforme appears as if it were the continuation upwards of the posterior column of the cord. Both constituents of the posterior column swell out somewhat in the region of the calamus scriptorius. The swelling of the funiculus gracilis, which is known here as the clava, *Cl* (or posterior pyramid), is more distinct than that of the funiculus cuneatus (tuberculum cuneatum), *Tbc*.

A number of nerves arise from the medulla. The origin of the first cervical nerve, *Rac. 1*, is also prolonged upwards into this region. Between the pyramid and the olive, and extending almost the whole length of the latter, come out the root-fibres of the hypoglossus (fig. 11), *XII*. Between the olive and the corpus restiforme, an uninterrupted series of roots take exit to join the n. accessorius Willisii, *XI*, the vagus, *X*, and the glossopharyngeal, *IX*.

The larger part of the n. accessorius arises in the spinal cord; the root-fibres which pass out through the lateral column extend as far downwards as the fifth pair of nerves. It is impossible to distinguish with certainty the upper roots of the n. accessorius which are attached to the medulla in the region of the olive from those of the vagus, or the vagus fibres from those of the glossopharyngeal. All one can do is to assign the more distal fibres to the accessorius, the more proximal to the glossopharyngeal. In the furrow between the pons and the pyramid, 2 mm. from the median line, the n. abducens arises in several bundles, quickly uniting together, which pass out in the transverse furrow between the pyramid and the pons.

Bundles of fibres, *Stm* (fig. 13), take origin in the floor of the fourth ventricle, and, encircling the corpus restiforme, just before it sinks into the cerebellum, join with another bundle, *VIII* (fig. 11), which comes out from the corpus restiforme itself, to form the n. acusticus. A little swelling at the edge of the fourth ventricle, the tæniola or fasciola cinerea, corresponds to one of the centres from which the eighth nerve takes origin (its accessory nucleus). Mesially to the auditory nerve and rather close to its ventral side the facial nerve takes exit in a strong bundle.

2. THE HIND-BRAIN.

The pons, *Po*, comprises an immense tract of crossing fibres, measuring about 3 cm. from before backwards, and 4 cm. from side to side. At either side the pons constitutes a more rounded column, the middle cerebellar peduncle (brachium cerebelli ad pontem), fig. 13, *Po +*, which passes dorsally into the cerebellum. This closes the

CEREBELLUM. 47

ring through which the columns of the hind-brain must pass on their road forwards.

The **cerebellum**, looked at from above (dorsally), presents a deep notch, incisura marsupialis, *Im*. On its ventral or under margin is a shallower, broader notch, incisura semilunaris, *Isl*. The former contains the process of the dura mater known as the falx cerebelli, while the latter is filled up with a portion of the mid-brain. On the dorsal surface a ridge extends from the one notch to the other, from which (as from the roof-tree of a house) the two surfaces of the cerebellum slope away. On either side of this ridge a shallow groove, sulcus longitudinalis superior cerebelli, *Slsp*, marks off the superior vermis, *Vrsp*, from the lateral lobes.

The dorsal surface of the cerebellum is completely covered in cortex-substance.

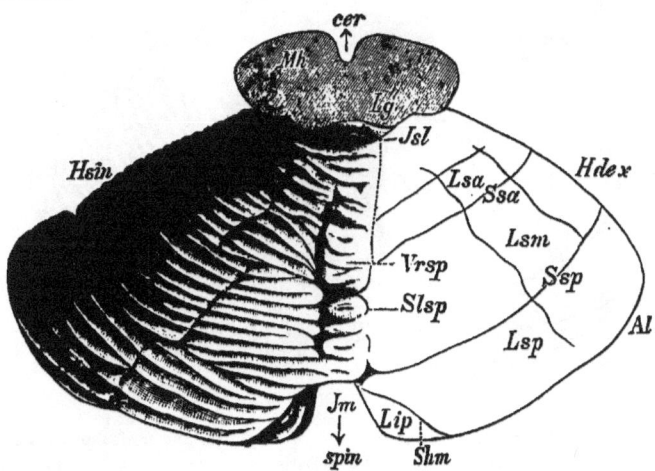

Fig. 14.—Cerebellum, dorsal view. *Natural size.* The mid-brain is cut across behind the corpora quadrigemina at *Mh*.—*Al*, Lateral angle of hemisphere; *H*, hemispheres of cerebellum, *dextera* et *sinistra*; *Jm*, incisura marsupialis; *Jsl*, incisura semilunaris; *Lg*, lingula; *Lsa*, lobus superior anterior; *Lsm*, lobus superior medius; *Lsp*, lobus superior posterior; *Shm*, sulcus horizontalis magnus; *Ssa*, sulcus superior anterior; *Ssp*, sulcus superior posterior; *Vrsp*, vermis superior; ↑ *cer*, pointing brainwards; and ↓ *spin*, spinewards.

The ventral side of the cerebellum can only be seen by cutting through the massive columns which unite it with the rest of the hind-brain. On this side the median part or vermis inferior, *Vrif*, is sharply cut off by deep furrows, sulci longitudinales inferiores, from the lateral hemisphere. The great lateral hemispheres arch over and hide the vermis inferior, shutting it up in the vallecula. It can only be brought into view by pressing the hemispheres aside.

The anterior part of the vermis inferior does not reach into the incisura semilunaris. A layer of white substance extends brainwards in front of it, the velum medullare anterius or roof of the front part of the fourth ventricle; on this is borne a recurved part of the superior vermis.

It follows from this that the superior vermis is much longer than the inferior. The ventral surface of the cerebellum is not entirely covered with grey matter.

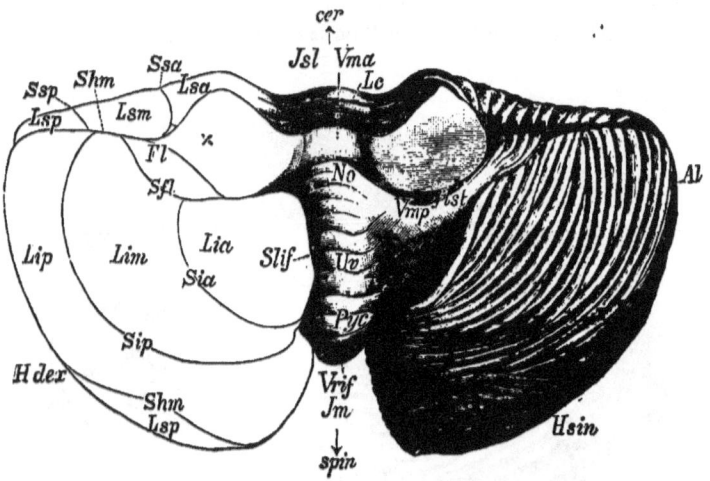

Fig. 15.—Cerebellum from the ventral side. *Nat. size.* The cerebellar peduncles are cut at ×. The anterior lamina tectoria is cut away from its attachment to the mid-brain. The lobus anterior inferior is broken away from the left hemisphere.—*Al*, Angulus lateralis; *Fl*, flocculus; *Flst*, pedunculus flocculi; *Hdex*, right hemisphere; *Hsin*, left hemisphere; *Jm*, incisura marsupialis; *Jsl*, incisura semilunaris; *Lc*, lobulus centralis; *Lia*, lobus inferior ant.; *Lim*, lobus inf. med.; *Lip*, lobus inf. post.; *Lsa*, lobus sup. ant.; *Lsm*, lobus sup. med.; *Lsp*, lobus sup. post.; *No*, nodulus; *Pyc*, pyramid; *Sfl*, sulcus flocculi; *Shm*, sulcus horizontalis magnus; *Sia*, sulcus inf. ant.; *Sip*, sulcus inf. post.; *Slif*, sulcus longitudinalis inf.; *Ssa*, sulcus sup. ant.; *Ssp*, sulcus sup. post.; *Uv*, uvula; *Vma*, velum medullare anterius; *Vmp*, velum medullare posterius; *Vrif*, vermis inferior.

The surface of the cerebellum is broken up in a characteristic way by a large number of **furrows.** They are not, as a superficial examination would lead one to think, of anything like equal depth. If the cerebellum is cut at right angles to the direction which these furrows assume, it will be seen that some of them extend so much deeper than others as to make it possible to divide the organ into lobes (fig. 16).

There is no uniformity in the classification of these lobes and their nomenclature. The lobes are divided by secondary furrows into lobules, which again bear convolutions. The greatest fissure is the sulcus horizontalis magnus, *Shm*, which divides the cerebellum into an upper [in lower animals anterior] and lower [or posterior] part. This deepest and most constant of the fissures of the cerebellum originates at the middle peduncle, and extends around the cerebellum almost parallel to its border. At first it lies a little on the under surface, it then extends over the border, and for a short distance before its termination belongs to the upper surface (fig. 14).

The fissures on the upper surface are thickly set. They form arches parallel to the hinder border of the cerebellum and the incisura semilunaris, the centre of their curvature being situate in the region of the corpora quadrigemina. Two of these are fairly constant and important; they divide the upper surface into three divisions. They are named sulci cerebelli superiores anterior, *Ssa*, et posterior, *Ssp*. The anterior fissure (fig. 15) commences on the middle peduncle and crosses the vermis to join the fissure on the opposite side. The upper surface of the vermis is divided by it into two almost equal segments (fig. 14). The posterior fissure commences in the great horizontal sulcus, a little in front of the postero-external angle of the cerebellum. Crossing the upper surface it almost reaches the horizontal sulcus again at the spot where the latter passes on to the vermis, without, however, actually joining it.

The fissures on the under surface of the cerebellum present the same want of regularity. Two principal fissures are, however, again to be recognised, sulci cerebelli inferiores, anterior et posterior, *Sia* and *Sip*. Another short but deep fissure leaves the great horizontal fissure in a gentle curve directed backwards, and ends in the groove between the cerebellum and the medulla, *Sfl*. The anterior inferior sulcus does not commence as does the posterior inferior in the great horizontal sulcus, but in the floccular sulcus.

The disposition of the sulci on the vermis is best seen in a median sagittal section (fig. 16). Both its superior and inferior surfaces are crossed by three chief fissures which are too short to need separate names. As a rule, no distinct furrow separates the upper from the under vermis; the continuation across the middle line of the great horizontal fissure may be used for this purpose.

The above-named fissures divide the surface of the cerebellum into

4

LOBES OF CEREBELLUM.

lobes and lobules. On the under surface at any rate they are not sufficiently constant to allow of a uniform nomenclature.

The hemispheres are divided into—on the upper surface :—

Anterior **lobe**, *Lsa*, *seu* lunatus anterior,
Middle „ *Lsm*, „ „ posterior, } Lobus quadrangulus.
Posterior „ *Lsp*, „ semilunaris superior.

On the under surface :—

Anterior **lobe**, *Lia*, *seu* amygdala, *seu* tonsil.
Middle „ *Lim*, „ { gracilis.
                      { cuneiformis, *seu* biventer.
Posterior „ *Lip*, „ semilunaris inferior.

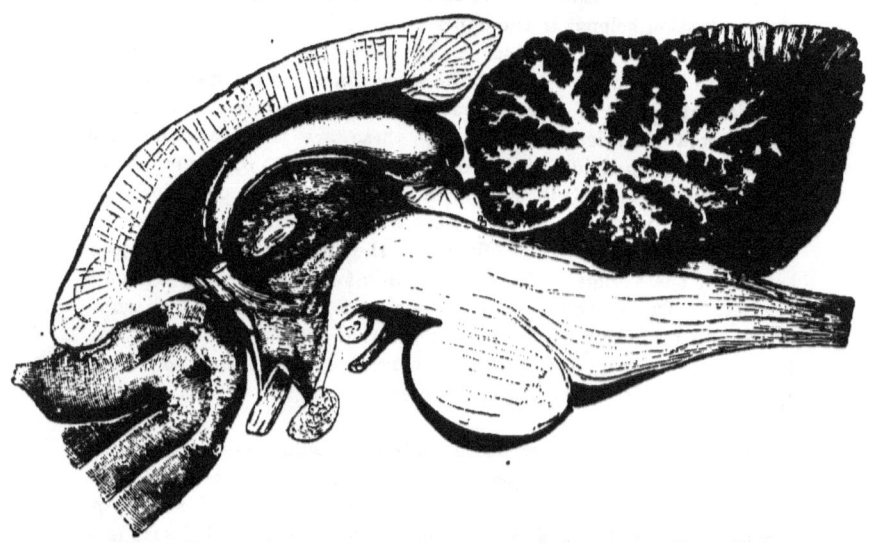

Fig. 16.—Sagittal section through the brain in the median line. *Right half. Nat. size.* Of the cortex of the cerebrum only a part of the frontal region is drawn. —*II*, Nervus opticus ; *III*, nervus occulomotorius ; *AAS*, aditus ad aquæductum Sylvii ; *AS*, aquæductus Sylvii ; *Cc*, canalis centralis ; *Ccll*, corpus callosum ; *Ch*, chiasma nervorum opticorum ; *Cm*, corpus mammillare ; *Coa*, commissura anterior ; *Coba*, commissura baseos alba ; *Com*, commissura mollis ; *Cop*, commissura posterior ; *Cscr*, calamus scriptorius ; *Cu*, culmen ; *Dc*, declive ; *Fcc*, folium cacuminis ; *Fcl*, columna fornicis cut across at ± ; *Fna*, funiculus ant. med. spinalis ; *Fnp*, funiculus post. med. spin. ; *Foca*, foramen cæcum ant. ; *Focp*, foramen cæcum post. ; *Gcc*, genu ; *Gh*, ganglion habenulæ ; *Glp*, glandula pinealis ; *Hy*, hypophysis ; *Lc*, lobulus centralis ; *Lg*, lingula ; *Lia*, lobus inf. ant. ; *Lim*, lobus inf. med. ; *Lip*, lobus inf. post. ; *Lsm*, lobus sup. med. ; *Lsp*, lobus sup. post. ; *Lt*, lamina terminalis ;

FLOCCULUS. 51

The three upper as well as the two posterior lobes on the under surface have a pronounced semilunar form. Only the anterior lobe or tonsil, which pushes itself in towards the middle line, has a more complicated shape. The two tonsils from opposite sides meet in the middle line above the medulla oblongata. The floccular sulcus cuts off a small, but very conspicuous, lobe, the flocculus or lobus vagi, *Fl*, which occupies the commencement of the great horizontal fissure resting on the middle peduncle. Some small accessory lobules, which lie beside the flocculus on the middle peduncles, are known as accessory flocculi.

Proceeding from the front of the vermis backwards along its upper surface, and continuing our course along its under surface (fig. 16), we find :—

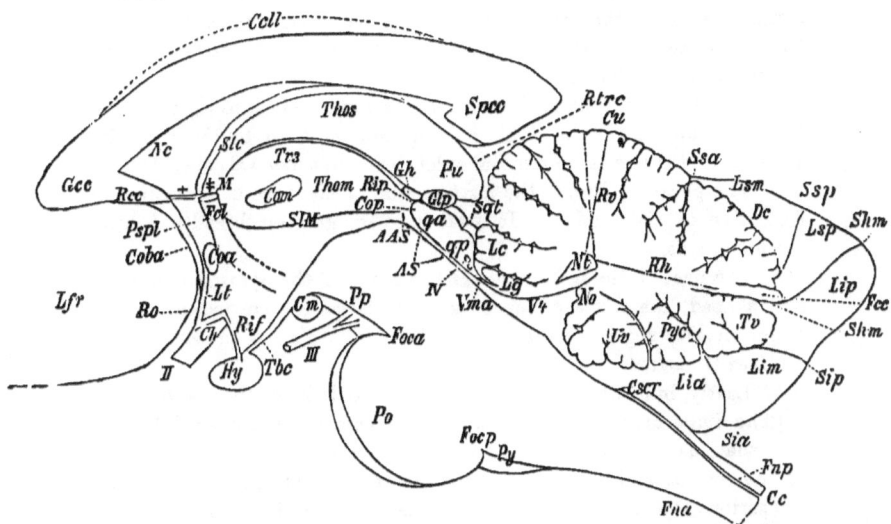

M, situation of foramen of Monro; *Nc*, nucleus caudatus; *No*, nodulus; *Nt*, nucleus tecti; *Po*, pons; *Pp*, pes pedunculi; *Pspl*, pedunculus septi pellucidi cut at + ; *Pu*, pulvinar thalami; *Pyc*, pyramis cerebelli; *Qa*, corpus quadrigeminum anterius; *Qp*, corpus quadrigeminum posterius; *Rcc*, rostrum ; *Rh*, ramus medullaris horizontalis; *Rif*, infundibulum ; *Rip*, recessus infrapinealis; *Ro*, recessus opticus; *Rtrc*, rima transversa cerebri; *Rv*, ramus medullaris verticalis; *Shm*, sulcus horizontalis magnus ; *Sia*, sulcus inf. ant. ; *Sip*, sulcus inf. post. ; *SlM*, sulcus Monroi ; *Spcc*, splenium ; *Sqt*, sulcus corp. quad. transversus; *Ssa*, sulcus sup. ant. ; *Ssp*, sulcus sup. post. ; *Stc*, stria cornea; *Tbc*, tuber cinereum ; *Thom*, thalamus opticus, mesial surface; *Thos*, thalamus opticus, upper surface; *Tv*, tuber valvulæ; *Tv3*, tænia ventriculi tertii; *Uv*, uvula ; *Vma*, velum medullare ant. ; *V4*, fourth ventricle.

1. The lingula, *Lg*, a tiny tongue-shaped lobule, made up of from five to eight minute convolutions, lying on the velum medullare anterius, *Vma*. Sometimes its under surface is free from the velum, in which case it also is marked by transverse convolutions. The lingula extends on either side in a little leaflet, the frenulum lingulæ, which represents an atrophied portion of the lateral hemisphere.

2. The central lobe, *Lc*, projects forwards until it touches the back of the corpora quadrigemina. To this piece of the vermis again belongs an inconspicuous portion of the hemisphere, the ala lobi centralis.

3. The upper lobe of the vermis (or monticulus) comprises by far the largest part of the vermis. It is again divided into two—(*a*) culmen (apex), *Cu*, reaching as far backwards as the union of the anterior superior sulci of the two sides, *Ssa*; (*b*) declive, *Dc*, reaching thence backwards to the posterior superior fissure (*Ssp*), belongs both to the upper and the under vermis.

4. The posterior lobe of the vermis; divided again into (*a*) the little folium cacuminis, *Fcc*, a single convolution bounded by the posterior superior and the great horizontal sulci; (*b*) the tuber valvulæ, *Tv*.

5. The pyramis, *Pyc*, is the next section of the vermis, consisting of from five to eight folia. It attains to its greatest breadth behind the amygdalæ.

6. The part of the inferior vermis in front of the pyramis is narrow and shaped like a steep house-roof. It is called, on account of its situation with regard to the tonsils, the uvula, *Uv*; it presents six to ten free transverse folia.

7. Lastly, in front of the uvula projects a little knob, the nodulus, *No*.

[The importance of these names is somewhat diminished by the fact that, as can be seen in fig. 16, they do not exhaust the lobes of the vermis, and also by a doubt as to their morphological value. No serious attempt has yet been made to trace the lobation of the cerebellum throughout the vertebrata.]

The **medullary centre** of the cerebellum consists of two egg-shaped masses of white substance belonging to the hemispheres, and united together by the medullary substance of the vermis. The white substance is, in the main, a repetition in miniature of the whole cerebellum; but the portion belonging to the vermis is not relatively so large as the rest.

Portions of the medullary centre are prolonged into the lobes and lobules, dividing repeatedly, and occupying the centres of all the folia. A special description of these divisions of the medullary substance is, therefore, unnecessary. Those of the vermis, as they are represented in fig. 16, may be mentioned. The central white substance of the vermis, called corpus trapezoides (a name which has given rise to mistakes), gives off two principal branches. One of these, the vertical branch, Rv, projects upwards into the monticulus. The other, or horizontal branch, Rh, is directed backwards into the central mass of the hinder lobes; quite near its origin this horizontal limb gives a considerable branch downwards into the pyramid. A less important branch passes in front of the vertical ramus into the central lobe, while another is continued in front of the horizontal one into the uvula. A still smaller branch enters the nodule, whilst the most minute of all forms the medullary substance of the lingula. The branches of the vermis taken together constitute, with their cortical covering, the arbor vitæ.

In connection with the cerebellum, although not really belonging to it, for it forms rather the embryonal roof of the fourth ventricle, the velum medullare posterius Tarini (*seu* valvula semilunaris), Vmp, must occupy our attention. To exhibit this structure it is necessary to cut off the medulla oblongata at the level of the hinder border of the pons, and then to break off the tonsils of the cerebellum. Fig. 15 shows the left velum medullare posterius on the right hand side of the picture. Each tonsil is now seen lying with its upper surface in a hemispherical depression, the floor of which is not formed by the substance of the cerebellum but by a delicate transparent membrane, which stretches from the uvula and nodule on either side as a semilunar leaflet attached along its posterior convex border to the cerebellum, but presenting a free concave border directed forwards. It is comparable in appearance to one of the semilunar aortic valves. Laterally the free edge is prolonged into a bundle of nerve-fibres, which can be followed as far as the flocculus, $Flst$; stalk of the flocculus.

Grey matter is also to be found in the medullary centre of the cerebellum. It can be shown by cutting the cerebellum horizontally, the section following the sulcus horizontalis magnus, or by making at right angles to this sulcus a section inclining obliquely outwards and backwards from the incisura semilunaris. In either case the corpus dentatum cerebelli, Ndt, appears as a narrow grey zigzag band.

The corpus dentatum, Ndt, fig. 17 (*seu* nucleus dentatus *seu* fimbriatus *seu* lenticulatus, corpus ciliare *seu* rhomboideum), is a puckered up bag of grey substance, the open mouth of which looks a little mesially and ventrally. It lies in the mesial half of the hemisphere, and so close to the ventricle that it is only separated from it by a thin layer of white substance. Its longest antero-posterior diameter (converging somewhat with that of the opposite side) is about 2 cm. The corpus dentatum is not seen in its greatest extension in a frontal section.

54   NUCLEUS OF ROOF.

Another not well-defined, light-grey or brownish mass, oval in shape, appears between the two corpora dentata in a frontal section. This is *Stilling's* roof-nucleus, *Nt* (nucleus tecti *seu* fastigii, substantia ferruginea superior). Between the corpus dentatum and the nucleus of the roof are some little scattered clumps of grey matter which *Stilling* names the nuclei emboliformis et globosus. *Meynert* calls them both the nuclei subdentati (gezackte nebenkerne). In fig. 17, taken from the brain of the monkey, these nuclei are not visible.

The presence of these nuclei inside the cerebellum depends upon the fact that from three directions large white bundles stream into this organ on either side. One of these bundles, the corpus restiforme, which skirts the margin of the fourth ventricle, has been already noticed, *Crst.*

The middle and largest peduncles of the cerebellum, which unite it with the pons, have also been noticed. They belong altogether to the hind-brain. *Henle* may be followed in considering the line joining the points of exit of the trigeminal and facial nerves (fig. 11) as the

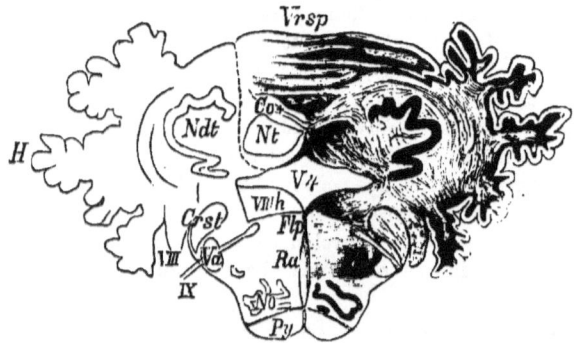

Fig. 17.—Frontal section through the medulla oblongata and cerebellum of an ape. *Twice natural size.*—*VIII*, Nervus acusticus; *VIIIh*, chief auditory nucleus; *IX*, nervus glossopharyngeus; *Co+*, superior commissure (and decussation); *Crst*, corpus restiforme; *Flp*, fasciculus longitudinalis post.; *H*, hemisphere of cerebellum; *Ndt*, nucleus dentatus; *No*, nucleus olivaris; *Nt*, nucleus tecti; *Py*, anterior pyramid of the medulla; *Ra*, raphe; *V4*, ventriculus quartus; *Va*, ascending root of trigeminus; *Vrsp*, vermis superior.

division between the middle peduncle and the pons proper. The greater number of fibres in the pons are directed transversely (*Foville* compares their appearance as seen from the mid-ventral line to a head of hair parted in the middle); a broad band of fibres is, however, conspicuous in the anterior half of the pons, which, starting in the usual direction, subsequently inclines backwards and outwards over the surface of the others towards the point of exit of the facial nerve. It is called the fasciculus obliquus (ruban fibreux oblique of Foville), fig. 11, *Fob*.

A bundle of fibres, the ponticulus, *Pol*, is usually to be seen along the hinder border of the pons, spreading over the pyramid.

The third pair of cerebellar peduncles, which have not yet been mentioned, pass from this organ towards the great brain converging in this direction much in the same way as the posterior peduncles converge spinewards. They and the posterior peduncles together bound a rhomboidal space; sinus rhomboidalis (or fourth ventricle). They look as if they went to the corpora quadrigemina, and hence have been named by mistake the processus cerebelli ad corpora quadrigemina. They are also named the brachia conjunctiva (*seu* conjunctoria, processus cerebelli ad cerebrum); compare figs. 12 and 18, *Brc*.

Between the mesial edges of the anterior peduncles lies a thin tongue-shaped membrane with its point turned brainwards, the velum medullare anterius, *Vma*, already described; on it lies the lingula, *Lg*.

The lateral borders of the anterior peduncles are not really visible, for as they converge forwards to slip under the posterior tubercles of the corpora quadrigemina they are overlapped, even from the moment that they leave the pons, by two other white tracts, the lemnisci, which converge more quickly than the anterior peduncles, and almost reach the middle line in front of the point of the velum medullare. The lemniscus (fillet, laqueus, ruban de Reil), *Lm*, has a triangular form, and is usually divided into two parts by a shallow furrow, which also runs brainwards and mesially. An isolated bundle can almost always be seen lying in the furrow between the anterior border of the pons and the superior cerebellar peduncles; passing on round the cerebral peduncles it sinks at last into the fissure between them. This bundle, the tænia pontis, *Tpo* (fig. 12), can sometimes be lifted up for a considerable distance as a free cord.

It has already been mentioned that the great fifth nerve takes its exit from the pons near its anterior border. The motor-root is attached in front of the larger sensory one, fig. 11, *Vm* and *Vs*.

The floor of the fourth ventricle (sinus *sive* fossa rhomboidalis) is exposed by cutting the cerebellum from all its connections with the rest of the brain. It is longest in its antero-posterior diameter (about 3 cm.). Its greatest breadth (about 2 cm.) is in the line connecting the attachments of the auditory nerves. The margins of the sinus rhomboidalis are formed in front by the superior cerebellar peduncles, and by the corpora restiformia behind.

Of the two diagonals of the sinus, the longitudinal is marked by

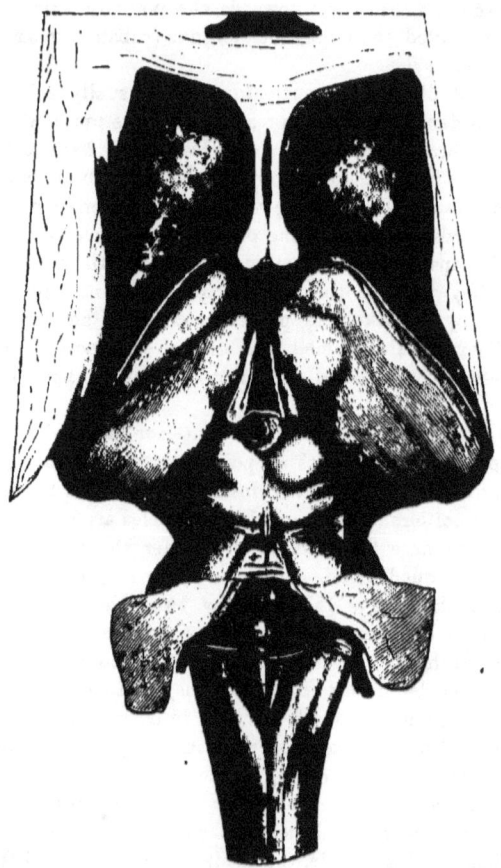

Fig. 18.—Hind-brain, mid-brain, and 'tween-brain from the dorsal side. *Nat. size.* The greater part of the secondary fore-brain is removed by three cuts made in the horizontal, sagittal, and frontal planes respectively.—*IV*, Nervus trochlearis; *VII*, nervus facialis; *VIII*, nervus acusticus; *Ac*, ala cinerea; *Brc*, brachium cerebelli ad corpus quadrigeminum; *Brqa*, brachium anterius; *Brqp*, brachium posterius; *Cgm*, corpus geniculatum mediale; *Cl*, clava; *Coa*, commissura anterior; *Com*, commissura mollis; *Crst*, corpus restiforme; *Cscr*, calamus scriptorius; *Et*, eminentia teres; *Fcl*, columnæ fornicis; *Fnc*, funiculus cuneatus; *Fng*, funiculus gracilis; *Fnl*, funiculus lateralis; *Foa*, fovea anterior; *Frv*, frenulum veli anterioris; *Fslp*, fissura longitudinalis posterior; *Gcc*, genu corporis callosi; *gh*, ganglion habenulæ; *Glp*, glandula pinealis; *K*, conductor sonorus; *Lc*, locus cœruleus; *Lg*, lingula; *Lm*, lemniscus; *M*, situation of foramen of Monro; *Nc*, nucleus caudatus; *Pdc*,

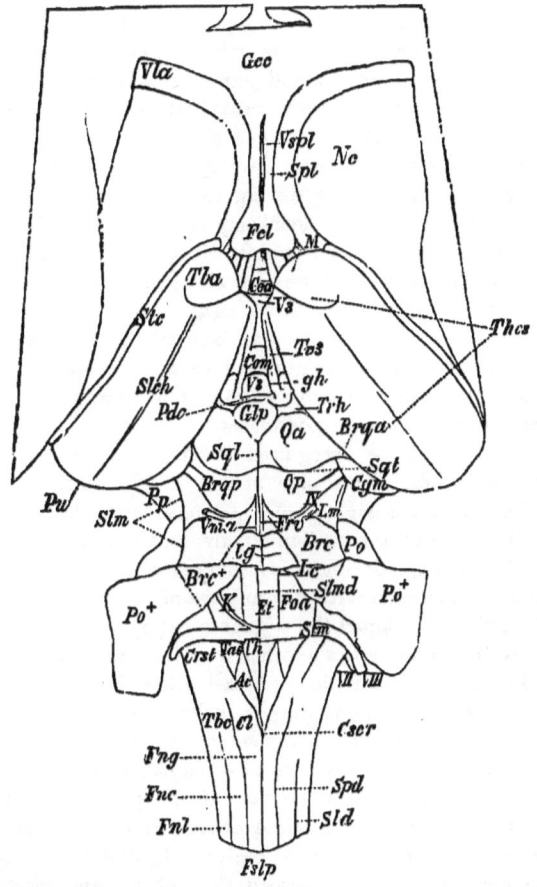

pedunculus conarii; *Po*, pons cut across at +; *Pp*, pes pedunculi cerebri; *Pu*, pulvinar; *Qa*, anterior corpora quadrigemina; *Qp*, posterior corpora quadrigemina; *Slch*, sulcus choroideus; *Sld*, sulcus lateralis dorsalis; *Slm*, sulcus longitudinalis mesencephali; *Slmd*, sulcus longitudinalis medianus ventriculi quarti; *Spd*, sulcus paramedianus; *Spl*, septum pellucidum; *Sql*, sulcus corp. quadrig. longitudinalis; *Sqt*, sulcus corp. quadr. transversus; *Stc*, stria cornea; *Stm*, striæ medullares *seu* acusticæ; *Tac*, trigonum acustici; *Tba*, tuberculum anterius; *Tbc*, tuberculum cuneatum; *Th*, trigonum hypoglossi; *Thos*, thalamus opticus; *Trh*, trigonum habenulæ; *Tv3*, tænia ventriculi tertii; *Vla*, anterior horn of the lateral ventricle; *Vma*, velum medullare anterius; *Vspl*, ventriculus septi pellucidi, *seu* quintus; *V3*, ventriculus tertius.

a conspicuous fissure (sulcus medianus longitudinalis sinus rhomboidalis), *Slmd*, while transverse bundles of fibres (striæ medullares, *seu* acusticæ), *Stm*, starting in the middle line, pass outwards to encircle the corpora restiformia, and join the auditory nerve. These striæ acusticæ are subject to great individual variations. They may in exceptional cases be absent on one side, or even on both sides. Occasionally they are very strongly developed. Sometimes individual bundles cross over one another during their course. Bundles lying quite free, not fused with the floor, may also be met with. Besides the usual tracts which cross the tæniola cinerea to join the auditory nerve, other bundles of white fibres are also generally to be met with in the floor of the fourth ventricles, such a bundle [the conductor sonorus *seu* tractus acusticus] or "Klangstab" of *Bergmann*, fig. 24, *K*, is often to be found originating in the median fissure, near the striæ medullares, and passing obliquely outwards and forwards towards the anterior cerebellar peduncles.

Three divisions can be recognised in the posterior half of the sinus rhomboidalis on either side of the middle line. The most mesial of these forms a right-angled triangle with the right angle bounded on one side by the median fissure, on the other by the striæ acusticæ. This triangular region is covered with white substance; it corresponds in the main with the nucleus of the hypoglossal nerve, and may, therefore, be named the trigonum hypoglossi, *Th*.

Laterally to this triangle lies another with its apex against the striæ acusticæ. Its surface is a little depressed below the rest of the floor of the fourth ventricle and is grey in colour. Since it corresponds very closely to the nuclei of the vagus (and glossopharyngeus) it may be called the trigonum vagi (it is more often named ala cinerea, hence the lettering *Ac*). The lateral portion of the posterior half of the floor of the fourth ventricle is raised above the general surface. It extends beyond the striæ acusticæ, brainwards, and there attains its greatest development. It is termed the tuberculum acusticum [a well-marked swelling in the child's medulla], since it corresponds to a group of nerve-cells which many regard as the nucleus of this nerve.

The proximal [anterior] half of the floor of the fourth ventricle is distinguished from the distal half by having a complete covering, albeit a thin one, of white substance, which gives to its lateral boundary a sharper definition than is exhibited by the distal half. A cylindrical eminence, about 4 mm. broad, lies on either side the middle line. Commencing, as the continuation upwards of the trigonum hypoglossi, it extends to the front of the fourth ventricle beneath the corpora quadrigemina; at their upper part, owing to the drawing together of the superior cerebellar peduncles, these eminences, *Et* eminentiæ teretes (wrongly called funiculi teretes), are somewhat contracted.

Laterally to the eminentia teres a depressed spot is visible, the fovea anterior, *Foa*, distinguished, as a rule, by the presence of a fairly large superficial vein.

Lastly, in the front of the floor, near the lateral angle, is to be noticed a dark-brown or bluish space, stretching forwards for from 4 to 6 mm. as far as the corpora quadrigemina, locus cœruleus, *Lc*. Its colour, which is not always visible until the surface has been scratched, is due to a strongly pigmented group of nerve-cells, substantia ferruginea, which shows through the upper layer of the medulla.

At its proximal extremity the ventricle has a breadth of 3 mm. Beneath the corpora quadrigemina it sinks into the aquæductus Sylvii, *Qp*.

The cerebellum cannot be looked upon as a portion of the roof of the hind-brain. It is a secondary formation which grows later from the two sides and arches over the fourth ventricle. The following structures cover in the ventricle :—

1. The front is covered by the velum medullare anterius.
2. The middle part by the vela medullaria posteriora.
3. The roof of the hinder part of the ventricle is formed by a thin vascular membrane, reduced for the greater part of its extent to a triangular layer of epithelium and pia mater, tela choroidea inferior ventriculi quarti. This is continuous in front with the vela posteriora. It is shown when the back of the cerebellum is lifted up from the medulla oblongata. Some other small and unimportant developments (little plates of white matter) are found in this part of the ventricle, namely, the obex (often absent) which fills in the angle between the diverging funiculi graciles and the tæniæ ventriculi quarti, *Alp* (ligulæ, alæ pontis, ponticulus), fig. 12, which skirt the outer margin of the ventricle as far forwards as the striæ acusticæ. These little plates of white matter are very delicate and easily torn off with the membranous roof of the ventricle with which they are intimately connected; in fig. 13 they are only partly visible. The base of the triangular membrane corresponds to the vermis, and fuses with its pia mater.

[In a section which cuts through the front of the fourth ventricle this group of pigmented cells appears as a round black spot on either side. It is as well to restrict the name substantia ferruginea to this clump of pigmented cells, which, although small, is as dark as the substantia nigra, and to use the term locus cœruleus for the grey-blue appearance which the floor of the fourth ventricle presents over the region beneath which the substantia ferruginea lies.]

A peculiar shaggy plexus of vessels, the plexus choroideus cerebelli medialis, hangs to the under surface of the tela choroidea on either side of the middle line. The depending fringes commence at the calamus scriptorius, and take a sagittal direction as far forwards as the back of the velum medullare posterius. Here they turn outwards, and lying on the under side of the cerebellum, run along the stalks of the flocculi to meet the auditory nerves, where they form a somewhat larger coil, the plexus choroideus cerebelli lateralis (ala, plexus nervi vagi). In the part of the roof of the ventricle, which is thinned out as the tela choroidea, three gaps are formed during the course of development, the only communications, perhaps, between the brain-ventricles and the pericerebral space. Between the two plexus choroidei mediales, and just in front of the calamus scriptorius, a large hole is pierced in the roof, easy to demonstrate although at one time its existence was much doubted, the foramen Magendii (apertura inferior ventriculi quarti, orifice commun des cavités de l'encéphale). There are also always to be found, as *Key* and *Retzius* have proved, two other openings, which lie at the lateral angles (recessus laterales ventriculi quarti) of the tela choroidea, just where the plexus choroideus lateralis comes out, aperturæ laterales ventriculi quarti. According to *Merkel* and *Mierzejewsky*, communications between the lateral ventricles of the great brain and the surface are also to be found in the form of elongated clefts above the gyri hippocampi.

3. THE MID-BRAIN.

Proceeding forwards we reach the mid-brain, or region in which the corpora quadrigemina are included. In connection with this region it will be necessary to describe structures which, although they belong properly speaking to the 'tween-brain, press themselves backwards into the mid-brain region.

The mid-brain is not more than a centimeter in length and is divided by a furrow, the sulcus lateralis mesencephali, *Slm*, (figs. 12, 18, and 19), into two easily distinguishable parts; the ventral (basal) and dorsal portions of the mid-brain.

The sulcus lateralis is seen whether the brain-stem is looked at from above or from the side. It commences at the front of the pons and bounds the structure already described under the name of fillet.

The great **cerebral peduncle** (pes pedunculi cerebri), *Pp* (figs. 11, 12, and 13), lies on the ventral side of this furrow, and projects, laterally, somewhat beyond it. As it comes out from the pons it has a breadth of 12 to 20 mm., which is increased during its short superficial course. Passing beneath the optic tract, TII, it disappears from view in the interior of the great brain. It consists of bundles of fibres visible as separate bundles from the surface, not following its main direction, but giving it the appearance of a twisted cord. Those which are most mesially placed as it leaves the pons, pass so abruptly to the outer side that they have an almost transverse course. These are the fibres, LmP, from the fillet to the peduncle, so called for reasons which will be presently explained. Each peduncle runs not straight forwards, but diverging from its fellow at an angle of 70° to 80°; a triangular space is thus left between the two, trigonum intercrurale, $Tric$ (fossa interpeduncularis [interpeduncular space]).

A deep furrow, from which the fibres of the oculomotor nerve emerge (sulcus oculomotorius), $SlIII$, marks the boundary between the pes pedunculi and the trigonum intercrurale. In the mid-line of the fossa is another well-marked furrow, sulcus substantiæ perforatæ posterioris, $Slpp$. The medial portion of the fossa, broad in front and pointed behind, constitutes the floor of the third ventricle. It is pierced by numerous blood-vessels, and hence is termed the substantia perforata posterior, $Sbpp$. The perforated space is bounded on either side by elongated swellings which really belong to the dorsal portion (the tegment) of the peduncle. They can only be seen when the peduncles are pushed aside, and hence are not visible in fig. 11.

The part of the mid-brain lying dorsally to the lateral sulcus presents two rounded swellings, the corpora quadrigemina (or in submammalian orders, bigemina), Qa and Qp. A median fissure rising abruptly from the velum medullare, and opening out in front into a little shallow triangular fossa, lodges the pineal gland, Glp, and separates the tubercles of the corpora quadrigemina of one side, from those of the other. The triangular fossa, just mentioned (trigonum subpineale), often presents a slight elevation in the centre (colliculus subpinealis of *Schwalbe*). At the back, where it sinks on to the valve of Vieussens, the fissure is bounded on either side by a little ridge of white substance (sometimes the two ridges are fused together), frenulum veli medullaris antici, Frv.

A transverse sulcus, sulcus corp. quad. transversus (*seu* frontalis), Sqt, crosses the median fissure at right angles, dividing the anterior pair of tubercles from the posterior. It is shallowest near the middle line.

The anterior tubercles, Qa, measure in the sagittal direction 8 mm., in the frontal direction 12 mm. The posterior tubercles, Qp, measure 6 mm. by 8 mm. The latter are distinguished by the abruptness of their posterior surfaces.

MID-BRAIN.

From each of the corpora quadrigemina a white bundle passes ventrally, laterally, and brainwards. These are the peduncles of the corpora quadrigemina (brachia conjunctiva). On each side the two brachia are separated by a continuation of the transverse furrow, which might well be called in this part of its course the sulcus interbrachialis.

The posterior brachium is soon divided into two by a shallow furrow. The posterior of these two divisions disappears in the sulcus lateralis. The anterior joins a spindle-shaped elevation of about 1 cm. in length, the internal geniculate body, Cgm (ganglion *seu* corpus geniculatum mediale *seu* internum) which is squeezed into the sulcus interbrachialis. It must be looked upon as a part of the 'tween-brain.

The anterior brachium, $Brqa$, continues its course, covered by the overhanging optic thalamus, almost to the optic tract. It is broadest where it comes out from the anterior tubercle, but loses a considerable portion of its substance beneath the lateral geniculate body.

A thin nerve tract, which is very visible in many animals but rarely distinctly seen in man, proceeds from in front of the anterior tubercles downwards, outwards, and backwards across the brachia, and then on across the pedunculus cerebri, tractus peduncularis transversus, fig. 12, Tpt. Its termination is never distinctly seen.

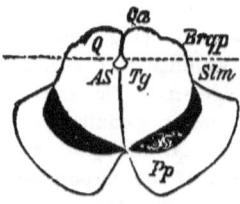

Fig. 19.—Transverse section through the anterior corpora quadrigemina (semi-diagrammatic).—AS, Aquæductus Sylvii; $Brqp$, brachium posterius corporum quadr.; Pp, pes pedunculi; Q, region of the corpora quadrigemina; Qa, anterior corpora quadrigemina; Slm, sulcus lateralis mesencephali; SS, substantia nigra Soemmeringi; Tg, tegmentum.

When the mid-brain is cut across at right angles to its long axis (fig. 19) the proximal continuation of the fourth ventricle, aquæductus Sylvii, is to be seen in the middle line. In the substance of this part of the brain-stem several strata are distinguishable :—

1. The region of the corpora quadrigemina is limited by a line drawn transversely through the aqueduct, Q.

2. A region of mixed grey and white substance, the tegment, Tg.

3. A stratum which is at once distinguished by its colour, due to the intensely black pigmentation of the cells of which it is composed, stratum nigrum, *Ss* (substantia nigra Soemmeringi, stratum intermedium).

4. Lastly, the lowest portion of the picture is occupied by the semilunar section of the pes pedunculi or crusta, *Pp*.

The attachment of the oculomotor nerve, fig. 11, *III*, has been already mentioned. Most of it comes out of the sulcus oculomotorius; detached bundles, however, pierce the mesial surface of the crusta. Not infrequently a detached portion of the nerve, isolated from the main bundle by a blood-vessel, takes exit from the peduncle considerably further outwards, *III'*.

The trochlear nerve, figs. 12 and 13, *IV*, arises as a thin thread, or in some cases as two separate roots, from the lateral angle of the velum medullare. It lies usually in the furrow between the posterior tubercles of the corpora quadrigemina and the superior cerebellar peduncles.

4. THE 'TWEEN-BRAIN.

It is very difficult to fix the boundaries between this portion of the brain and the parts in front of and behind it, namely, the secondary fore-brain and the mid-brain.

The most important structures which it includes are the optic thalami, *Tho*, with the two corpora geniculata, *Cgl* and *Cgm*, the optic tracts, *TII*, and the corpora mammillaria, *Cm* (figs. 11, 12, and 13). By some people the outer part of the thalamus is looked upon as belonging to the secondary fore-brain; an opinion based upon phylogenetic considerations, an explanation of which would lead us too far.

One part of the optic thalamus, the pulvinar, *Pu*, has already been shown to press backwards against the corpora quadrigemina. If all the rest of the great brain which lies superficial to the thalami is cut away so that these are left uncovered, a view into the ventricles of the great brain is obtained. We will content ourselves, in the first place, with a description of the external appearances of the parts included in the 'tween-brain, reserving the study of their intimate structure until sections through the whole of the great brain are under discussion.

The OPTIC THALAMUS (couche optique), fig. 18, is a large oval body which lies upon the pedunculus cerebri, with its long axis inclined forwards and inwards. Its lateral portion, which is continued into the optic tract (fig. 13), arches outwards over the peduncle towards the base of the brain. The upper surface of the thalamus, *Thos*, appears white owing to a thin covering of fibres (stratum zonale), while its mesial surface is grey. The two are separated by an angular border.

The fairly flat mesial surfaces (fig. 16) of the two thalami are very near together, and at one part are actually fused in the middle line forming the [soft, grey, or] middle commissure, *Com* (commissura mollis, trabecula cinerea). It is a short band, usually flattened from above downwards, and easily broken. This commissure is not seldom completely wanting. In some cases, on the other hand, it is double. In hydrocephalous distension of the ventricle, it may be drawn out to a considerable length (17 mm.—*Anton*).

The cavity of the 'tween-brain is named the middle or third ventricle, *V3*. The aqueduct of Sylvius opens out on reaching its oblique posterior wall into the aditus ad aquæductum Sylvii, fig. 16, *AAS*. From this a fissure runs along the middle of the posterior wall and floor of the ventricle. In the floor it opens out into a funnel-shaped depression, recessus infundibuli, *Rif*. This recess produces a grey conical swelling on the surface of the basis cerebri projecting behind the optic chiasm, tuber cinereum (fig. 11), *Tbc*. To the point of the infundibulum, *If*, hangs an ellipsoidal body, the hypophysis cerebri [pituitary body], *Hy*.

The upper surface of the thalamus is bounded laterally by a furrow (fig. 21) which contains a large vein, as well as a thickened ridge of ependyma. At the bottom of the furrow there lies also a bundle of fibres. Thickened ependyma and fibres together make the stria cornea, *Stc* (stria terminalis, tænia cornea), figs. 16, 18, and 20. The furrow begins at the front of the thalamus, runs outwards and backwards, and can be followed into the inferior horn of the lateral ventricle.

In addition to the general rounding of its upper surface the optic thalamus presents certain minor elevations (fig. 18). A distinct rounded elevation, about as large as a bean, constitutes its anterior end, tuberculum anterius, *Tba*. A shallow furrow, sulcus choroideus, *Slch*, starts at the back of this tubercle, and divides the rest of the surface into a mesial and a lateral portion. The back of the thalamus is elevated into the considerable rounded pulvinar, *Pu*. Beyond this the thalamus bends downwards and outwards, and narrows into a swelling, somewhat smaller than a bean, corpus geniculatum laterale (*seu* externum), in which the optic tract terminates. The optic tract encircles the peduncle, and meets on the basis cerebri with the tract of the opposite side in the chiasma nervorum opticorum, *Ch*. The lateral corpus geniculatum does not lie immediately on the peduncle, for the mesial portion of the optic tract, which is directed towards the mesial geniculate body, insinuates itself between the two. A white tract, which unites the two geniculate bodies together, and is best seen in the new-born child, is named by *Rauber*, ansa intergenicularis.

The boundary between the upper and mesial surfaces of the thalamus is rendered more evident by a ledge of white matter which is generally continued into a plate of gelatinous substance projecting towards the middle line, tænia thalami (T. ventriculi tertii) figs. 16 and 18, *Tv 3*. This overhanging ledge swells just in front of the trigonum subpineale into a club-shaped body, the ganglion habenulæ, *Gh*. Between this and the back of the thalamus lies a small triangular region, the trigonum habenulæ, *Trh*. Provided the membranes of the brain have not been dragged off roughly, a little conical body, the glandula pinealis (conarium), *Glp*, is seen lying in the horizontal fissure between the corpora quadrigemina. It is 8 to 12 mm. long. Short peduncles pass from the anterior end of the pineal gland, which forms part of the posterior wall of the third ventricle, to the ganglion habenulæ on either side, pedunculi habenulæ, *Pdc*. The posterior part of the third ventricle presents a little pit beneath the pineal body, recessus infrapinealis (ventriculus conarii), *Rip* (fig. 16).

Below this a well-formed tract of white matter, visible when the pineal gland has been removed, crosses above the anterior opening of the aqueduct of Sylvius, the posterior commissure, fig. 16, *Cop*. It bounds the trigonum subpinealis in front.

We are already acquainted with most of the structures which are found on the ventral surface of the 'tween-brain. First comes the optic chiasm, *Ch* (fig. 11), with the tuber cinereum, *Tbc*, in the angle made by the posterior edges of the optic tracts. Behind this again lie two white rounded eminences about the size of peas, corpora mammillaria (*seu* candicantia), *Cm*. They form the proper anterior border of the trigonum interpedunculare.

5. THE GREAT BRAIN.

The total fore-brain is split by the great longitudinal fissure (see fig. 7) into two equal halves, the hemispheres.

The surface of the hemispheres is almost everywhere covered with grey matter, the cortex. It is depressed into fissures and raised into convolutions, but certain grey masses found in the interior of the great brain will be referred to first.

If the method recommended for exhibiting the optic thalamus has been followed, a rounded swelling freely projecting into the ventricle, the **caudate nucleus** (corpus striatum, intra-ventricular portion of the corpus striatum), *Nc* (fig. 20), is seen to its outer side and separated from it by the stria cornea. It is largest in front of the thalamus, and thins away behind into a narrow riband. This riband, or tail of the nucleus, lies parallel to the stria cornea. It curves backwards, downwards, and, finally, forwards, and can be followed as far as the tip of the temporo-sphenoidal lobe (fig. 19a). It thus comes about that the nucleus

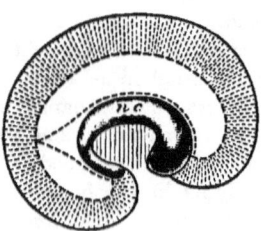

Fig. 19a.—Diagram to show the cortical relations of the nucleus caudatus, *nc* (*after Wernicke*). The head of the caudate nucleus is in continuity with the cortex of the frontal lobe, the extremity of the tail with that of the temporal lobe.

caudatus describes an arch, the anterior limb of which is formed by the massive head, the posterior limb by the tail. The latter part of the caudate nucleus lies in that portion of the ventricle called the descending horn. If a horizontal section is made through the hemisphere parallel to the surface of the optic thalamus and nucleus caudatus, and only cutting off their domes, a rounded body, about half a centimeter in transverse diameter and projecting (as a rule) backwards into a point, is seen in the anterior part of the thalamus. It is the anterior nucleus (upper nucleus, centre antérieur), *Na*. The fairly-distinct capsule which invests the anterior nucleus is prolonged backwards as a plate of white substance, lamina medullaris medialis thalami optici, *Lmm*. This lamina, therefore, divides the thalamus into two pieces of almost equal breadth. The lateral nucleus, *Nl*, projects beyond the mesial nucleus (nucleus internus), *Nm*, both in front and behind. The lateral boundary of the thalamus is marked by a white lamella, lamina medullaris lateralis thalami, *Lml*.

If a second horizontal section is made through the hemisphere about ½ cm. below the surface of the head of the nucleus caudatus (fig. 21), an idea is obtained of the depth to which this grey mass reaches. The nucleus anterior of the thalamus is no longer visible, but the lamina medullaris medialis, *Lmm*, and the nucleus lateralis, *Nl*, are clearly distinguished. The lateral boundary of the thalamus is formed by the feebly-developed lamina lateralis, *Lml*.

Another grey mass appears in this section which nowhere reaches to the surface, but is embedded in white matter, the **lenticular**

NUCLEUS LENTICULARIS. 67

nucleus (nucleus lenticularis, extra-ventricular portion of the corpus striatum), *Nlf*. It lies like a blunt wedge with its angular border thrust in between the nucleus caudatus and optic thalamus, separated from each of these by the white substance of the internal capsule, *Ci*.

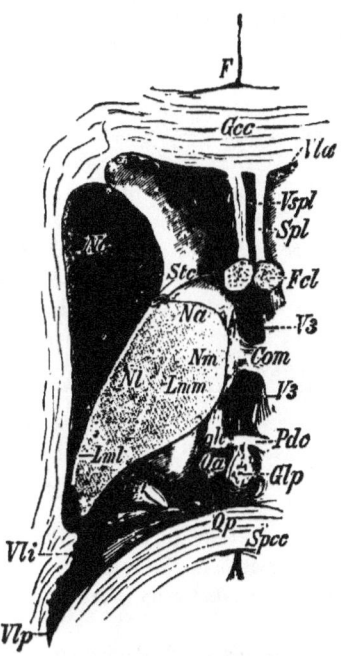

Fig. 20.—Section through the 'tween-brain and the neighbouring part of the forebrain half a centimeter beneath the upper surface of the optic thalamus and nucleus caudatus. *Nat. size*. Only the thalamus, the nucleus caudatus, and the parts immediately surrounding them are represented.—*Com*, Commissura mollis; *Fcl*, columna fornicis; *Fcr*, crus fornicis; *Gcc*, genu corporis callosi; *gh*, ganglion habenulæ; *Glp*, glandula pinealis; *Lml*, lamina medullaris lateralis; *Lmm*, lamina medullaris medialis; *Na*, nucleus anterior; *Nc*, nucleus caudatus; *Nl*, nucleus lateralis; *Nm*, nucleus medialis; *Pdc*, pedunculus pinealis; *Qa*, anterior corpora quadrigemina; *Qp*, posterior corpora quadrigemina; *Spcc*, splenium corporis callosi; *Spl*, septum pellucidum; *Stc*, stria cornea; *Vla*, anterior horn of lateral ventricle; *Vli*, descending horn of lateral ventricle; *Vlp*, posterior horn of lateral ventricle; *Vspl*, ventriculus septi pellucidi; *V3*, third ventricle.

Two thin white laminæ traverse the nucleus lenticularis, dividing its substance into three segments, which may be named, proceeding from the inner angle outwards, the first, second, and third pieces of the nucleus lenticularis, Nlf_1, Nlf_2, Nlf_3. The mesial and second segments (which together constitute the globus pallidus) of the lenticular nucleus are pale, like the thalamus, while the outer segment or putamen is as dark as the nucleus caudatus.

The lateral surface of the nucleus lenticularis corresponds in situation to the portion of the cortex of the great brain called the island of Reil, *I*. The nucleus and the cortex are separated by one grey and two white layers. Next to the nucleus lenticularis comes a thin layer of white matter, the outer capsule (capsula externa), *Ce*, to the outer side of which is applied the grey sheet of the claustrum (nucleus tæniæformis *seu* lateralis), *Cl*. Between the claustrum and the cortex of the island of Reil lies the sheet of white matter termed lamina fossæ Sylvii (capsula extrema). The mesial surface of the claustrum corresponds to the outer surface of the nucleus lenticularis, its lateral surface adapts itself to a certain extent to the cortex of the island of Reil, exhibiting similar small elevations and depressions.

The anterior angle of the nucleus lenticularis is situate somewhat farther back than the front of the nucleus caudatus. The posterior angle lies a little behind the thalamus.

To get a complete picture of the nucleus lenticularis, a frontal (transverse vertical) section must be made through the hemisphere at the level of the front of the thalamus (fig. 22). It now appears as a wedge, with its convex base resting on the cortex of the island of Reil, its angle—more acute than in a horizontal section—is directed beneath the thalamus. Between the nucleus and the cortex is again seen the claustrum shut in between the external capsule and the lamina fossæ Sylvii.

The region which lies below the thalamus (regio subthalamica), seen in this section, and sections carried more posteriorly, contains both grey and white matter; and can only be treated of when we are dealing with the minute structure of the brain. If a section is carried a little farther forward than in fig. 22, so that it traverses the optic chiasm, the lateral segment of the nucleus lenticularis is seen, better than in this section, to be in direct relation to another grey mass, the nucleus amygdaleus (nucleus amygdaliformis), *Am*. This nucleus is understood to be a thickened portion of the cortex of the temporo-sphenoidal lobe. [The claustrum also is fused at its anterior end with the nucleus amygdaleus.]

The tractus opticus, *II*, in its course around the cerebral peduncle pushes itself in between the nucleus lenticularis and nucleus amygdaleus.

The **white matter of the hemisphere** reaches its greatest development above the central ganglia (or basal ganglia, a term

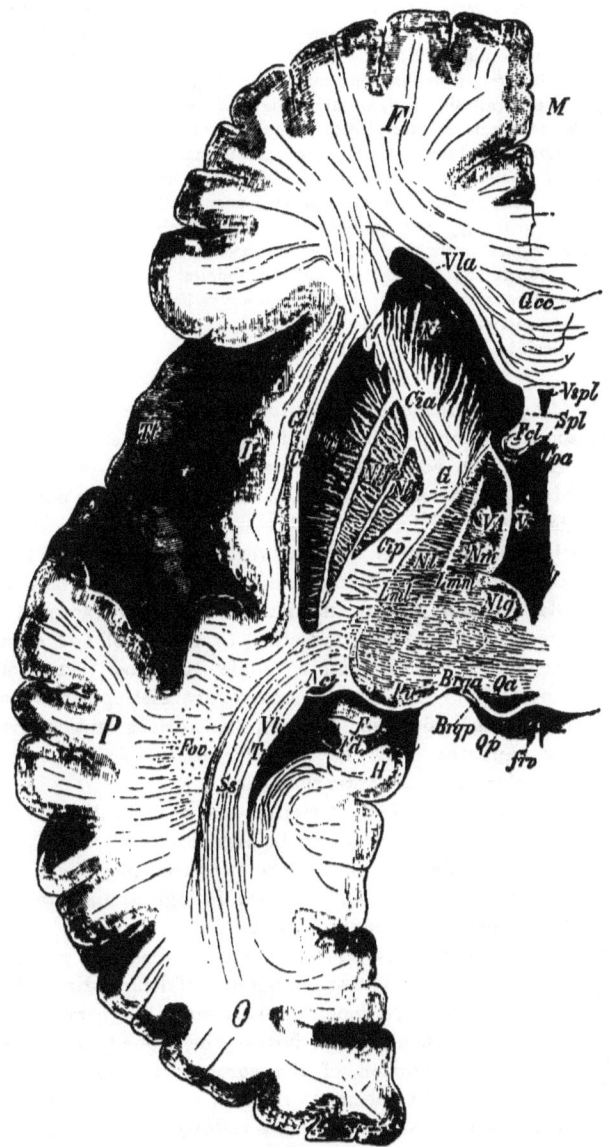

Fig. 21.—Horizontal section, one centimeter deeper than in fig. 20. *Nat. size.* The operculum, which had been detached from its connections by the section, is removed.—*Brqa*, Brachium anterius; *Brqp*, brachium posterius; *Ce*, capsula

externa; *Cia*, capsula interna, anterior limb; *Cip*, capsula interna, posterior limb; *Cl*, claustrum; *Coa*, commissura anterior; *F*, fimbria; *F*, frontal lobe; *Fcl*, columna fornicis; *Fd*, fascia dentata; *Fov*, fasciculus occipitalis verticalis of *Wernicke; frv*, frenulum veli anterioris; *G*, genu capsulæ internæ; *Gcc*, genu corporis callosi; *H*, gyrus hippocampi; *I*, island of Reil; *Lml*, lamina medullaris thalami lateralis; *Lmm*, lamina medullaris thalami medialis; *M*, great longitudinal fissure; *Nc*, nucleus caudatus (head); *Nc'*, nucleus caudatus (tail); *Nl*, nucleus lateralis thalami; *Nlf*, nucleus lenticularis; *Nlf 1 and 2*, globus pallidus; *Nlf 3*, putamen; *Nm*, nucleus medialis thalami; *Ntg*, nucleus tegmenti ruber; *O*, occipital lobe; *P*, parietal lobe; *Pu*, pulvinar; *Qa*, anterior corpora quadrigemina; *Qp*, posterior corpora quadrigemina; *Spl*, septum pellucidum; *Ss*, sagittal fibres of occipital lobe; *Tp*, tapetum; *Tt*, gyrus temporalis transversus; *VA*, Vicq d'Azyr's bundle; *Vla*, anterior horn of lateral ventricle; *Vli*, inferior horn of lateral ventricle; *Vspl*, ventriculus septi pellucidi; *V3*, ventriculus tertius.

comprehending nuclei caudatus et lenticularis and the optic thalamus). In a horizontal section parallel to, but above, the upper surface of the corpus callosum, the whole central mass of the hemisphere appears white (centrum semiovale Vieussenii). Such a section is not figured, but a frontal section of the centrum is seen in fig. 22, *CsV*. In deeper sections, which pass through the included grey masses (fig. 21), the white matter is seen to be broken up into limited tracts, which invest the lenticular nucleus, as the internal and external capsules. The former is seen in a horizontal section to consist of two segments meeting one another at an oblique angle, the knee (genu) of the internal capsule, *G*. The two segments are distinguished as the anterior limb, *Cia*, compressed between the nucleus lenticularis and the nucleus caudatus, and the posterior limb, *Cip*, between the nucleus lenticularis and the optic thalamus.

Certain special masses of white matter—the corpus callosum, fornix, and anterior commissure—must now be described.

(*a.*) **The Corpus Callosum.**—If the two hemispheres are pressed apart, a white structure of from 7 to 9 cm. in sagittal diameter is seen crossing the bottom of the longitudinal fissure.

Its fibres exhibit a transverse arrangement, *Ccll*. In addition to the transverse fibering, some distinct bundles of longitudinal fibres lie on its upper surface, striæ longitudinales mediales (nervi Lancisii), *NL* (fig. 25), and between them a furrow, the raphe (sutura corporis callosi). From that portion of the corpus callosum seen at the bottom of the fissure, its substance radiates outwards on either side into the hemispheres (radiatio corporis callosi). Above the corpora quadrigemina, the posterior edge of the corpus callosum is, as seen in a sagittal section, roundly thickened and rolled over, splenium corporis callosi, *Spcc*. In front it turns over in the knee (genu corporis

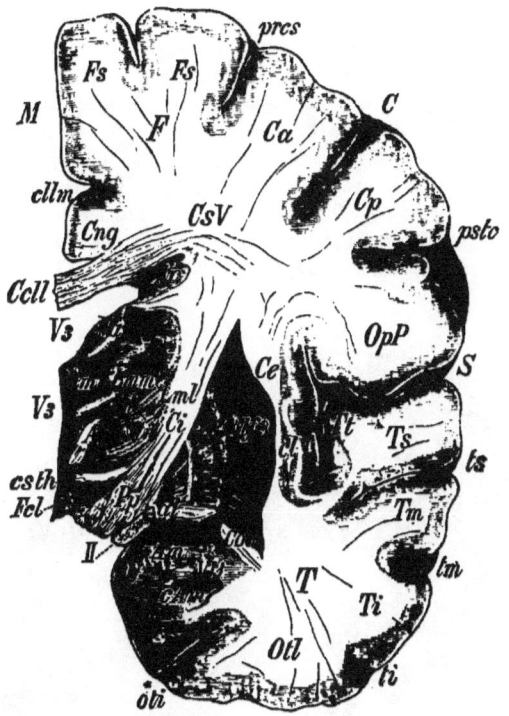

Fig. 22.—Frontal section through the human cerebral hemisphere. Left side, posterior portion. *Natural size.*—*II*, Tractus opticus; *al*, situation of ansa lenticularis; *Am*, amygdala; *C*, central fissure; *Ca*, gyrus centralis anterior; *cAm*, anterior end of cornu Ammonis; *Ccll*, corpus callosum; *Ce*, capsula externa; *Ci*, capsula interna; *cl*, claustrum; *cllm*, sulcus calloso-marginalis; *Cng*, gyrus fornicatus; *Coa*, commissura anterior; *Cp*, gyrus centralis posterior; *csth*, corpus subthalamicum; *CsV*, centrum semiovale Vieussenii; *F*, frontal lobe; *Fcl*, anterior pillar of fornix; *Fs*, gyrus frontalis superior; *I*, island of Reil; *Lmm*, *Lml*, lamina medullaris medialis et lateralis; *M*, great horizontal fissure; *Na*, *Nm*, *Nl*, nucleus anterior, medialis et lateralis, thalami optici; *Nc*, nucleus caudatus (tail); *Nlf 3, 2, 1*, the three portions of the nucleus lenticularis; *OpP*, opercular portion of inferior parietal lobule; *oti*, sulcus occipito-temporalis inf.; *Otl*, gyrus occipito-temporalis lateralis; *Pp*, pes pedunculi; *prcs*, sulcus præcentralis, pars superior; *pstc*, sulcus postcentralis; *S*, fissura Sylvii; *Str*, stratum reticulare; *T*, temporal lobe; *Tt*, gyrus temporalis transversus; *Ts*, *Tm*, *Ti*, gyrus temporalis superior, medius et inferior; *ts*, *tm*, *ti*, sulcus temporalis superior, medius et inferior; *U*, uncus gyri hippocampi; *VA*, bundle of Vicq d'Azyr; *Vli*, anterior end of the inferior horn of the lateral ventricle; *V3*, ventriculus tertius.

72   FIFTH VENTRICLE.

callosi), *Gcc*, and becoming quickly thinner folds downwards and backwards to form the rostrum, *Rcc*. The radiation outwards of the corpus callosum will be described later on. Between the rostrum and genu of the corpus callosum in front, and the fornix behind, is shut in a thin triangular plate of nerve-substance, septum pellucidum, *Splc*. This plate consists of two layers which contain between them a median cleft, the fifth ventricle (ventriculus septi pellucidi), *Vspl* (figs. 20 and 21). The size of this ventricle varies to a not inconsiderable extent in different individuals. The lower angle of the septum is continued on to the base of the brain, between the rostrum and the fornix, as the pedunculus septi pellucidi, *Pspl* (figs. 23 and 25). In fig. 23, the

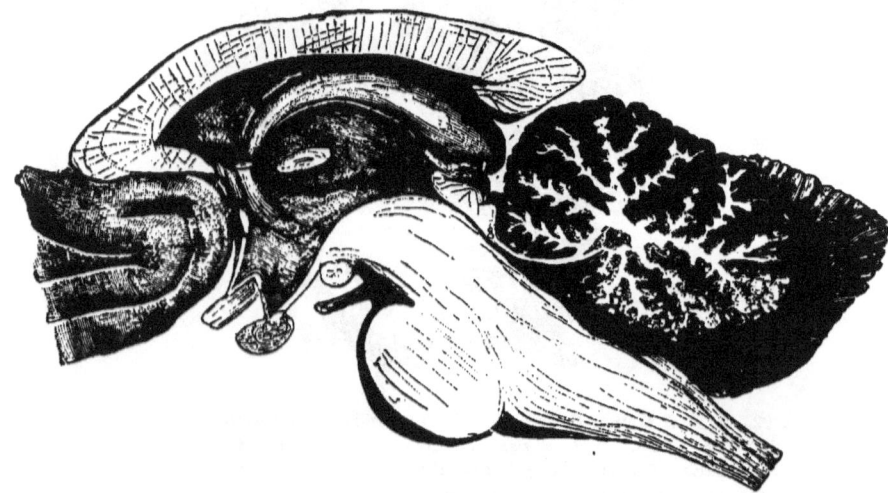

Fig. 23.—Sagittal section through the brain in the median line, right half. *Natural size.* Of the convolutions on the mesial surface of the hemisphere, only a part of those in the frontal region are visible (*Lfr*).—*II*, Nervus opticus; *III*, nervus oculomotorius; *IV*, crossing of the n. trochlearis; *AAS*, aditus ad aquæductum Sylvii; *AS*, aquæductus Sylvii; *Cc*, canalis centralis; *Ccll*, corpus callosum; *Ch*, chiasma nervorum opticorum; *Cm*, corpus mammillare; *Coa*, commissura ant.; *Coba*, commissura baseos alba; *Com*, commissura mollis; *Cop*, commissura posterior; *Cscr*, calamus scriptorius; *Cu*, culmen; *Dc*, declive; *Fcc*, folium cacuminis; *Fcl*, columna fornicis cut across at ±; *Fna*, funiculus ant. med. spinalis; *Fnp*, funiculus post. med. spin.; *Foca*, foramen cæcum ant.; *Focp*, foramen cæcum post.; *Gcc*, genu; *Gh*, ganglion habenulæ; *Glp*, glandula pinealis; *Hy*, hypophysis; *Lc*, lobulus centralis; *Lg*, lingula; *Lia*, lobus inf. ant.; *Lim*, lobus inf. med.; *Lip*, lobus

FORNIX.

septum pellucidum and fornix, are marked with the signs + and ±. The principal part of both has been removed in this preparation.

(b.) **Fornix** (voute à trois ou quatre piliers, trigone cerebral) appears as a paired structure composed of longitudinal fibres, which lies on the under surface of the corpus callosum, and arches over the thalamus (fig. 24). The fornix comes up out of the descending horn of the lateral ventricle as a flat band attached to the brain-wall by one of its edges only. In this situation it is termed the fimbria, *Fi*. Converging towards its fellow of the opposite side it reaches the under surface of the corpus callosum a little in front of the splenium. The free portions constitute the crura fornicis (posterior pillars of the

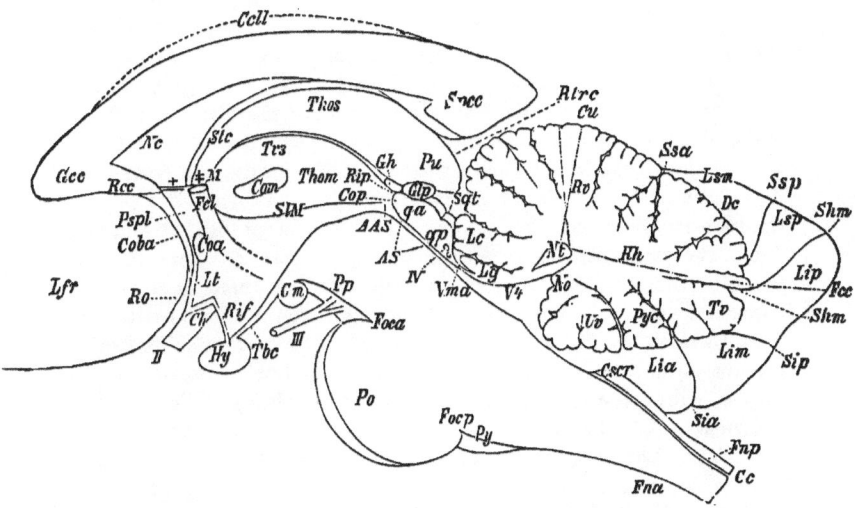

inf. post.; *Lsm*, lobus sup. med.; *Lsp*, lobus sup. post.; *Lt*, lamina terminalis; *M*, situation of |foramen of Monro; *Nc*, nucleus caudatus; *No*, nodulus; *Nt*, nucleus tecti; *Po*, pons; *Pp*, pes pedunculi; *Pspl*, pedunculus septi pellucidi cut at +; *Pu*, pulvinar thalami; *Pyc*, pyramis cerebelli; *Qa*, corpus quadrigeminum anterius; *Qp*, corpus quadrigeminum posterius; *Rcc*, rostrum; *Rh*, ramus medullaris horizontalis; *Rif*, infundibulum; *Rip*, recessus infrapinealis; *Ro*, recessus opticus; *Rtrc*, rima transversa cerebri; *Rv*, ramus medullaris verticalis; *Shm*, sulcus horizontalis magnus; *Sia*, sulcus inf. ant.; *Sip*, sulcus inf. post.; *SlM*, sulcus Monroi; *Spcc*, splenium; *Sqt*, sulcus corp. quad. tranversus; *Ssa*, sulc. sup. ant.; *Ssp*, sulcus sup. post.; *Stc*, stria cornea; *Tbc*, tuber cinereum; *Thom*, thalamus opticus, mesial surface; *Thos*, thalamus opticus, upper surface; *Tv*, tuber valvulæ; *Tv3*, tænia ventriculi tertii; *Uv*, uvula; *Vma*, velum medullare ant.; *V4*, fourth ventricle.

fornix), *Fcr.* The two pillars are separated from the thalamus by the interval of the third ventricle.

The two crura of the fornix unite a little in front of the posterior commissure. From this spot forwards they constitute a single band, the body of the fornix, *Fcp*, about 20 to 25 mm. long, firmly united to the corpus callosum. In front the septum pellucidum is pushed in between the fornix and the corpus callosum. Anteriorly, the fornix splits into two rounded columns (columnæ fornicis, anterior pillars), *Fcl* (figs. 20, 21, 23, and 24), which pass backwards as well as downwards, being covered with a thin layer of grey substance belonging to the thalamus. If this grey investment is removed the columns of the fornix can be followed as well-defined white tracts as far as the corpora mammillaria (radix descendens fornicis of *Meynert*), fig. 23. Another bundle, ascending from each corpus mammillare to the optic thalamus, can be laid bare by a little manipulation of the grey matter. *Meynert* regards this as the real continuation of the fornix, which he looks upon as looping over in the corpus mammillare. *Forel* and *Gudden* deny that it has such a relation to the fornix, and hence one does not usually term it the ascending root of the fornix, but prefers the indeterminate expression, bundle of *Vicq d'Azyr* (figs. 21 and 22), *VA*. The two posterior crura (pillars) of the fornix, where they lie beneath the corpus callosum, include between them a structure, triangular and equilateral, with an angle pointing forwards, psalterium (lyra Davidis), *Ps* (fig. 26). It consists of white matter exhibiting a transverse fibrillation, and is often not completely united to the corpus callosum, the space left between them being called *Verga's* ventricle, *VV*. The total length of the fornix is about 10 cm.

(c.) **The Anterior Commissure,** *Coa* (figs. 13, 21, 23, and 24), appears in median section as a very conspicuous bundle, cut transversely in front of the anterior pillars of the fornix, a small portion of it only being free in the middle line; on either side it plunges into the substance of the hemisphere. The anterior commissure is very easily followed by simple dissection or by making a series of frontal sections. It is a well-defined bundle which, after crossing the median line, passes laterally, and then arches backwards under the nucleus lenticularis.

[In a hardened brain, owing to the shrinking of its fibres, the anterior commissure lies almost free in the channel which it occupies in its course through the great brain. With any blunt instrument it can be followed beneath the neck of the nucleus caudatus, across the internal capsule and through the nucleus lenticularis, which it pierces at the back of its middle segment, missing the internal, but traversing the back of the external segment. It exhibits a remarkable torsion, its fibres being twisted upon one another in such a manner as to suggest that while the middle portion was fixed the two ends of the bundle have been rotated upwards and backwards (*see* Appendix A, Rotation of Great Brain). This commissure makes its appearance at an early stage in the development of the brain. It appears to be

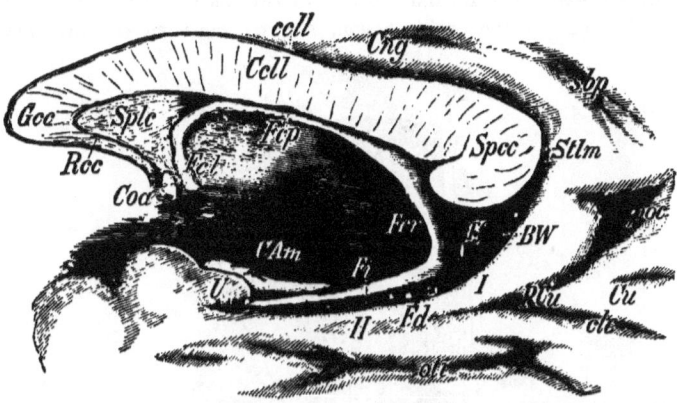

Fig. 24.—Part of a median section through the great brain. The optic thalamus is broken away; the parts of the temporal lobe are somewhat separated from one another. *Natural size.*—*BW*, Gyrus subcallosus; *CAm*, cornu Ammonis; *Ccll*, corpus callosum; *ccll*, sulcus corporis callosi; *clc*, fissura calcarina; *Cng*, gyrus cinguli; *Coa*, commissura anterior; *Cu*, cuneus; *Fcl*, columna fornicis; *Fcp*, corpus fornicis; *Fcr*, crus fornicis; *Fd*, fascia dentata; *Fi*, fimbria; *Gcc*, genu corporis callosi; *H*, gyrus hypocampi; *I*, isthmus gyri fornicati; *oti*, sulcus occipito-temporalis inferior; *PCu*, pedunculus cunei; *poc*, fissura parieto-occipitalis; *Rcc*, rostrum corporis callosi; *sbp*, sulcus subparietalis; *Spcc*, splenium corporis callosi; *Splc*, septum pelucidum; *Stlm*, stria longitudinalis medialis; *Tf*, tuberculum fasciæ dentatæ.

present in all vertebrates, attaining a greater relative importance amongst the lower members of the sub-kingdom, in which the cortex of the great brain and its special commissure, the corpus callosum, are rudimentary, than in Man.]

The part of the **base of the fore-brain,** which lies in front of the optic chiasm, must now be treated of in further detail. The lateral and mesial portions of this surface will be separately described. On either side lies a light grey area, bounded behind by the optic tract, in front by the frontal convolutions, and laterally by the temporo-sphenoidal lobe, *T;* this is known as the substantia perforata anterior (lamina cribrosa), fig. 25, *Spa*. Numerous apertures for vessels are seen in this region, especially in its antero-lateral portion. It is these holes which have given the area its name. Separate white bundles, emerging from the side of the temporo-sphenoidal lobe, cross this area, as well as the transverse orbital convolution, to reach the free white column of the olfactory tract, *Trol.* The olfactory tract passes forwards and slightly inwards for a distance of about 3·5 cm. At its anterior end it bears a yellowish grey swelling, the olfactory bulb (bulbus nervi olfactorii), *Bol.* The median portion of this part of the base of the brain, in front of the optic chiasm, is narrower than the substantia

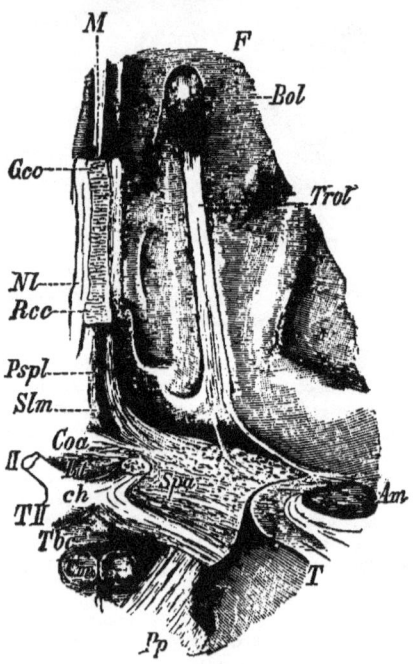

Fig. 25.—Part of the base of the brain, the left hemisphere in front of the optic chiasm. *The apex of the temporal lobe is cut away.*—*II*, Nervus opticus; *Am*, nucleus amygdaleus; *Bol*, bulbus olfactorius; *ch*, chiasma; *Cm*, corpus mammillare; *Coa*, elevation in the grey commissure of the floor of the third ventricle caused by the underlying anterior commissure; *F*, frontal lobe; *Gcc*, genu corp. callosi; *Lt*, lamina terminalis; *M*, longitudinal fissure; *NL*, nervus Lancisii; *Pp*, pes pedunculi; *Pspl*, pedunculus septi pellucidi; *Rcc*, rostrum corporis callosi; *Slm*, sulcus medius subst. perf. ant.; *Spa*, substantia perforata anterior; *T*, temporal lobe; *Tbc*, tuber cinereum; *Trol*, tractus olfactorius; *TII*, tractus opticus; *U*, uncus.

perforata, but reaches farther forwards; it constitutes the most anterior portion of the floor of the third ventricle (or grey floor-commissure).

The part which lies immediately in front of the chiasm is very easily torn, it is named the lamina terminalis, *Lt*, fig. 25 (see also fig. 23). A slight elevation is produced by the anterior commissure, here covered by a thin layer of grey matter.

In front of this grey elevation is seen a furrow, sulcus medius substantiæ perforatæ anterioris, *Slm*, which extends to the rostrum corporis callosi, *Rcc*. On either side of the median furrow is seen a thin longitudinal swelling which emerges from under the rostrum; pedunculus septi pellucidi, *Pspl*. Its hinder end turns outwards towards the perforated space on which it is lost.

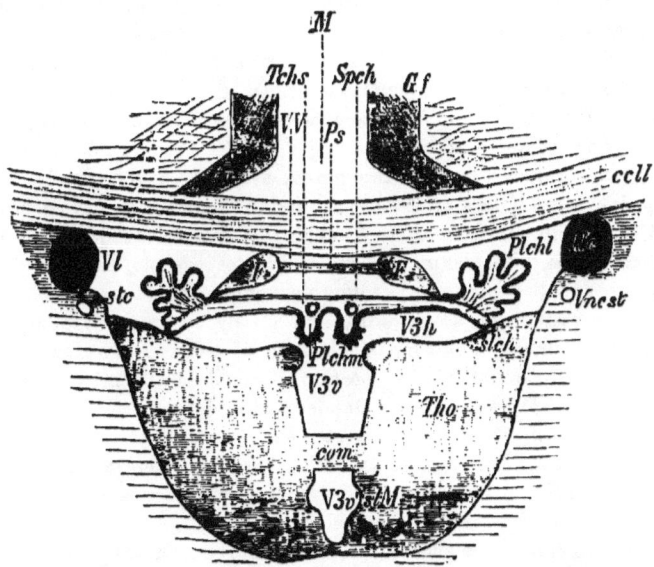

Fig. 26.—Diagram of the cerebral ventricles and the plexus choroideus.—*ccll*, Corpus callosum; *com*, commissura mollis; *F*, fornix; *Gf*, gyrus fornicatus; *M*, great longitudinal fissure; *Nc*, nucleus caudatus; *Plchl*, plexus choroideus lateralis; *Plchm*, plexus choroideus medialis; *Ps*, psalterium; *slch*, sulcus choroideus; *slM*, sulcus Monroi; *Spch*, suprachoroidal space; *stc*, stria cornea; *Tchs*, tela choroidea superior; *Tho*, thalamus opticus; *Vl*, ventriculus lateralis; *Vncst*, vena striæ corneæ; *V3*, third ventricle; *V3h*, its horizontal portion; *V3v*, its ventricle portion; *V3v′*, the same portion below the grey commissure; *VV*, Verga's ventricle.

6. THE VENTRICLES OF THE GREAT BRAIN.

Although the anatomical disposition of the ventricles of the great brain seems simple, their morphological relations to the nervous substance are only to be made out by careful ontogenetic study.

The ventricles of the brain can be entered from behind beneath the splenium corporis callosi. However much the transverse slit which here exists (fissura transversa cerebri anterior, rima transversa) is closed up by the membranes of the brain, it yet affords this opening. The first thing which meets the view when one removes the back of the brain, with the corpus callosum and body of the fornix (on one side at any rate the fornix should be left for further study) is not the optic thalamus but a vascular membranous fold which covers it. Seen in its whole extent at once this fold has the form of an equilateral triangle. The base of the triangle corresponds with the transverse fissure; its apex reaches the anterior pillars of the fornix; its antero-lateral borders lie parallel to the striæ corneæ and somewhat mesial to them, and are attached to the surface of the thalamus (fig. 26), this fold of membrane is the tela choroidea superior (velum triangulare [*seu* interpositum]), *Tchs*. The lateral margin of the tela choroidea carries a convoluted system of blood-vessels, more extensive behind than in front, the choroid plexus of the great brain, *Plchl*. At the junction of the lateral and posterior borders of the tela choroidea, the plexus attains its greatest development swelling into the so-called glomus. From this angle of the tela the choroid plexus is continued downwards, and finally forwards [around the cerebral peduncle], following the course of the crus fornicis as far as the anterior point of that portion of the lateral ventricle, which we shall learn presently to call the descending horn.

If the fornix, *F*, has not been removed one notices that its sharp lateral border is attached to the tela choroidea along a line parallel to, but slightly on the mesial side of, the stria cornea and the line of attachment of the tela to the thalamus.

The situation of the choroid plexus is marked on the thalamus by a shallow groove, sulcus choroideus, *slch* (figs. 18 and 26).

On the under side of the tela choroidea, near the middle line, are attached two narrow strips of choroidal plexus, plexus choroidei medii, *Plchm*. They extend from the anterior angle of the tela to its base.

The whole hollow space in the interior of the great brain is divided into three portions by the attachments of the tela to the thalami, a middle, *V3*, and two symmetrical lateral ventricles, *Vl*. Another space is left between the fornix with the psalterium, and the tela (below *Verga's* ventricle therefore), spatium supra-choroideum, *Spch*. On the diagram (fig. 26) its extent is purposely exaggerated.

The **third** (or **middle**) **ventricle** is made up of two portions, one vertical, the other horizontal, united as seen in cross-section in a T. The vertical limb of the *T*, bounded by the mesial surfaces of the thalami, is the principal portion of the ventricle (fig. 26, *V3v* and

$V3v'$, and figs. 20, 21, and 22). At its hinder end the aquæductus Sylvii opens into the ventricle at the aditus ad aquæductum, AAS (fig. 23). From this spot its floor sinks downwards somewhat quickly to the apex of the infundibulum. The anterior wall is formed by the lamina cinerea terminalis, Lt, already described. The lowest part of this surface is so pushed into the ventricle by the optic chiasm, that a pouch is formed above the chiasm, recessus chiasmatis (*seu* opticus), Ro. The upper edge of this vertical portion of the ventricle is formed by the stria medullaris thalami, to which, for the most part by the intervention of the tænia ventriculi tertii, $Tv3$, the plexus choroideus medius is attached. In front, where the stria medullaris approaches quite close to the anterior pillar of the fornix, a space, foramen of Monro, M, is left between the thalamus and the fornix. The plexus choroideus lateralis with a vein passes out of the lateral ventricle into the third ventricle through this hole, and bends backwards in the plexus choroideus medius. The foramen of Monro constitutes the only direct connection between the middle and lateral ventricles (see also fig. 7).

A shallow groove is to be noticed on the mesial face of the thalamus, which passes in a gentle curve beneath the commissura mollis, from the foramen of Monro to the aditus ad aquæductum Sylvii, sulcus Monroi, SlM. (fig. 23).

The horizontal portion of the third ventricle, $V3h$ (not always included in the third ventricle in descriptions of this part of the brain), comprises that portion of the ventricle which is shut in between the tela choroidea and the upper surface of the optic thalamus. It extends from the stria medullaris outwards to the line of attachment of the tela choroidea. This cleft narrows towards the front, where it ends in a point. It corresponds obviously to the tela choroidea media in form.

The paired LATERAL VENTRICLES (ventriculi laterales *seu* tricornes), Vl, lie in the interior of each hemisphere, and communicate through the foramen of Monro with the middle ventricle; they are not directly in connection with one another.

Just as the whole hemisphere is to be looked upon as presenting an arch open in front with a posterior prolongation, the occipital lobe, so the cavity which it contains is an arched space with a special occipital prolongation backward.

In each lateral ventricle (figs. 18, 20, and 21) is distinguished a central or principal part (cella media), from which a horn (anterior horn, Vla) passes forwards, a diverticulum is continued backwards (posterior horn, Vlp), and, lastly, the ventricle ends in the inferior limb of the arch, the inferior [or descending] horn, Vli.

The **anterior horn** is the part of the lateral ventricle which corresponds to the head of the nucleus caudatus, and reaches still farther forwards into the frontal lobe. Its mesial wall is formed by the septum pellucidum. The corpus callosum constitutes its front wall and roof.

The **cella media** begins at about the level of the foramen of Monro. Its roof is formed by the middle portion of the body of the corpus callosum. In the floor of the cavity lie, in order from without inwards, the tail of the nucleus caudatus, the stria cornea, the lateral portion of the optic thalamus, and the plexus choroideus lateralis (figs. 18 and 26). The upper surface of the fornix may also be included in the floor of the cella media, since this structure lies with only its mesial edge resting against the corpus callosum.

The **posterior horn** of the lateral ventricle (fig. 27) begins at about the level of the splenium corporis callosi, and reaches usually nearly to

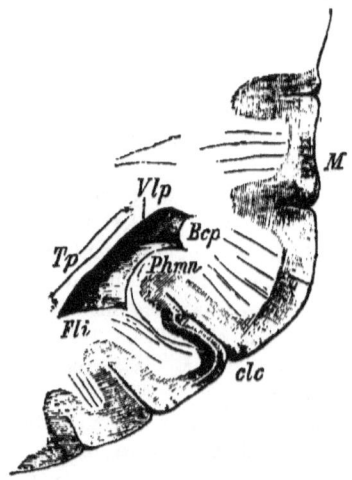

Fig. 27.—Frontal section through the right cerebral hemisphere, behind the splenium corporis callosi. *Posterior segment. Natural size.*—*M*, Mesial surface of the hemisphere; *clc*, fissura calcarina; *Vlp*, posterior horn of the lateral ventricle; *Bcp*, bulbus cornu posterioris; *Phmn*, pes hippocampi minor; *Fli*, fasciculus longitudinalis inf.; *Tp*, tapetum.

the occipital pole of the hemisphere. For upper and outer walls the posterior horn has the continuation of the corpus callosum or tapetum, *Tp*. The mesial and lower wall is formed, as shown in a frontal section, by three distinct elongated elevations. The upper corresponds to the margin of the corpus callosum, forceps posterior corporis callosi (bulbus cornu posterioris), *Bcp*. The middle swelling, calcar avis (pes hippocampi minor), *Phmn*, is formed by a fissure (fissura calcarina), *clc*, which, cutting deeply into the mesial surface of the hemisphere, pushes in front of it the wall of the ventricle. In some brains in which this swelling is strongly developed, its surface is somewhat indented transversely, faintly recalling a bird's claw. The lowest of the three swellings is produced by a thickening in the mass of longitudinal white fibres, fasciculus longitudinalis inferior, *Fli*. The choroid plexus does not enter the posterior horn.

The **inferior (descending) horn** (fig. 28), *Vli*, extends far forwards into the temporal lobe, but terminates about 2 cm. behind its pole. It is apparently open to the mesial surface through the hippocampal fissure (fissura cornu Ammonis, *h*). For the greater part of its extent the inferior horn is roofed in by the tapetum; the tail of the caudate nucleus and the stria cornea also extend to the front of this horn. Near the anterior end of the cornu Ammonis the tail of the caudate nucleus, which by this time is reduced to a thin grey band, begins to swell suddenly, and passes over into the nucleus amygdaleus (figs. 22 and 25).

Let us enter the inferior horn from the mesial side through the fissure hippocampi, *h*, with a view to explore its inferior wall. A succession of structures are met with all arranged longitudinally; first, a broad convolution, gyrus hippocampi (subiculum cornu Ammonis), *H*, on the surface of which, in the fresh brain, a reticulated white layer is recognisable, substantia reticularis Arnoldi; secondly, a frequently notched grey cord, almost hidden at the bottom of a furrow, fascia dentata, *fd;* thirdly, a flattened white triangular column, the fimbria, *Fi*, covering up the fascia dentata, which is only distinctly visible after the fimbria has been pushed aside; fourthly, a considerable white swelling, the pes hippocampi majoris (cornu Ammonis, *CAm*), which is greatly enlarged and distinctly indented in front; fifthly, in the depth of the inferior horn is to be found not infrequently a swelling, eminentia collateralis Meckelii, *EcM*, which, like the pes hippocampi minoris of the posterior horn, is simply due to the deep indentation of the surface by a fissure (fissura collateralis *seu* occipito-temporalis inferior), *oti*. The eminentia collateralis is separated from the cornu Ammonis by a furrow, which I will call the fissura subiculi interna, so deep that it almost splits the subiculum. It is not distinctly marked off from the tapetum on the outer side.

Of the several structures just mentioned, the subiculum and fascia dentata, as well as part of the fimbria, lie outside the inferior horn proper. The fimbria presents a sharp edge, to which the plexus choroideus lateralis, *Pc*, is attached. Only the portion of the fimbria which lies laterally to this edge enters the inferior horn. The cornu Ammonis and eminentia collateralis properly form its floor.

Followed back to the splenium corporis callosi, one sees that the subiculum cornu Ammonis is continued over the corpus callosum as the gyrus fornicatus (*seu* gyrus cinguli, *Cng*); also, that the fascia dentata is the termination of the free edge of the cortex. Above the

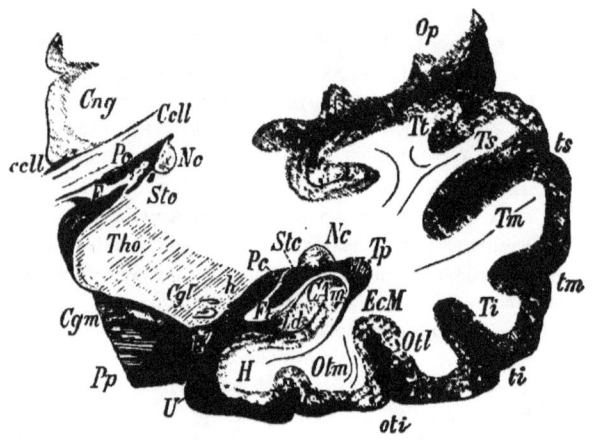

Fig. 28.—Frontal section through the right hemisphere behind the uncus. *Anterior segment*. The upper part is not represented.—*II*, Tractus opticus; *CAm*, cornu Ammonis; *Ccll*, corpus callosum; *ccll*, sulcus corporis callosi; *Cgl*, corpus geniculatum laterale; *Cgm*, corpus geniculatum mediale; *Cng*, cingulum; *EcM*, eminentia collateralis Meckelii; *F*, fornix; *fd*, fascia dentata; *Fi*, fimbria; *h*, fissura hippocampi; *H*, gyrus hippocampi; *Nc*, nucleus caudatus; *Op*, operculum; *oti*, sulcus occipito-temporalis inferior; *Otl*, *Otm*, gyri occipito-temporalis lateralis et medialis; *Pc*, plexus choroideus lateralis; *Pp*, pes pedunculi; *S*, fissura Sylvii; *Stc*, stria cornea; *Tho*, thalamus opticus; *Tp*, tapetum; *Ts*, *Tm*, *Ti*, gyri temporalis superior, medius et inferior; *ts*, *tm*, *tc*, sulci temporalis sup. med. et inf.; *Tt*, gyrus temporalis transversus; *U*, uncus; *Vli*, inferior horn of the lateral ventricle.

corpus callosum it constitutes a thin layer of grey substance hardly distinguishable from the cortex of the gyrus fornicatus, indusium griseum. The free mesial edge of the indusium is thickened, and forms (without other addition) certain recognisable longitudinal striæ, striæ longitudinales mediales (*seu* nervi Lancisii, *Stlm*—figs. 25 and 29). Just before the fascia dentata, having reached the splenium corporis callosi, begins (much diminished in size) to ascend on to its upper side, it swells out into a tubercle looking as if the great splenium pressed it down, tuberculum fasciæ dentatæ (*Zuckerkandl*), *Tf*.

CONNECTION OF LATERAL VENTRICLE WITH SURFACE. 83

Between this tuberculum and the ascending gyrus hippocampi lie certain minute convolutions, better seen in many animals than in Man, which *Zuckerkandl* calls callosal convolutions, *BW*. Exceptionally in Man, these convolutions form a cord-like body, which stretches on to the upper surface of the corpus callosum beneath the gyrus fornicatus.

The fimbria becomes the crus fornicis. The splenium corporis callosi squeezes itself in between the crus and the continuation

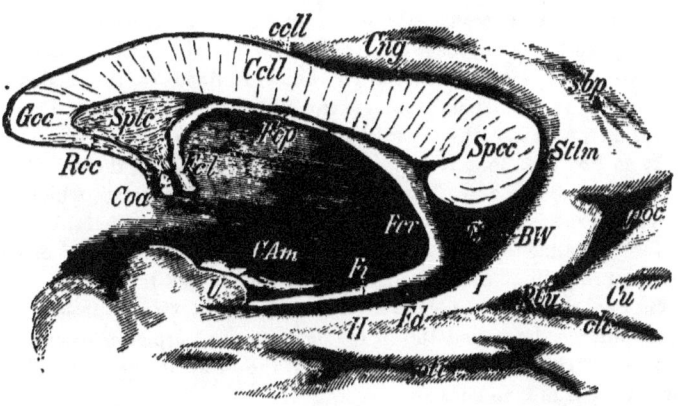

Fig. 29.—Portion of a median section of the cerebrum. The optic thalamus has been pulled away. The structures of the temporo-sphenoidal lobe are a little separated from one another. *Natural size.*—*BW*, Gyrus corporis callosi; *CAm*, cornu Ammonis; *Ccll*, corpus callosum; *ccll*, sulcus corporis callosi; *clc*, fissura calcarina; *Cng*, gyrus cinguli; *Coa*, commissura anterior; *Cu*, cuneus; *Fcl*, columna fornicis; *Fcp*, corpus fornicis; *Fcr*, crus fornicis; *Fd*, fascia dentata; *Fi*, fimbria; *Gcc*, genu corporis callosi; *H*, gyrus hippocampi; *I*, isthmus gyri fornicati; *oti*, sulcus occipito-temporalis inferior; *PCu*, pedunculus cunei; *poc*, fissura parieto-occipitalis; *Rcc*, rostrum corporis callosi; *sbp*, sulcus subparietalis; *Spcc*, splenium corporis callosi; *Splc*, septum pellucidum; *Stlm*, stria longitudinalis medialis; *Tf*, tuberculum fasciæ dentatæ; *U*, uncus.

upwards of the fascia dentata. Between these two structures and the corpus callosum a triangular area is left.

Starting at the foramen of Monro and travelling along the concave border of the fornix, the plexus choroideus finds entrance from the mesial surface into the lateral ventricle through a curved cleft (figs. 26 and 28). Development teaches us, however, that no true gap in the brain-wall exists, for the foramen of Monro is all that is left of a much larger passage from the primary to the secondary forebrain found in the fœtus. The velum interpositum, which springs from the primitive falx cerebri, is, with its choroid plexus, developed early in fœtal life. By its further growth on either side the velum interpositum pushes the inner walls of the primitive prosencephalic (cerebral) vesicles before its margin into their cavities (the future lateral ventricles), and so makes in each of them a cleft which extends backwards from the foramen of Monro; the transverse fissure of the cerebrum (fissura choroidea *seu* transversa cerebri). The involuted portion of the wall of the ventricle is thinned down to a mere layer of epithelium, which covers the choroid plexus, but still, along the whole extent of the transverse fissure, closes the ventricle in.

In fig. 26 the space between the lateral margin of the fornix and the line along which the tela choroidea is fixed to the thalamus, corresponds to the transverse fissure. Through this gap the choroid plexus advances into the lateral ventricle.

[It is not common in English text-books to extend the use of the term "transverse fissure" to all parts of the cleft through which the velum interpositum gains admittance to the lateral ventricles, but the custom is rather to limit the term to the "transverse fissure of *Bichat*" or incisura pallii. *See* Macalister's *Anatomy*, p. 702.]

Fig. 28 shows the plexus choroideus, *Pc*, in two situations, one in the inferior horn and the other beneath the corpus callosum. In this latter situation the pitting in of the wall of the ventricle by the plexus is also to be seen.

7. THE FISSURES AND CONVOLUTIONS ON THE SURFACE OF THE GREAT BRAIN.

The great brain may be regarded as a single almost globular body, divided by the great longitudinal fissure into two hemispheres, each of which presents a convex outer (lateral) and a flat mesial surface, which meet at an edge, sharp for the greater part of its extent.

On the surface of the adult brain a large, although variable, number of fissures are visible. Between them the surface is raised into convolutions.

It must be allowed that the fissures and convolutions of the cortex are not constant in arrangement; most of them, however, follow a definite type, and much trouble has been taken to determine the laws of their topographical distribution. We cannot yet regard the investigations into their developmental history and arrangement in different animals as complete.

In the following account, *Ecker's* nomenclature will be adopted on the ground that, being accepted by most anatomists, and being understood in all lands, his classification has come, in a sense, to be an international one.

The question is often discussed whether greater attention should be paid to the convolutions or the fissures. The proper way to look at the matter is to regard the fissures as cut into the surface of the brain, the convolutions as the portion of tissue left between adjoining fissures.

If an embryonic human brain is examined at the fifth or sixth month, or if we look at the brain of a rodent animal, certain fissures are seen cutting into the flat surface in regions where no convolutions have yet appeared; the latter only make their appearance as the fissures become numerous and approach near together.

Fissures may be arranged in order of importance in the three following groups:—

1. Principal or total fissures (fissuræ, scissuræ, sulci primarii).
2. Typical or secondary fissures (sulci secundarii).
3. Atypical or tertiary fissures (sulci tertiarii).

The chief fissures are the first to appear and permanently the deepest. They are called total fissures, because in early embryonic life, when the wall of the ventricle is thin, they involute it into the ventricular cavity. An example of this condition persists in the adult brain in the posterior horn of the ventricle, the calcar avis, *Phmn*, being formed in this way (fig. 27). The later subsidiary fissures sink into the surface only; they are divisible into those which are present in every normal brain (secondary fissures), and those which are subject to individual variations in number and direction (tertiary fissures).

The portions of the brain marked off by fissures are distinguished as lobes, lobules, and gyri.

The chief divisions are distinguished as lobes. This delimitation applies not to the cortex only, but also to the underlying mass of the brain. Each lobe comprises convolutions of which some in ordinary parlance are termed lobules. Typical convolutions, it goes without saying, are those bounded by typical fissures. Atypical fissures bound atypical convolutions. We only recognise as convolutions those which appear on the surface, and often forget that little convolutions are to be found in the bottom of certain fissures; deep or bridging convolutions. The superficial connections between adjoining convolutions are named by *Merkel*, gyri transitori [annectant convolutions]. The amount of cortex hidden away in the fissures is in the human brain about double that which appears on the surface.

Chief Fissures.—1. Fissura Sylvii (fossa Sylvii,* fissura lateralis),

* [A term better restricted to the open depression on the fœtal brain, which precedes the closed-in fissure of Sylvius.]

fig. 30. It is essentially distinguished from all other fissures by the manner of its origin. Its appearance is due to the fact that the great brain, during its growth, curves round its central stem-connection, making on its surface an arch open in front and below, which closes in an area, at first oval and later triangular in form, the "island."

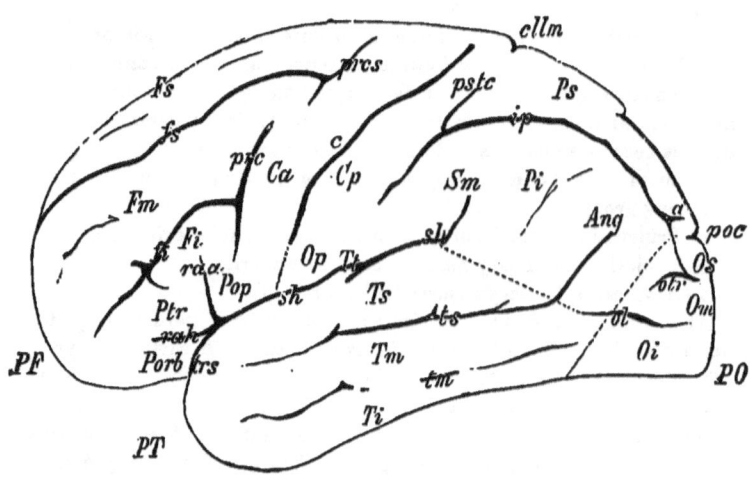

Fig. 30.—Left hemisphere from the side. *Half natural size.*—*Ang*, Gyrus angularis; *c*, central fissure; *Ca*, gyrus centralis anterior; *cllm*, sulcus calloso-marginalis; *Cp*, gyrus centralis posterior; *fi*, sulcus frontalis inferior; *Fi*, gyrus frontalis inferior; *Fm*, gyrus front. medius; *Fs*, gyrus frontalis superior; *fs*, sulcus frontalis superior; *ip*, fissura interparietalis; *Oi*, gyrus occipitalis inferior; *ol*, sulcus occipitalis lateralis; *Om*, gyrus occipitalis medius; *Op*, operculum; *Os*, gyrus occipitalis superior; *otr*, sulcus occipitalis transversus; *PF*, frontal pole; *Pi*, lobulus parietalis inferior; *PO*, occipital pole; *poc*, fissura parieto-occipitalis, pars lateralis; *Pop*, pars opercularis; *Porb*, pars orbitalis; *prc*, sulcus præcentralis inferior; *prcs*, sulcus præcentralis superior; *Ps*, lobulus parietalis superior; *pstc*, sulcus postcentralis, a constant little side branch of the interparietal fissure in front of the parieto-occipital fissure; *PT*, temporal pole; *Ptr*, pars triangularis; *raa*, ramus anterior ascendens; *rah*, ramus anterior horizontalis; *sh*, pars horizontalis; *Sm*, gyrus supramarginalis; *Ti*, gyrus temporalis inferior; *tm*, sulcus temporalis medius; *Tm*, gyrus temporalis medius; *trs*, truncus fissuræ Sylvii; *Ts*, gyrus temporalis superior; *ts*, sulcus temporalis superior; *Tt*, gyrus tempor. transversus. The boundaries between the four lobes when not made by fissures are marked with dotted lines.

During the further growth of the brain, the island is, in a sense, fixed to the stem portion of the hemisphere, while the rest of the great brain is free; hence surrounding parts bulge over the island,

and, closing it in from three sides (from the front, from above, and from below), leave it lying at the bottom of a [V-shaped] cleft, the fissura Sylvii. The island is seen only after the neighbouring convolutions have been pulled aside.

The form of the Sylvian fissure is determined by this growth of the hemisphere from three sides. It consists of a short commencing portion, *trs* (truncus fissuræ Sylvii), which ascends abruptly from the substantia perforata anterior on to the lateral surface of the hemisphere, and then bends over into the principal or horizontal portion of the fissure, *sh* (ramus horizontalis posterior); this ramus runs, slightly ascending, far backwards. Two short lateral fissures usually ascend from the anterior portion of the horizontal ramus; of these, the first runs horizontally forwards, *rah* (ramus anterior horizontalis); the other ascends vertically, *raa* (ramus anterior ascendens).

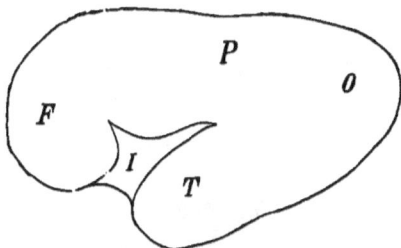

Fig. 31.—Left hemisphere of a human embryo at the fifth month.—*F*, Frontal; *P*, parietal; *O*, occipital; *T*, temporal lobes; *I*, island of Reil.

2. Sulcus centralis (sulcus Rolandi, fissura transversalis, scissura perpendicularis), *c*. This fissure also runs its course on the convex surface. It begins at about the level of the centre of the mesial cortex-border, but without quite reaching the edge, and is thence directed obliquely forwards and downwards towards the horizontal limb of the Sylvian fissure, into which, however, it seldom extends. Its lower end lies not quite 3 cm. behind the ramus ascendens of the Sylvian fissure. Since the central fissure does not cut deeply enough into the surface to produce a bulging of the ventricle wall, it ought not, strictly speaking, to be treated as a chief fissure, but its early origin, depth, and constancy, justify us in assigning this rank to it.

3. Fissura parieto-occipitalis (fissura occipitalis, f. occipitalis perpendicularis), *poc*, belongs in its principal part to the mesial, in its smaller part to the lateral surface. Hence two divisions are distinguished, and often called by separate names. A mesial portion (fissura perpendicularis interna), fig. 33, and a lateral portion (upper

part or fissura perpendicularis externa), fig. 32. The fissure on the mesial surface is distinguished by its depth and extent. Commencing

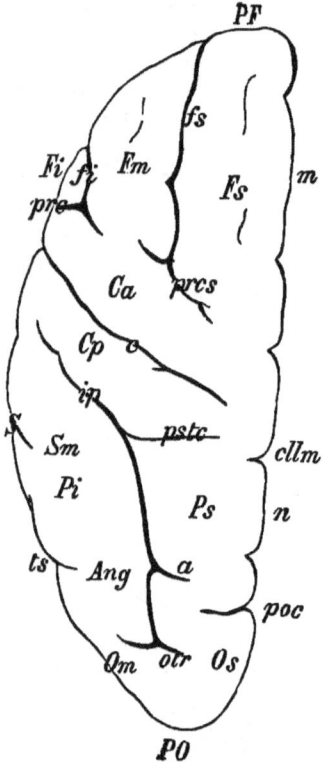

Fig. 32.—Left hemisphere from above. *Half nat. size.*—*a*, Side branch of the intraparietal fissure in front of the parieto-occipital; *Ang*, gyrus angularis; *c*, central fissure (fissure of Rolando); *Ca*, gyrus centralis ant.; *cllm*, sulcus calloso-marginalis; *Cp*, gyrus centralis posterior; *fi*, sulcus frontalis inf.; *Fi*, gyrus frontalis inf.; *Fm*, gyrus frontalis medius; *Fs*, gyrus frontalis sup.; *fs*, sulcus frontalis sup.; *ip*, fissura intraparietalis; *Om*, gyrus occipitalis medius; *Os*, gyrus occipitalis superior; *otr*, sulcus occipitalis transversus; *PF*, frontal pole; *Pi*, lobulus parietalis inf.; *PO*, occipital pole; *poc*, fissura parieto-occipitalis; *prc*, sulcus præcentralis inferior; *prcs*, sulcus præcentralis superior; *Ps*, lobulus parietalis sup.; *pstc*, sulcus centralis post.; *S*, fissura Sylvii; *Sm*, gyrus supramarginalis; *ts*, sulcus temporalis superior. The antero-posterior diameter of the corpus callosum, as it lies in the great longitudinal fissure, is indicated by the distance between *m* and *n*.

at the cortex-border some 4 or 5 cm. in front of its posterior angle, it runs downwards and sharply forwards, joining another fissure (fissura calcarina, about to be described) at an acute angle. As already mentioned, the parieto-occipital fissure extends over the border, and runs a short course (1 to 2 cm.) on the convex surface. Exceptionally, it reaches a long way down.

4. Fissura calcarina (fissura occipitalis horizontalis, pars posterior fissuræ hippocampi), *clc* (fig. 33), belongs exclusively to the mesial surface. It commences usually in two very short limbs, runs horizontally forwards, joins the parieto-occipital fissure, and terminates not far below the splenium corporis callosi.

Two fissures which are not to be recognised in the fully-developed brain must be added to the "total" fissures. They bound the gyrus arcuatus, which lies on the mesial surface of the fœtal brain.

1. The arcuate fissure, which in its upper portion bounds the corpus callosum (sulcus corporis callosi, *ccll*), but which below corresponds to the fissure (fissura hippocampi) which causes the hippocampus major to bulge into the descending horn of the lateral ventricle (figs. 28 and 33), *h*.

2. Fissura choroidea, which in the developed brain is no longer obviously present, nor is it any longer in relation to the cortex proper. It is represented by the folding of the choroid plexus into the lateral ventricle already frequently mentioned (*cf.* p. 84).

The Separate Lobes of the Great-Brain.—The attempt has been made to use the chief fissures as boundary lines for the lobes of the brain, but these fissures only constitute portions of such border lines, and for the rest the division must always remain an arbitrary one. The frontal lobe (lobus frontalis) is the piece in front of the central fissure and above the fissure of Sylvius. The parietal lobe (lobus parietalis) begins behind the central fissure, and reaches backwards as far as the parieto-occipital fissure and downwards to the fissure of Sylvius, but it is not completely separated by means of these fissures, either from the occipital lobe which lies behind it or the temporal lobe which lies below. Hence these boundaries are artificial, and differently understood by different authors. Remaining as far as possible true to *Ecker's* typical classification of the convolutions, we will employ for the purpose of distinguishing the parietal from the occipital lobe a shallow impression on the under side of the hemisphere, corresponding to the upper angle of the petrous portion of the temporal bone; an impression often to be seen only just after the brain is taken from the skull. A line continuing the direction of the parieto-occipital fissure as far as this depression separates the two lobes. It is still more difficult to define the temporo-sphenoidal (*seu* temporal) lobe. Usually the Sylvian fissure ends by turning upwards at a somewhat acute angle. From this angle we can draw

ISLAND OF REIL.

a line backwards and downwards towards the fissure, which will presently be described as occipitalis lateralis (fig. 30), *ol.* Below and in front of this line lies the temporo-sphenoidal lobe.

It remains to mention the island of Reil which lies at the bottom of the Sylvian fissure (insula Reili, lobus caudicis, *seu* intermedius

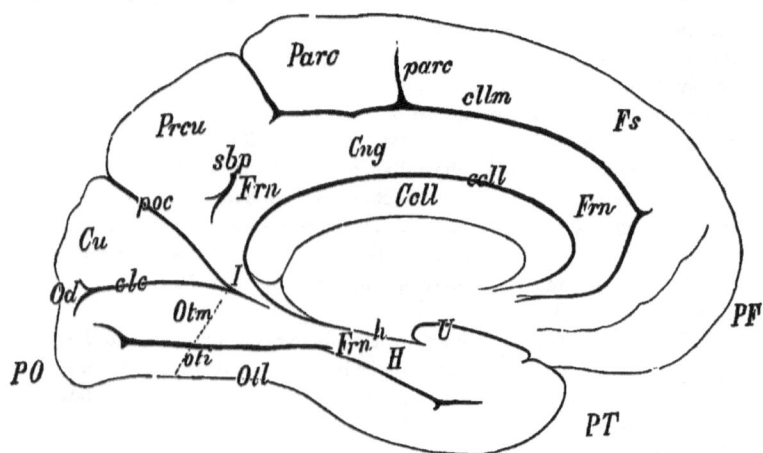

Fig. 33.—Left hemisphere, mesial surface. *Half nat. size.*—*Ccll,* Corpus callosum; *ccll,* sulcus corporis callosi; *clc,* fissura calcarina; *cllm,* sulcus calloso-marginalis; *Cng,* gyrus cinguli; *Cu,* cuneus; *Frn,* gyrus fornicatus; *Fs,* gyrus frontalis sup.; *H,* gyrus hippocampi; *h,* fissura hippocampi; *I,* isthmus gyri fornicati; *Od,* gyrus descendens; *oti,* sulcus occipito-temporalis inferior; *Otl,* gyrus occipito-temporalis lateralis; *Otm,* gyrus occipito-temporalis medialis; *Parc,* lobulus paracentralis; *parc,* sulcus paracentralis; *PF,* frontal pole; *PO,* occipital pole; *poc,* fissura parieto-occipitalis; *Prcu,* præcuneus; *PT,* temporal pole; *sbp,* sulcus subparietalis; *U,* uncus. The boundary between occipital and temporal lobes is shown by the dotted line, as also in fig. 30.

seu opertus *seu* centralis, lobus insulæ). Its boundaries are easily defined.

The following points are to be attended to in marking out these lobes:—

The convolution lying in front of the central fissure is sometimes reckoned to the parietal lobe.

On the mesial surface the boundary between frontal and parietal lobes is not marked. Supposing the central fissure were prolonged over the border on to this surface, it would divide a small, but most characteristic, lobule (the paracentral lobule) into two parts—a similar artificial division of the long convolution (gyrus fornicatus) which surrounds the corpus callosum would be necessary. The gyrus fornicatus is sometimes looked upon, as we shall presently see, as a special lobe.

By some (*Eberstaller* especially) it is denied that the occipital lobe reaches on the convex surface so far downwards and forwards as we have described.

It must not be forgotten that any division of the hemispheres into lobes is artificial, valuable only as a help to localising spots on its surface; we can easily overlook the faults which are inseparable from any method of delimitation.

[Although the classification of the lobes of the brain just discussed is highly artificial, and like all other attempts at mapping out the brain into lobes, has no object other than to enable one to indicate with precision localities upon its surface, it yet appears to the

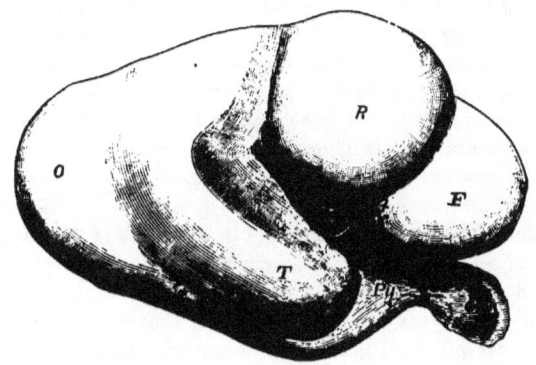

Fig. 34.—Diagram showing the lobation of the cerebrum.—*F*, Frontal lobe; *R*, Rolandic lobe; *O*, occipital lobe; *T*, temporal lobe; *I*, island of Reil; *Py*, pyriform lobe (uncinate gyrus); *OB*, olfactory bulb.

translator that the brain during its growth exhibits a well-marked tendency to bulge into defined lobes. A survey of all accessible mammalian brains leads him to the conclusion that these natural lobes have a distinct morphological, and, therefore, presumably also a distinct physiological significance. As shown in the accompanying diagram, the anterior end of the cerebral hemisphere is the part which has the appearance of greatest stability. The appearance of the fossa of Sylvius on the outer surface seems to be due to an intimate relation which exists between the nucleus lenticularis and the overlying cortex, whereby a portion of the surface, afterwards known as the island of Reil, is fixed and prevented from participating in the free growth of the rest of the hemisphere. The result of this fixation of the floor of the fossa of Sylvius is a bulging of the general surface of the hemisphere over the fossa, by which it comes at last to be covered in at the bottom of the "fissure" of Sylvius. In its overgrowth the surface exhibits a lobar conformation. The frontal lobe swells backwards; but the growth of this region is less exuberant than that of the rest of the convex surface. The portion of the brain which surrounds the crucial or central (Rolandic) sulcus—the sigmoid gyrus of animals—the ascending frontal and parietal convolutions or operculum of Man, constitutes the most distinct of all the lobes of the brain. Examination of a large number of brains leads to the conviction that the crucial and central sulci are homologous; but the marking out of the lobe is not affected by the view taken upon this question. The sigmoid gyrus, or operculum, as it may well be called, grows downwards as a lappet which overhangs the fossa of Sylvius. The development of this lobe varies distinctly as the force, rapidity, and specialisation of movement exhibited by the animal. The inferior and posterior part of the hemisphere bulges forwards as the temporo-sphenoidal lobe, across or below the fossa of Sylvius, its position depending upon the extent to which this fossa is overhung by the lobes already mentioned. The prolongation backwards of the posterior and superior part of the hemisphere as a natural lobe is obvious in many animals. In the rabbit, for example, it assumes a rounded form, the surface between the lobe and the rest of the brain being somewhat depressed. In this respect the brain of the rabbit contrasts remarkably with that of the mole. These four bulgings, frontal, opercular, occipital, and temporo-sphenoidal, are the largest and most distinct, but they include other less obvious elevations. It is very difficult to say how that portion of the surface which lies behind the Rolandic and temporo-sphenoidal and in front of the occipital lobe should be allocated, although there are sufficient indications of the existence upon it of other less pronounced swellings. Doubtless each region of the cortex, in which a variable function is localised, is liable to variations in size (*cf.* figs. 149, 150, 151)].

1. Frontal Lobe.—Three surfaces are to be distinguished—lateral, mesial, and basal. Since the basal surface lies on the roof of the orbit, it is often termed "orbital." Three constant fissures are found on the lateral surface—

(1) Sulcus præcentralis, *prc* + *prcs*, fig. 30 (vertical frontal fissure, sulcus prærolandicus), lies in front of and almost parallel with the central fissure.

(2) Sulcus frontalis superior, *fs;* and

(3) Sulcus frontalis inferior, *fi*, runs forwards from the præcentral sulcus, parallel with the inner border of the hemisphere.

The præcentral sulcus, which begins a short distance above the

Sylvian fissure, does not, as a rule, reach so far as the posterior end of the superior frontal fissure; a short fissure, *prcs*, running in the same direction is, however, always to be found at the hinder end of the superior frontal fissure, and it may be regarded as the continuation upwards of the præcentral which is then divided into two, sulci præcentrales inferior et superior. One of the longitudinal frontal fissures starts from each of these divisions. Usually the superior præcentral fissure runs a little downwards as well as upwards from the superior frontal.

Four convolutions are marked out by these fissures:—

(1) Gyrus centralis anterior, *Ca* (ascending frontal, præcentral, premier pli ascendant); a convolution which, running parallel with the central fissure, of which it forms the anterior boundary, traverses the whole of the lateral surface of the hemisphere from the fissure of Sylvius upwards. From it there extend forwards—

(2) Gyrus frontalis superior, *Fs* (upper, first or third (*Meynert*), frontal convolution, gyrus frontalis marginalis).

(3) Gyrus frontalis medius, *Fm* (middle or second frontal convolution).

(4) Gyrus frontalis inferior, *Fi* (inferior, third, or first (*Meynert*) frontal convolution, pli surcilier; *on the left side only*, Broca's convolution). [The region of the cortex, injury to which produces aphasia, was localised by Broca as the back of this convolution on the left side at its junction with the ascending frontal.]

The superior frontal convolution includes the border of the hemisphere, for it extends over on to the mesial surface. Its lateral surface is often, like that of the gyrus frontalis medius, complicated with a number of shallow inconstant fissures.

The inferior frontal convolution running forwards from the lower end of the ascending frontal winds round both the anterior ascending and the anterior horizontal limbs of the fissure of Sylvius. Hence it is divided into three parts—(*a.*) pars opercularis, *Pop*, between the sulcus præcentralis and the ramus ascendens fissuræ Sylvii; (*b.*) pars triangularis, *Ptr* (cap de la circonvolution de Broca), between the ascending and anterior horizontal rami; (*c.*) pars orbitalis, *Porb*, in front of the horizontal ramus; this latter properly belongs to the orbital surface of the frontal lobe.

"Connecting-convolutions" between the several frontal convolutions commonly complicate the survey.

All three frontal convolutions are continued on to the orbital surface of the lobe. Here the arrangement of fissures and convolutions is very inconstant (fig. 35). It often happens that the superior (here mesial) and inferior (here lateral) frontal convolutions run backwards as far as the anterior perforated space, *Spa*, where they are united together by a connecting piece. The middle convolution is thus shut off from the perforated space by the folding together of the

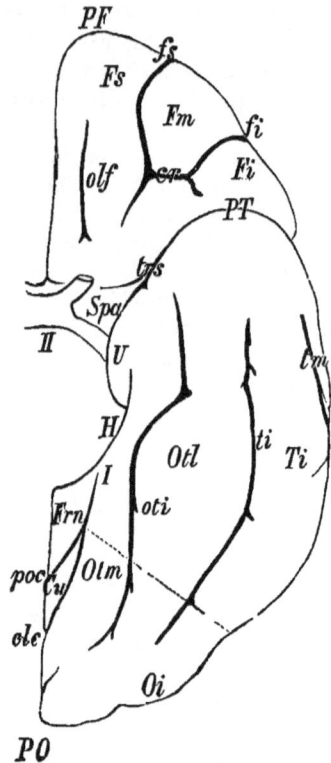

Fig. 35.—Left hemisphere from the base. *Half nat. size.*—*II*, Chiasma nervorum opticorum; *clc*, fissura calcarina; *cr*, sulcus cruciatus; *Cu*, cuneus; *Fi*, gyrus front. inferior; *fi*, sulcus frontalis inferior; *Fm*, gyrus front. medius; *Fn*, gyrus fornicatus; *Fs*, gyrus frontalis superior; *fs*, sulcus frontalis superior; *H*, gyrus hippocampi; *I*, isthmus; *Oi*, gyrus occipitalis inferior; *olf*, sulcus olfactorius; *oti*, sulcus occipito-temporalis inferior; *Otl*, gyrus occipito-temporalis lateralis; *Otm*, gyrus occipito-temporalis medialis; *PF*, frontal pole; *PO*, occipital pole; *poc*, fissura parieto-occipitalis; *PT*, temporal pole; *Spa*, substantia perforata anterior; *Ti*, gyrus temporalis inf.; *ti*, sulcus temporalis inferior; *tm*, sulcus temporalis medius; *trs*, truncus fissuræ Sylvii; *U*, uncus.

other two. All the fissures on the orbital surface unite together to form an H or an X, sulcus cruciatus, *cr* (orbitalis, cruciformis, triradiatus). A fissure which runs parallel with the great horizontal fissure is always to be seen on the orbital portion of the superior frontal convolution. In this lies the olfactory tract. It is termed, therefore, the olfactory fissure (*seu* sulcus rectus), *olf.*

It is quite unjustifiable to look upon the orbital surface of the frontal lobe as a lobe proper (lobus orbitalis). The mesial surface of the frontal lobe is better described later on, in connection with the mesial surface of the other lobes.

The most anterior point of the frontal lobe is termed the frontal pole, *PF*.

The inferior frontal convolution hardly ever reaches, without interruption, on to the orbital surface. Usually a short transverse fissure from 3 to 5 cm. long, sulcus fronto-marginalis (*Wernicke*), is found in the vicinity of the frontal pole. It is not shown in fig. 30.

2. **Parietal lobe** presents a lateral and a mesial surface, of which only the former will be considered now. A single typical fissure indents this lobe, the intraparietal, *ip*, fig. 30 (sulcus parietalis, fissura parietalis cum f. paroccipitalis of *Wilder*). It commences behind the central fissure and above the fissure of Sylvius. At first it ascends parallel to the central fissure; it then sweeps backwards in a great curve, and finally crosses the imaginary boundary between the parietal and occipital lobes to end in the latter. A continuation of its first portion ascends parallel to the central fissure towards the border of the hemisphere, which it does not, however, reach; so that, in a sense, a third transverse fissure is formed (the sulci præcentralis et centralis being the other two), which may be called sulcus centralis posterior (*seu* postrolandicus), *pstc*. Interruptions to the course of the intraparietal fissure are very common, especially on the right side. A short lateral branch, *a*, which passes towards the border of the brain, running at the same time backwards, in front of the parieto-occipital fissure, is almost constant.

Three convolutions are to be distinguished in the parietal lobe:—

(1) Gyrus centralis posterior, *Cp* (ascending parietal, postrolandicus, deuxième pli ascendant), is bordered in front by the central fissure. Around the upper end of this fissure it becomes continuous with the anterior central convolution with which it has been running parallel. Its upper part is usually narrow, and therefore markedly different from the broad upper part of the anterior central convolution.

(2) Gyrus parietalis superior, *Ps* (lobulus parietalis superior, præcuneus, gyrus parietalis primus), is that portion of the parietal lobe which lies behind the posterior central convolution and above the intraparietal fissure. It extends over the border of the hemisphere on to the mesial surface. This portion of the convolution is known as præcuneus, *Prcu*, fig. 36.

(3) Gyrus parietalis inferior, *Pi*, fig. 30 (lobulus parietalis inferior, lobulus tuberis, gyrus parietalis secundus), lying beneath the intraparietal fissure skirts around the hinder end of the Sylvian fissure (this portion is called the supramarginal gyrus, *Sm*), and then in a similar manner, encloses the superior temporal fissure which lies below and parallel to the Sylvian fissure. This second portion of the inferior parietal lobule is known as the gyrus angularis (pli courbe), *Ang*. The inferior parietal convolution is by no means sharply bounded on its occipital side. [There is no other part of the hemisphere the definition of which, as a whole, and its division into parts is so difficult as the inferior parietal lobule. Either of its constituent convolutions may be cut into two by the fissure which it normally confines. Supramarginal and angular convolutions may be simple as in fig. 30 or folded so much that their outline is difficult to trace.]

The inferior frontal convolution, with the exception of its orbital portion, together with the united lower ends of the two central convolutions and the inferior parietal, so far as it lies over the island of Reil, constitute the operculum, *Op*. If the operculum be lifted up, or a frontal section made through the brain, it is seen that its deep surface, which looks towards the fossa Sylvii and the temporo-sphenoidal lobe, is marked by several inconstant fissures.

The mesial surface of the temporal lobe will be described later on.

3. The Occipital Lobe.—The occipital lobe has as the whole the form of a three-sided pyramid with its base resting upon the parietal and temporo-sphenoidal lobes, and its apex projecting as the occipital pole, *Po*. Hence we have to distinguish three surfaces, lateral, mesial and inferior (or basal). At present the lateral only will be dealt with.

The fissures on the lateral surfaces are very inconstant, the following being more easily found than the rest :—

(1) Sulcus occipitalis transversus, *otr* (hinder transverse portion of the intraparietal fissure). It lies behind the parieto-occipital fissure, and is, as a rule, continuous with the intraparietal. It runs transversely across the occipital lobe for a variable distance, and is to be regarded as the analogue of the conspicuous fissure which occupies this situation in the monkey's brain.

(2) Sulcus occipitalis lateralis, *ol* (sulcus occipitalis longitudinalis inferior). The fissure looks as if it were the prolongation backwards of the principal portion of the upper temporal fissure. It lies in the line which this fissure would follow if prolonged backwards, on the lower part of the occipital lobe, nearly to the occipital pole. *Eberstaller* looks upon it as the inferior boundary of the occipital lobe.

Three not always equally well-defined convolutions converge towards the occipital pole—

(1) Gyrus occipitalis superior, *Os* (gyrus occipitalis primus *seu* parieto-occipitalis medialis).

(2) Gyrus occipitalis medius, *Om* (*seu* secundus).

(3) Gyrus occipitalis inferior, *Oi* (*seu* tertius, *seu* temporo-occipitalis).

The superior occipital passes into the superior parietal convolution through the medium of a connecting convolution which curves round the lower end of the parieto-occipital fissure (gyrus paroccipitalis of *Wilder*, premier pli de passage of *Gratiolet* [first annectant convolution of *Turner*]). The middle convolution is the continuation of the gyrus parietalis superior (gyrus angularis). The inferior occipital ends by joining with the middle, and in part also with the inferior, temporal convolution.

4. The Temporal Lobe.*—It presents a lateral and an inferior surface which are continuous with one another around the outer margin of the brain. Four fissures, all sagittal in direction, are to be distinguished. From the Sylvian fissure downwards they are as follows :—

(1) Sulcus temporalis superior, *ts* (parallel fissure [superior temporo-sphenoidal fissure], sulcus temporalis primus), a very constant and obvious fissure. Its chief portion is directed straight backwards towards the occipital lobe; its hinder end, which turns upwards, is surrounded by the gyrus angularis.

(2) Sulcus temporalis medius, *tm* (sulcus temporalis secundus), very often interrupted by bridging convolutions.

(3) Sulcus temporalis inferior, *ti* (*seu* tertius).

(4) Sulcus occipito-temporalis inferior, *oti* (inferior longitudinal fissure, fissura collateralis).

Of these four the two first are visible on the lateral surface of the brain; the remaining two belong to its under surface.

The convolutions on the lateral surface are arranged in three parallel folds like those of the frontal lobe, only more simply. In front, at the tip of the temporo-sphenoidal lobe (extremitas temporalis, temporal pole, *PT*), these three convolutions, as well as some of those which lie on the under surface, unite in a rounded dome.

(1) Gyrus temporalis superior, *Ts* (gyrus inframarginalis, parallel convolution, gyrus temporalis primus). This convolution forms the lower boundary of the Sylvian fissure; it is continued behind into the inferior parietal lobule.

If the lobes of the brain are pulled apart, so that a view is obtained of the fossa Sylvii in all its depth, it will be seen that, just as in the

---

* In *English text-books* termed usually *temporo-sphenoidal lobe*, a somewhat cumbrous papellation.

case of the under side of the operculum, so also with regard to the upper surface of the temporo-sphenoidal lobe (figs. 21 and 22), a considerable portion of cortex, hitherto hidden in the fissure, is brought to light. Three, and even in some cases four, convolutions are thus exposed. They originate in the superior temporal convolution and converge backwards towards the hinder angle of the island of Reil. The most constant and longest of these gyri temporales transversi (*Heschl*) is the anterior, *Tt* (fig. 21).

(2) Gyrus temporalis medius, *Tm* (*seu* secundus).

(3) Gyrus temporalis inferior, *Ti* (*seu* tertius), forms the transition from the lateral to the under surface of the temporal lobe.

(4) Gyrus occipito-temporalis lateralis, *Otl* (gyrus *seu* lobulus fusiformis), lying between the sulcus temporalis inferior and the sulcus temporo-occipitalis inferior, is usually broadest in the middle, and, therefore, more or less spindle-shaped. It can almost always be followed as far backwards as the occipital pole, hence it is also an essential constituent of the under surface of the occipital lobe.

(5) Gyrus occipito-temporalis medialis, *Otm* (gyrus *seu* lobulus lingualis), between the inferior occipito-temporal and the calcarine fissures, also originates on the under surface of the occipital lobe, of which it occupies the greater portion. It narrows in front and passes on into the gyrus hippocampi, which convolution, not mentioned in our enumeration, is really the last or sixth of the sagittally-running temporal convolutions; as in a certain sense it represents the mesial aspect of the lobe, it will be described in connection with the other convolutions on the mesial surface of the hemisphere.

The Mesial Surface of the Hemisphere.—On this surface the arched form of the hemisphere is most conspicuous, not only in the arrangement of the whole structure, but also in the configuration of its several constituents.

The sagittal section of the corpus callosum, *Ccll*, fig. 36, has the form of an arch, around which curves a convolution which commences beneath the rostrum of the corpus callosum on the frontal portion of the mesial aspect. It is continued backwards over and beneath the splenium, and runs forwards even as far as the apex of the temporo-sphenoidal lobe. This convolution is the gyrus fornicatus, *Frn*. It is separated from the corpus callosum by the sulcus corporis callosi, *ccll*. It comprises two portions—(1) the part lying close to the corpus callosum, gyrus cinguli, *Cng* (gyrus corporis callosi; often this portion alone is reckoned as gyrus fornicatus); and (2) a free-lying portion, gyrus hippocampi, *H* (subiculum cornu Ammonis). The portion of the gyrus fornicatus in which these two segments are united is strikingly constricted, *I*, isthmus gyri fornicati. Here the middle occipito-temporal convolution, *Otm*, becomes superficial, while another constituent of the mesial surface of the hemisphere, the cuneus, *Cu*, is insinuated between it and the gyrus fornicatus. It is connected with the latter by the stalk of the cuneus, *PCu*, fig. 29 (pedunculus cunei). The gyrus hippocampi swells out considerably at the anterior part of the temporo-spenoidal lobe, forming a hook-like curve, *U* (uncus, gyrus uncinatus). The inner boundary of the arch which forms the gyrus fornicatus corresponds with the embryonic sulcus arcuatus [of *Arnold*]. It is represented in the region of the gyrus cinguli by the sulcus corporis callosi, *ccll;* in its lower portion it corresponds to the fissura hippocampi, *h* (*cf.* p. 89).

Broca describes the gyrus fornicatus (with addition of the olfactory tract) as a special lobe, lobus limbicus. Similarly *Schwalbe*, on genetic grounds, institutes his lobus falciformis, which comprises the gyrus fornicatus, septum pellucidum, and fascia dentata.

[In all animals, the lower and anterior part of the hemisphere is distinguished from the rest of the cortex-covered cerebrum by a profound difference in appearance. The roundly-swelling convoluted part of the hemisphere terminates at the rhinal fissure. The portion of the hemisphere below this fissure, known as the pyriform lobe, is flatter than the rest, indented by blood-vessels, but not fissured, and usually whiter in colour. In front it tapers off into the olfactory tract and bulb. Behind it suddenly narrows into the gyrus fornicatus, or first convolution embracing the corpus callosum. The size of the pyriform lobe varies as the development of the olfactory apparatus. On the other hand, the extent to which it is overlapped by the temporal lobe depends upon the size of the latter, which is large in osmatic, small in anosmatic, animals. In Carnivora the temporal lobe projects far forwards over the pyriform, the rhinal fissure being bent at an acute angle. In Herbivora the rhinal fissure is almost straight, and the pyriform lobe is exposed on the lateral surface of the hemisphere. In Man, although this broad anatomical distinction between the pyriform lobe and the rest of the hemisphere is not visible, it is easy to recognise in the uncinate gyrus (including the gyrus hippocampi) the pyriform lobe of animals. The extreme anterior end of the gyrus fornicatus, the "terrain désert" of *Broca*, resembles to some extent the gyrus uncinatus in its superficial aspect, and is also connected, apparently, with one of the roots of the olfactory tract. It was quite unjustifiable, however, to introduce the gyrus fornicatus into the connection, making one large "lobe limbique" in the "form of a racquet," since the ground upon which this was done, the supposed connection of the limbic lobe with the sense of smell, is at once shown to be untenable by the fact that in animals in which the sense of smell is totally absent (marine mammalia) the gyrus fornicatus is well developed.]

The portion of the mesial surface of the hemisphere, to which the gyrus fornicatus does not lay claim, is occupied by cortical formations belonging to the convolutions already described. These are the convolutions, which, lying on the border of the hemisphere, belong as well to its mesial as to its lateral and under surfaces.

One fissure, sulcus calloso-marginalis, *cllm* (sulcus fronto-parietalis internus), which commences below the genu corporis callosi and

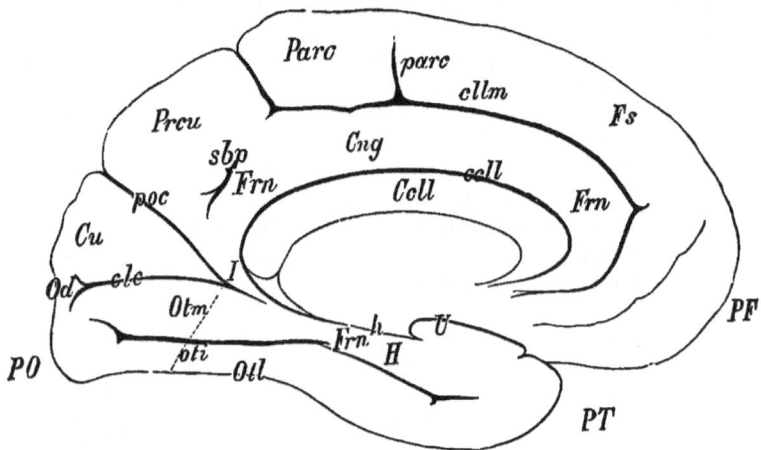

Fig. 36.—Left hemisphere, mesial surface. *Half nat. size.*—*Ccll*, Corpus callosum; *ccll*, sulcus corporis callosi; *clc*, fissura calcarina; *cllm*, sulcus calloso-marginalis; *Cng*, gyrus cinguli; *Cu*, cuneus; *Frn*, gyrus fornicatus; *Fs*, gyrus frontalis sup.; *H*, gyrus hippocampi; *h*, fissura hippocampi; *I*, isthmus gyri fornicati; *Od*, gyrus descendens; *oti*, sulcus occipito-temporalis inferior; *Otl*, gyrus occipito-temporalis lateralis; *Otm*, gyrus occipito-temporalis medialis; *Parc*, lobulus paracentralis; *parc*, sulcus paracentralis; *PF*, frontal pole; *PO*, occipital pole; *poc*, fissura parieto-occipitalis; *Prcu*, præcuneus; *PT*, temporal pole; *sbp*, sulcus subparietalis; *U*, uncus. The boundary between occipital and temporal lobes is marked with a dotted line.

forms an arch parallel with the arch of the corpus callosum, about midway between it and the border of the hemisphere, constitutes the upper boundary of the gyrus cinguli. A little in front of the splenium this fissure cuts its way up to and over the border of the hemisphere, appearing, for a short distance, on the lateral surface, behind the central fissure. Above the centre of the corpus callosum the calloso-marginal fissure sends upwards a short lateral branch, sulcus paracentralis, *parc*. After the calloso-marginal fissure has turned upwards towards the margin, its original arched direction is only continued by a shallow depression or an inconstant fissure, sulcus subparietalis, *sbp*.

Apart from the gyrus fornicatus a number of different named areas are met with on the mesial surface of the hemisphere, commencing at the frontal end :—

1. Gyrus frontalis superior, *Fs*.

2. The loop of communication between the upper ends of the two transverse convolutions, lobulus paracentralis, *Parc*, which reaches backwards to the ascending portion of the sulcus calloso-marginalis, while its anterior boundary is formed by the sulcus paracentralis.

3. Præcuneus, *Prcu* (lobulus quadratus, mesial surface of the gyrus parietalis superior), is a plump four-sided piece of cortex, of about the same size as the lobulus paracentralis, which extends as far backwards as the internal parieto-occipital fissure. Exceptionally, the superior parietal lobe, which, undoubtedly, is continuous with the præcuneus around the border of the hemisphere, is included under this name.

4. The cuneus, *Cu* (lobulus triangularis, mesial surface of the gyrus occipitalis superior), a conspicuous portion of the cortex lying between the parieto-occipital and calcarine fissures. The front of the wedge runs forwards as far as the isthmus fornicatus. It is known as the pedunculus cunei, *PCu* (fig. 29, not shown in fig. 36).

5. The gyrus occipitalis descendens, *Od*, is seen as a narrow convolution lying behind the bifid end of the calcarine fissure near the occipital pole. It unites the cuneus with

6. The gyrus occipito-temporalis medialis, *Otm*.

The Island of Reil.—This region is hidden by other portions of the hemisphere which grow over it from three sides. The result is, that the island assumes the form of a triangular pyramid, the base of which rests on the brain-stem, while its apex (pole of the island of Reil, the point on its surface which projects farthest laterally) is cut off from the rest of the cortex by the sulcus circularis Reili. It is only wanting in front and on the ventral side, where the island is continuous with the lamina perforata anterior (limen insulæ) *trs* (fig. 35). The island is divided by a constant fissure, sulcus centralis insulæ, which runs almost parallel with the sulcus Rolandi, into a larger anterior and a smaller posterior part. The posterior part is split up into a number of convolutions, gyri recti (*seu* breves insulæ, *Guldberg* and *Eberstaller*).

The simple arrangement of convolutions described above is usually masked by the presence of numerous atypical furrows and convolutions; hence it is often difficult for an inexperienced observer to find his way about the surface of the hemisphere and pick out the typical convolutions from the apparent chaos. The first fissure to look for is always the Sylvian, which cannot be confused with any other. Next the central fissure [of *Rolando*] is found. In looking for this a mistake might in the first moment be made. A good guide, however, is afforded by the posterior transverse convolution rapidly narrowing in its upper part and bounded at the top by the portion of the calloso-marginal fissure which turns over on to the lateral surface. If now the parieto-occipital fissure is found as it turns over the border of the hemisphere, a sufficient number of starting points are determined to enable the student to orient the rest of the surface.

The beginner is strongly recommended to study the convolutions and fissures in as many brains as possible in order that he may learn to recognise them quickly. As long as the surface is covered by pia mater the fissures are extremely difficult to find.

VARIETIES AND ANOMALIES OF CONVOLUTION.

We have so described the principal and chief subordinate fissures as to make it possible to find them with more or less ease in any normal brain.

Non-essential differences in the arrangement of the primary and secondary fissures, as well as wide variations in the course of the tertiary fissures, characterise individual brains. There are always differences between the two hemispheres, and the richer the brain in convolutions the greater is the difference between its two sides. Similar individual variations in the arrangement of the convolutions in different individuals and differences between the two hemispheres are found in animals. In them, too, variation increases with development.

The form of the skull is of great importance in determining the type of brain-convolution. In dolichocephalic heads the brains incline towards an exaggeration of the longitudinally arranged convolutions and fissures, while in people with brachycephalic skulls the transverse fissures and convolutions are dominant. Early synostosis alters the arrangement of the convolutions in a similar manner (*Zuckerkandl*).

One is often tempted to assert that complexity in convolution keeps pace with intellectual power; such a connection is not, however, demonstrable in every individual case. Often it seems possible to recognise a conspicuous development of convolutions to which a definite physiological purpose can be assigned in individuals remarkable for the preponderance of the corresponding faculties.

The best example of this is the inferior frontal convolution [of the left side] which is in intimate connection with the faculty of speech. *Rüdinger* asserts that in the brains of great orators this convolution is strongly developed; in the otherwise exceptionally small brain of Gambetta,* the pars triangularis was large, strongly twisted and, in a certain sense, doubled (*Duval*). On the other hand, the pars opercularis may be so slightly developed, that a portion of the island of Reil is uncovered and exposed from the surface. Great pains have been taken to discover a difference in type of convolution in the two sexes, but only slight and inconstant differences have been discovered (*Huschke, Wagner, Rüdinger*). It has been pointed out that the frontal lobe is better developed in man; the sulcus centralis longer than in women (*Passet*).

All the principal fissures are present in human brains at the time of birth, but secondary and tertiary fissures are still some time from their complete development (only a month according to *Sernoff*). The relation of the fissures to one another changes during the growing period, some parts of the brain developing quickly, others lagging behind; thus it comes about that the angle, open in front, which the two central fissures make with one another, averages in the child 52°, in adults 70° (*Hamy*).

If, as the result of senile marasmus, or owing to other causes (chronic mental disease, for example), atrophy of the brain sets in, the convolutions become narrower, the fissures broader. On the contrary in hypertrophy of the brain, the convolutions are pressed up against the bones of the skull and flattened.

When the fissures are conspicuously increased, although they are only superficial, the condition is known as polygyry. Occasionally little knobs of cortical substance project from the surface of convolutions, especially the superior frontal.

The frequency of **anomalies in convolution** is variously estimated; some people describing as an anomaly what is only regarded by others as a variation.

For example, the sulcus centralis may descend into the Sylvian fissure, or it may, as more often happens, end at some distance from it. Neither case can be properly regarded as an anomaly. Rather might one regard as anomalies those cases in which this fissure is divided into an upper and a lower half (*Heschl*), owing to the exceptional development of a convolution, which is almost always present although out of sight (the posterior end of the middle frontal convolu-

* [The report which went the round of the newspapers at the time of Gambetta's death that his brain was phenomenally small, weighing no more than 1,100 grms., seems to have been a complete mistake. *Duval's* estimate of 1,241 grms. has been confirmed by *Rüdinger* and others.]

tion); or again, a condition in which the gyrus centralis is double (*Giacomini*). Anomalies, in the strict sense of the word, are extremely rare. The sulcus postcentralis, or the anterior end of the sulcus intraparietalis, not rarely cuts into the Sylvian fissure.

More striking anomalies of the brain-surface are found as purely teratological conditions; for example, cyclopia, associated in microcephalic brains, with absence of certain parts, as, for example, the corpus callosum, the occipital lobes (inoccipitia, *Richter*), or the olfactory lobes (arrhinencephalia, *Kundrat*). Also as the result of certain destructive pathological processes occurring either in intra- or extra-uterine life; for example, in porencephalia a certain portion of the surface of the brain is absent, and the ventricle is only separated from the surface by the meninges. A very rare, but interesting, anomaly in the arrangement of the convolutions, is that condition in which the two hemispheres are not completely separated from one another, but certain convolutions bridge across the great longitudinal fissure (*Hadlich*, *Wille*, *Kundrat*, *Arnold*, *Turner*).

PHYSIOLOGICAL MEANING OF THE CONVOLUTIONS.

Our exact knowledge of the topography of the cortex has been acquired since the time when it was first realised that the different regions into which it is divided are endowed with separate functions. Some physiologists still either disbelieve in **localisation** or only allow the application of the law in a modified sense; but a long series of successful clinical diagnoses now place beyond the reach of contradiction the fact that certain regions of the cortex are to a greater extent than the rest associated with certain functions. A full agreement as to the division of functions in the cortex does not yet obtain amongst the followers of the localisation theory, on which account we shall content ourselves with stating those points which may be considered as definitely settled. We shall take our stand upon a moderate localisation, such as was first enunciated by *Exner*.

Individual centres and cortex-fields are not to be considered as sharply outlined and definitely marked off from neighbouring regions; the so-called centres are rather the spots of maximal relation to functions which fade away into neighbouring areas. Hence it follows that the cortex-fields to a certain extent overlap one another.

In the following summary we shall speak of the centres in this sense as comprehending the spots of maximal physiological relation.

The functions of the gyrus frontalis superior and medius are not sufficiently well known. Attempts have often been made to associate them with the higher psychical functions—"intelligence." They suffer most in dementia paralytica. The gyri frontalis inferior, centrales anterior et posterior, the lobulus paracentralis, and the anterior part of the superior parietal lobule, together constitute the motor region (motor-field or -sphere). Here the motor activity of the cortex is localised to the greatest extent; it chiefly controls the muscles of the opposite side of the body; to a subordinate degree those of the same side also.

The groups of muscles represented in this region may be classified as follows:—

Tongue muscles—gyrus frontalis inferior, especially its pars opercularis, and probably also its pars triangularis (left-side; motor speech-centre).

Face muscles—under part of the gyrus centralis anterior.

Muscles of the upper extremity—middle part of the gyrus centralis anterior extending over on to the gyrus centralis posterior.

Muscles of the lower extremity—upper part of both gyri centrales, lobulus paracentralis, and some of the anterior part of the lobulus parietalis superior.*

[Muscles of the trunk—marginal gyrus in front of the paracentral lobule—*Schäfer* and *Horsley*.]

A safe localisation of the remaining voluntary muscles is not yet possible. The involuntary muscles stand, in all probability, in no direct dependence upon the cortex of the brain.

Munk's view, that sensation for the regions corresponding to the muscles thus represented in the cortex is also localised in these centres, cannot yet be maintained on clinical evidence.

The functions of a large part of the parietal lobe have not yet been cleared up.

The occipital lobe, its cuneus certainly, and, perhaps, the neighbouring portion of the parietal lobe (that is to say, the gyrus angularis, *Ferrier*), are connected with the sense of sight. This region is the seat of sight-perceptions for the temporal half of the retina of the same side, and the nasal half of the retina of the opposite side. Whether or not the motor centres for the extrinsic eye-muscles lie in this region, extending somewhat over to the neighbouring parts of the parietal lobe must be left undecided.

The temporal lobe probably stands in the same relation to perceptions of sound as the occipital lobe to perceptions of sight; this holds good for the upper convolution only, or at the most for this and the middle

* For a detailed account of the anatomical situation of the motor centres, see *Schäfer* and *Horsley—Proc. Royal Society*, vol. xxxvi., p. 437; *Phil. Trans.*, vol. clxxix., B, p. 1, and several papers in *Brain*.

convolution. The anterior part of the temporo-sphenoidal lobe, especially the region of the uncus, is intimately connected with the olfactory apparatus. All the rest of this very important region is, in its physiological relations, as yet unexplained.

From what has been said it will be evident that a sharp delimitation of the cortical centres does not in effect exist. Probably individual variations are present in no slight degree. As far as is known the portions of the cortex hidden away in the fissures join in function with the superficial parts.

With regard to the fissures, it is not yet decided whether they simply serve the purpose of increasing the superficial spread of the cortex or also at the same time serve to mark out territories physiologically distinct. [No question in the comparative anatomy of the brain concerns the neurologist more closely than the question as to the morphological value of the fissures. Are they merely plaitings of a shifting surface, or are they the boundaries of morphologically distinct organs? A study of the development of the fissures in the brains of animals which stand far apart in phylogeny teaches that they appear with such regularity as to sequence and progressive extension, and obey such definite rules as to relative depth as would, in the case of other parts of the body, justify us in considering them as the divisions between structures of separate function.] According to the views above set forth the second object could, owing to the partial overlapping of the centres, be effected to a limited extent only by the fissures.

There can be no manner of doubt that skull-case and skull-contents mutually influence one another's growth, but it would be quite a mistake to trace the arrangement of the convolutions to the resistance offered by the wall of the skull.

This much may be taken as certain—the fissuring of the cortex of the great brain effects an increase in its superficial area. The same holds good for the cerebellar cortex, the convolutions of the inferior olive and the corpus dentatum cerebelli. Plaits of the vascular pia mater extend into the fissures of the cerebrum and cerebellum, carrying thus to its substance the greatest amount of nutriment possible. From this point of view the fissures are nutrient in function (*J. Seitz*).

For all convolutions the law holds good that the thinner the cortex the narrower the convolutions. Hence the occipital convolutions are the narrowest of those of the great brain. The cerebellar convolutions are still narrower.

Comparative anatomy could, until recently, offer but few points of view in the consideration of the fissures of the great brain. The larger the animal the more convoluted is its brain; but other factors besides size and intelligence determine its richness in convolution (*Krueg*).

SECTION III.—HISTOLOGICAL ELEMENTS OF THE CENTRAL NERVOUS SYSTEM.

For a right apprehension of the structure of the central nervous system a knowledge of all the elements of which it is composed is absolutely necessary.

It is not exclusively composed of nervous elements; indeed, these are enfolded in a network of other structures which serve for their nutrition and support.

The following is a summary of all the different kinds of tissue met with in the central nervous system, of each of which we shall have later on to give a detailed account:—

    A. Nervous constituents.
        1. Nerve-fibres.
        2. Nerve-cells.

    B. Non-nervous constituents.
        1. Vessels.
        2. Epithelia.
        3. Supporting tissues:
            (a.) Connective tissues;
            (b.) Neurogleia.

A. NERVOUS CONSTITUENTS.

1. NERVE-FIBRES.

We shall soon see that we must distinguish several kinds of nerve-fibre; there is, however, one histological constituent, **the axis-cylinder,** common to them all, for its presence is characteristic of a nerve-fibre. The axis-cylinder can only with difficulty be seen in fresh peripheral nerve-fibres. It is only distinct after the action of various reagents; so that for a time people doubted its existence in the living fibre, and took it to be an artificial product.

There are various methods for bringing out the axis-cylinder. A fresh nerve should be taken from a recently killed animal—for example, the sciatic nerve of a frog serves particularly well on account of its large fibres, a part is finely teased as quickly as may be without the addition of anything, except a little serum placed on an object glass, care being taken that the individual fibres are long and spread straight out. Afterwards collodion is put round the preparation, and it is covered with a cover-slip. The axis-cylinders are then seen as darker bands traversing the centres of the fibres; such a preparation can only be kept a short time.

If a coarsely teased piece of fresh nerve is laid for twenty-four hours in a weak solution of perosmic acid (0·1 per cent.), and then, after washing, more finely teased, a preparation is made which shows the axis-cylinder as a central clear band, as well as various other details of structure presently to be described. Such fibres can afterwards be coloured with picrocarmine, fuchsin, and other reagents.

The following good method for isolating fresh nerve-fibres—often a matter of some difficulty—is given by S. *Mayer*. A piece, about half a centimeter long, is cut out of a larger nerve in such a way that it is left lying on the subjacent muscle, upon which it can be raised

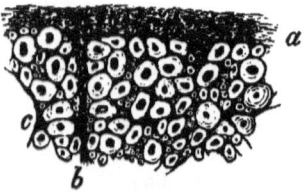

Fig. 37.—Cross-section from the anterior columns of the spinal cord, stained in carmine. *Magn*. 150 *diam*.—*a*, Peripheral grey cortex layer; *b*, smaller septum. In the medullary substance, besides the cross-sections of larger and smaller medullated fibres, three distinct stellate connective-tissue cells are seen, one of these is indicated by the letter *c*.

without being touched and placed upon a dark background. It will soon be seen that the strong nerve-sheath has retracted a little from its contents at both cut ends, the nervous substance projecting with but a small admixture of non-nervous element. It glistens like satin. If one end be now fixed with a needle, a small bundle can be easily drawn out from the other end with another needle, and allows itself to be teased without further resistance.

A section, cut transversely from a spinal cord hardened in bichromate of potassium and stained in carmine, is shown in fig. 37.

By these methods the axis-cylinder appears as an almost homogeneous band, which presents numerous curves and irregularities due

PRIMITIVE FIBRILLÆ.

to the methods employed for hardening. *Fleischl* says that the axis-cylinder is better preserved in fibres hardened in alcohol.

By employing other reagents, further details in the structure of the axis-cylinder may be brought to light. It can be resolved into a hollow membrane "axis-cylinder-sheath" filled with stiff protoplasm or, according to others, with fluid protoplasm, in which a number of exceedingly fine fibres are included (primitive fibrillæ). The number of these fibrillæ depends upon the thickness of the axis-cylinders. *Kupffer* counted over a hundred in the thicker fibres of the frog's sciatic nerve; their diameter is nearly always so inconsiderable that it is impossible to measure them, even when very highly magnified. In the centre of the axis-cylinder these fibrillæ are mostly closely packed, while the peripheral part often appears destitute of fibrillæ. The large nerve-fibres in the abdominal cord of the crayfish, examined in a drop of the animal's blood, show these bundles of fibrillæ as a central fasciculus (*Remak, Freud*). After maceration of fresh nerve-fibres in weak chromic acid, the fibrillar structure of the axis-cylinder becomes sometimes distinctly visible. For permanent preparations *Kupffer* recommends the following method. The nerve is fixed in a condition of physiological extension, which can be accomplished (as suggested by *Ranvier*) in the following manner:—The nerve is fastened by means of a ligature at either end to a little rod of wood (*e.g.*, a match) which has previously been cut thin at the middle. Nerve and wood are then laid for two hours in a 0·5 per cent. osmic acid solution, washed for two hours in distilled water, and then placed from one to two days in a strong watery solution of acid fuchsin. *Jacobi* finds that a concentrated solution of Bismarck-brown used in the same way yields better results. After that it is washed for six to twelve hours in

Fig. 38.—Medullated peripheral fibre. *Hardened in potassic chromate, stained in carmine, teased. Magn.* 200.—*a*, Axis-cylinder; *b*, Ranvier's node; *c*, nucleus of Schwann's sheath.

absolute alcohol, cleared in clove-oil, embedded in paraffin (maintained for twenty-four hours just above its melting point), and after that cut in longitudinal and cross-sections. Such preparations show the primitive fibrillæ in an otherwise uncoloured axis-cylinder. At those particular spots in the nerve, which we shall know later on as Ranvier's nodes, the individual fibrillæ come close together (*Boveri*).

Nansen looks upon the axis-cylinder as made of a large number of closely packed "primitive tubes;" these cylindrical tubes consist of an extremely fine connective-tissue sheath (spongioplasm) and viscous contents (hyaloplasm).

If small pieces of the spinal cord of a recently-killed animal are left for eight to fourteen days in a weak solution of silver nitrate (1 in 400) in the dark, washed in water, and then, after teasing, exposed for a short time to daylight in a drop of glycerin and distilled water, many axis-cylinders are met with amongst its fibres, which for some part

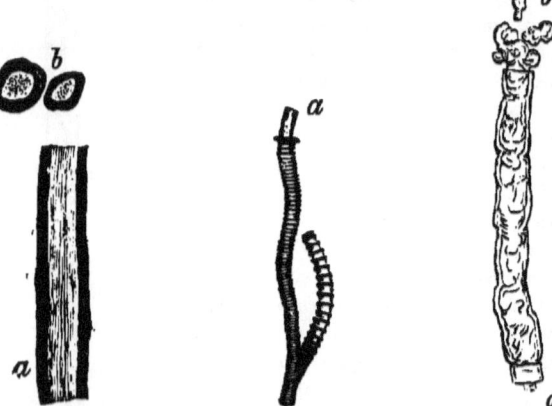

Fig. 39.—Peripheral nerve-fibres of the frog. *Perosmic acid and Bismarck-brown. Magn.* 1,000.— *a*, Longitudinal section; *b*, cross-section.

Fig. 40.—Axis-cylinder of a fibre from the white substance of the spinal cord. Its sheath stained with silver nitrate exhibits Frommann's stripes. At *a* the axis-cylinder is naked. *Magn.* 400.

Fig. 41.—A fresh medullated nerve-fibre from the sciatic of the frog. *Magn.* 200. Commencing coagulation of the myelin.—*a*, The axis-cylinder projecting free; *b*, escaping drops of myelin.

of their course appear coloured brown by the silver, owing to their being divested of their medullary sheath. On close observation it is seen that this colouration is not continuous for the most part, but made up of a succession of darker and lighter cross-bands. The width of these bands varies from 1 to 4 μ; on any small piece of the axis-cylinder, it is, however, regular, the cross stripes being in strong contrast with the longitudinal striation which the former methods have shown us.

This brown colouration with silver nitrate looks as if it affected the axis-cylinder itself; as a matter of fact, however, it is a delicate membrane investing the tube in which the axis-cylinder substance, with its primitive fibrillæ, is contained—**axis-cylinder-sheath**. [Not the axis-cylinder-sheath or "rind" in *Schiefferdecker's* sense; but a coagulum from the lymph which accumulates between the elastic rind and the medullary sheath on the shrinking of the former in the silver nitrate. According to *Schiefferdecker*, the delicate elastic rind encloses very fluid watery albumen in which it is possible, but not probable, that fibrillæ lie.] Often one sees the coagulated axis-cylinder projecting some distance out of its sheath. Upon what the peculiar nature of this silver impregnation depends is still uncertain; but it would, nevertheless, be quite a mistake to look upon it as an artificial product which could be passed over. The thickness of the axis-cylinder does not always stand in a direct relation to the breadth of the transverse striæ. This striation, first described by *Frommann*— hence known as Frommann's cross bands—is found not only in peripheral medullated nerves, but also in the not yet medullated fibres of the spinal cord of new-born animals. The axis-cylinder-sheath is in most nerve-fibres surrounded by still other investments, by the medullary sheath, Schwann's sheath, and the fibrillar sheath [Henle's sheath]. These three sheaths are found in most peripheral nerves, and hence may next afford us material for observation.

The **medullary sheath** which surrounds the axis-cylinder-sheath soon begins to coagulate, especially its outer layers, in freshly prepared nerves, producing the appearance characteristic of doubly-contoured nerves (fig. 41). Later on the medulla coagulates right up to the axis-cylinder in globular masses, which greatly alter the appearance of the fibre. At the cut ends of the nerves, these coagulation-products escape from the sheath of Schwann as peculiar rounded variously-shaped drops.

The greater number of stains, such as carmine for example, are absorbed but little by the medullary sheath, so that, in most methods of staining, this sheath is left nearly or quite uncoloured. A transverse section of a peripheral nerve, or of the white columns of the spinal cord shows colourless rings of medulla, usually stratified concentrically, surrounding coloured axis-cylinders. Sometimes some of these rings are stained (fig. 37) in such a way that either the cross-sections of certain nerves are coloured while others remain untouched; or certain nerves show several concentric rings of colour. This really depends, however, upon the coagulation-process and not upon any histological differences which might have physiological bearings. In longitudinal sections many nerve-fibres appear irregularly tinted, since for a distance the medulla is colourless and then again coloured. The object of several methods, as for instance the acid-fuchsin staining and Weigert's hæmatoxylin method, is to colour the medullary sheath.

Nerve-fibres which have been placed for twenty-four hours in a weak solution of perosmic acid (0·1 to 0·2 per cent.), examined in glycerin, show that the medullary sheath is not continuous. It is best to maintain them during this immersion in a condition of physiological extension. The medullary sheath is broken at regular intervals (1 to 2 mm. apart in the frog), leaving a space between the segments. The myelin sheath is somewhat enlarged as it approaches the interruption, at either side of which it ends while the axis-cylinder extends across (fig. 44). These gaps were first described by *Ranvier*, and hence are known as Ranvier's nodes (étranglements annulaires). They are to be seen, although less clearly than in specimens prepared as above, not only in freshly-prepared nerves, but also in the living nerve in the frog's lungs (*Rawitz*). This proves that they are not artificial products.

In osmic preparations it is further seen that the medullary sheath is formed of segments which, arranged like overlapping funnels, surround the axis-cylinder. Such a segmentation was known to *Stilling*, and was again described, simultaneously, by *Schmidt*, *Lantermann*, and *Zawerthal*. These medullary segments may be products of the method; but under any circumstances their regular arrangement depends upon some pre-existing quality in the medulla which demands our attention. Both the nodes of Ranvier and the cones of Lantermann present considerable variety in form. The two must not be confused together.

Schwann's sheath (membrana limitans, neurilemma) is a thin, delicate, but yet firm membrane, which closely invests the medulla; hence it is, as a rule, no more visible than the axis-cylinder-sheath. Schwann's sheath is exhibited when by slight pressure on the glass covering freshly-isolated nerves, some of the medulla is squeezed out from the tubes which contain it.

A thin fresh nerve fasciculus is washed in water and then placed in a solution of nitrate of silver (1 in 300) for from ten minutes to an hour at most. Again it is washed in water and then examined in glycerin. After exposure for a short time to light such a nerve-bundle shows, under a low power, numbers of black crosses occurring at intervals (fig. 43). The meaning of these crosses is not made out until after teasing the fasciculus with needles. Now it is seen that each cross consists of a vertical and a horizontal bar; the latter extends from the sheath of the axis-cylinder to the sheath of Schwann, and corresponds to a silver-impregnated diaphragm which occupies the situation of Ranvier's node; it may hence be termed the intermedullary sheath. Through its central aperture passes the axis-cylinder. Others (*e.g.*, *Kuhnt*) term certain very delicate membranes which are supposed to extend across from the axis-cylinder-sheath to the sheath of *Schwann* between Lantermann's cones, "intermedullary sheaths." [If a name is needed they might be distinguished as "intramedullary" sheaths.] The vertical bar of the cross depends upon the impregnation with silver of the ends of the axis-cylinder, where it adjoins the intermedullary sheath. The farther from this diaphragm the fainter becomes the staining of the axis-cylinder. The intermedullary sheath, which is unstained by osmic acid constitutes a connection between Schwann's sheath and the axis-cylinder-sheath.

Fig. 42.—Nerve-fibre from the sciatic of the frog, after treatment with osmic acid. *Magn.* 400.—*a*, Ranvier's node; *b*, nucleus of Schwann's sheath. The darkly-stained medullary sheath exhibits the clefts which separate Lantermann's cones.

Fig. 43.—A short piece of a thin peripheral nerve treated with nitrate of silver. *Magn.* 30. It shows numerous nodes of Ranvier.

Fig. 44.—An isolated medullated fibre treated with silver nitrate. *Magn.* 200. —*a*, a Ranvier's node.

Between each two nodes of Ranvier, Schwann's sheath presents an oval nucleus, at either pole of which lies a little granular protoplasm. The nucleus occupies a depression in the medullary sheath. Only in fishes are several nuclei found in a single internode. The nuclei of Schwann's sheath are best seen in preparations stained with carmine or aniline colours or in osmic preparations; in the latter they appear greenish-grey (figs. 38 and 42).

Lastly, each fibre is invested by a delicate coat, the **fibrillar sheath** [sheath of Henle]. It consists of a thin membrane in which delicate longitudinally running fibres seem to lie (presumably plaits in the membrane). The nuclei which lie on this sheath and stain out conspicuously with fuchsin are regarded by *Key* and *Retzius* as belonging to endothelial cells which clothe the sheath. A lymph space lies between the fibrillar sheath and Schwann's sheath; but probably the fibrillar sheath is not closed in completely on all sides.

[The endothelial sheath, which invests the fibres and provides a bath of lymph in which each fibre lies, is a specially modified layer of the immediately contiguous connective-tissue cells. It is seen best where it surrounds isolated fibres, whether medullated or not (*e.g.*, olfactory nerves). Within a nerve-fasciculus it forms part of the endoneurium. In the central nervous system its place is taken by neurogleia (connective) tissue.]

Of the numerous views with regard to the constitution of medullated fibres, we will mention only the opinion of *Ewald* and *Kuhne*, founded upon their digestion-method, that the whole myelin-sheath is traversed by a close-set network of horny substance (neuro-keratin). *Stilling* had long ago so represented the constitution of the fibres. *Rezzonico*, *Golgi*, *Cattani*, and others have detected in Lantermann's constrictions peculiar spirally arranged fibres, which also correspond to Kuhnt's cones.

Nerve-fibres which possess all the elements above mentioned occur exclusively in peripheral nerves.

Axis-cylinders, the essential elements in the conception of nerve-fibres, are found without further investment in the central nervous system, and in end organs. They are, for the most part, very fine, and may consist of but a single primitive fibril.

Axis-cylinders surrounded by sheaths of Schwann,* but without medulla, compose the greater portion of the sympathetic commissural cords. In these cords, however, as well as in the sympathetic plexus, not a few medullated fibres are to be found. Usually their medullary sheaths are thin. Numerous oval nuclei are disposed around the non-medullated fibres with their long axes in the direction of the fibre. These nuclei belong to the sheath of Schwann, which lies so close on the axis-cylinder that its membranous nature can be scarcely recog-

*[The membranous sheath which surrounds non-medullated fibres and carries numerous nuclei on its internal surface has been regarded usually as the homologue of the similar sheath which invests isolated medullated fibres; the sheath of Henle. It has been customary, on the other hand, to look upon Schwann's sheath as a membrane, possibly the investing membrane of the myelin cells, which cannot exist apart from the medullary sheath.]

nised. It is characteristic of non-medullated (grey or Remak's) fibres that they appear to divide and unite again in plexuses. Possibly the plexiform appearance is illusive, depending upon the difficulty in isolating non-medullated fibres. They are supposed by *Boveri* to possess a delicate analogue of a medullary sheath. To see these fibres well the sympathetic nerve should be taken from the neck of a living or recently killed animal, placed in a weak solution of bichromate of potassium (1 in 200), stained in carmine, and teased out (fig. 45).

Fig. 45.—Remak's fibres from the sympathetic of the neck of a rabbit. *Carmine staining. Magn.* 200.

Fig. 46.—Central medullated nerve-fibre from the brain, without sheath of Schwann. *Magn.* 200.

Fig. 47. — Very fine varicose axis-cylinders from the bulbus olfactorius of the dog. *Magn.* 400.

Scattered grey fibres are also to be found in peripheral nerves. The olfactory nerves consist exclusively of such axis-cylinders, naked except for the investing sheath of Schwann [and Henle's sheath]. Both motor and sensory nerves at their peripheral ends lose first their sheath of myelin and then their sheath of Schwann.

Medullated fibres destitute of Schwann's sheath are also to be found; all medullated fibres of the central nervous system belong to this class.

They are best seen after a small piece of brain or spinal cord has been placed for twenty-four hours in a weak solution of perosmic acid (1 in 1000) and then teased. Such medullated fibres, having no supporting investment, do not show a distinct border when they are teased out; but the medulla bulges ("varicose fibres") or breaks away from the axis-cylinder, leaving it free for a longer or shorter distance (fig. 46).

The fine non-medullated, as well as the finest of the medullated, fibres which have no sheath of Schwann, show the axis-cylinder beset with little swellings (varicose axis-cylinders). It might be concluded from this that the finest axis-cylinders are not invested with an axis-cylinder-sheath (fig. 47).

No difference in structure between motor and sensory fibres has yet been detected; nor is it, as was formerly supposed, correct to say that motor fibres are uniformly larger than sensory; rather has *Schwalbe* proved that in general those fibres are thickest which have the longest course. So far as mammals are concerned, it appears that the fibres are larger in the larger animals, a conclusion in harmony with the statement just made as to the relation of diameter to length.

[The size of a nerve-fibre depends upon other factors besides its length. Up to a certain point it varies (if a motor fibre) as the size of the muscle-fibre which it innervates, although in considering cases bearing upon this rule the fact that a motor nerve-fibre usually divides to supply a number of muscle-fibres must be taken into account. The relation in number between the fibres of a muscle and the fibres of the nerve supplying it is constant for each case, but varies widely for different muscles, from 1 nerve-fibre to 40 muscle-fibres in the leg, to 3 nerve-fibres to 7 muscle-fibres in the extrinsic muscles of the eyeball. Many more observations and numerous corrections are necessary before the laws which govern the size of nerve-fibres can be formulated. The quality of its action as well as the size of the muscle-fibre appears to influence the size of the nerve. The fibres going to the slowly-acting red muscles of the rabbit are smaller than the fibres supplying the more highly differentiated white muscles. Spinal motor and sensory fibres reach a diameter of about 20 μ in the dog (*Gaskell*). Medullated sympathetic fibres are usually from 2 μ to 2·5 μ in diameter. At the first moment it might be thought that the fibres supplying plain muscle do not support the generalisations just made; but it must be remembered that their disposition is very different to that of the motor-fibres of striated muscle; instead of each fibre running straight from its nutritive cell in the spinal cord to its destination, it ends in a sympathetic ganglion in a "distributive" cell from which a large number of fine non-medullated fibres proceed to the muscle. It is possible that in Auerbach's and Meissner's plexuses the process of subdivision is carried still further.]

In the spinal cord of many fishes a single conspicuous fibre is to be seen in the anterior column on either side, *Mauthner's* "colossal fibre" with a diameter of nearly 0·1 mm. So, too, in Malapterurus electricus the fibre in the spinal cord destined for the electric organ is marked out by its considerable size. The medullated fibres of the brain are distinguished from those of the spinal cord by their tenuity. No medullated fibres are found in any invertebrate, nor are they present in cyclostomes or lophobranchii. *Rawitz* has found, however, in the nervous system of acephalæ a substance which may be ranked with myelin. So, too, in embryos of early date all fibres are non-medullated; and the fact that they acquire their medullary sheaths by degrees (in

DEVELOPMENT OF MEDULLARY SHEATH.

some fibres only after birth) is the basis of one of the most important of anatomical methods.

The peripheral fibres in the puppy, for example, are already medullated at birth; but so unequally is the medullary sheath disposed that the fibres look as if varicose (fig. 48). Perhaps the cause of Lantermann's cones is to be sought for in this bead-like deposition of medullary matter.

The histological interpretation of the medullary sheath has suffered several changes in recent years; one of the opinions with regard to its constitution which most deserves attention is that of *Boveri*, that it consists of a succession of hollow cells so placed together as to form a continuous tube through which passes, without interruption, the axis-cylinder. Each segment lying between two of Ranvier's nodes answers to such a cell. Each possesses a single nucleus. At the node the sheath of Schwann dips in towards the axis-cylinder-sheath. This latter sheath does not belong in any sense to the axis-cylinder, and, therefore, would be better named the inner neurilemma. The sheath of Schwann is not supposed to extend without a break across the node, and so the independence of each cell is kept up. In spite of this *Jacobi* seems to be correct when, as the result of his exact investigations, he says that Schwann's sheath presents no gaps at Ranvier's nodes, even in the earliest stages of its development.

Fig. 48.—Peripheral nerve-fibre from a new-born puppy, partially surrounded with myelin. *Magn.* 200.—*a*, nucleus of Schwann's sheath.

Vignal is doubtless right in supposing that the bundle of fibrils, of which the embryonic axis-cylinder is made up, is embraced and surrounded by soft amœboid cells which encircle it much in the way in which an amœba draws itself around a foreign body.

From this point of view each node expresses the plane of contact of two neighbouring cells—an intercellular space in fact (*Boveri*).

[The account of the structure of nerve-fibres given above makes it quite clear that the essential part of a nerve-fibre, the axis-cylinder, is an unbroken process of an epiblastic cell, the nerve-cell in the spinal cord or spinal ganglion. Spinal cord and ganglia were originally portions of the epiblast. The cord is formed by the involution of the neuro-epithelial tube. The ganglia are derived by delamination from rudiments which lie outside the rudiment for the cord. The epithelial cells of the embryonic cord throw out processes which seek, down through the mesoblast, for the muscle-fibres they are destined to supply, or else are elongated strands uniting the cell in the cord with a sister-cell in a sympathetic ganglion. The cells of the spinal ganglion throw out a process on either side. The distal process seeks a sensory cell. The proximal process works its way into the cord (*Beard*). It is almost impossible to suppose that the nerve-filament finds its muscle-fibre without a guide; perhaps the junction between nerve-cell and muscle-cell is effected very early, at a time when they are almost or quite in contiguity, and the subsequent elongation of the nerve-fibre is due to the change in situation of the muscle; at present the subject is beset with difficulties which will only clear up when fresh facts are brought to light. What is the origin of the myelin and other cells by which the axis-cylinder is ensheathed? In all probability the primitive neuro-epithelial cells fall into two groups. Some become nerve-cells; others, less favoured, serve for their support. These latter, again, exhibit a subdivision of labour. Some of them acquire a large amount of the fatty metabolite myelin, and, applying themselves to the nerve-fibres within the cerebro-spinal system, invest them with their cylindrical medullary segments (*Schiefferdecker* is convinced that intra-axial fibres show Ranvier's nodes). Other supporting cells constitute the connective-tissue of the brain and cord, the cells of Deiters, or neurogleia-cells of various kinds. Have the myelin-cells of peripheral nerves a similar epithelial origin? *Vignal* is inclined to answer this question in the affirmative. He believes that they emigrate with the axis-cylinder from the central system. The sheath of Schwann would seem to be the outer cell-membrane of the myelin-cell; the single nucleus indicating that each segment is formed from one cell only. It is difficult, however, to account for the absence of any such membrane around the myelin-cells within the cerebro-spinal axis, if both within and without the axis they are derived from similar epithelial elements. It is possible that the axis-sheath is the same cell-membrane on the incurved side of the myelin-cell.]

Pathological Changes in Nerve-fibres.—But few alterations in nerve-fibres due to disease are known as yet; methods are certainly needed to enable us to recognise such changes.

The most important form of **degeneration of a medullated fibre** yet observed, and the one most studied hitherto in detail, is that which results when a peripheral fibre is cut off from the nerve-centres (*cf.* p. 19). The first investigations in which this method was used were made by *A. Waller*, and hence the process is known as "Wallerian degeneration." If a peripheral nerve is cut across (the sciatic nerve is convenient for the purpose) one finds during the interval between the second and the fourteenth day in a mammal (in cold-blooded animals

SECONDARY DEGENERATION.

the changes occur more slowly) after it has been hardened and coloured in osmic acid the following changes—at first the myelin breaks up irregularly; later it shows only scattered blackened drops and numerous dark granules, Schwann's sheath being filled out in places only. The protoplasm and the nuclei appear to be increased. The axis-cylinder is broken into a succession of detached pieces of different length, often twisted or coiled; finally it disappears (fig. 49). In the last stage Schwann's sheath alone survives to represent the nerve-fibre. It looks like a thin cord of connective-tissue; scattered groups of granules, and a few myelin drops alone reminding us that we have to do with a tubular membrane. Multiplication of nuclei occurs by caryomitosis (*Tangl*).

As the first result of section of peripheral nerves the degeneration advances, in the peripheral as well as in the central segment, only as far as the next node of Ranvier where it seems to be stopped (*Engelmann*). While, however, in the central stump, despite its functional inactivity, no further changes occur for months, degeneration advances rapidly throughout the whole length of the peripheral portion of the nerve.

Now, recent observations have shown that even in the central stump many nerve-fibres degenerate (in Man) in the manner just described, while, on the other hand, in the peripheral portion a certain number of fibres remain intact (*F. Krause*). In most animals the number of fibres which degenerate in the central stump is small. Obviously these are sensory fibres whose trophic centres are situate towards the periphery.

Degenerative processes in non-medullated fibres are known in but a few cases, as for instance in the fine fibres of the cornea. *Ranvier* has observed in the Remak's fibres of peripheral nerves a multiplication of nuclei and the appearance of peculiar highly-refracting granules as well as fat granules.

The progress of degeneration within the central nervous system when fibres are cut off from their centres appears to differ from that occurring in peripheral nerves in certain points not yet sufficiently understood.

Fig. 49. — Two fibres from the anterior roots springing from a softened spinal cord. *Magn.* 200.

Later, in two to three months, the process of **regeneration** may be observed in cut nerve-fibres; on the whole the central idea of the process consists in this—the axis-cylinders of the central portions of the cut nerves constitute the foundations for the newly-created nerve-fibres, which grow through the scar and enter the old sheaths of Schwann, several nerves often entering a single sheath, along which their course to the periphery is directed. At first these newly-formed fibres are finer than the old ones, and the distance between the nodes is less. A direct union of a cut nerve by "primary intention" is impossible (*Krause*).

S. Mayer has proved that a continual replacement of nerve-fibres goes on normally in peripheral nerves; since fibres are always to be met with in conditions of de- and re-generation. This is especially the case in the nerves of the brown rat (mus decumanus).

Neuritic degeneration is usually described in connection with the form which results from the cutting off of the trophic influence of a centre; the processes were supposed to be identical, whereas they are, as a matter of fact, quite different.

This second form of nerve-change is met with in many infectious diseases (especially in the paralysis which follows diphtheria), as well as in certain paralyses due to poisons, as in lead paralysis (*Gombault*). In these cases the nerve is not diseased throughout, but only in certain segments, which alternate with others that remain quite normal. Further, the disease affects only the myelin-sheath and the sheath of Schwann, while the axis-cylinder appears at first, at any rate, to remain intact. Lastly, the products of the destroyed myelin are not large drops, but little groups of fatty granules. Just like this, too, is (according to *Gombault*) the degeneration of peripheral nerves which accompanies alcoholic paralysis; without, however, the same segmental localisation. The pathological changes in the central nerve-fibres which occur in disseminated sclerosis may be ranged here; for the axis-cylinder, despite the changes in the medullary sheath, is long preserved.

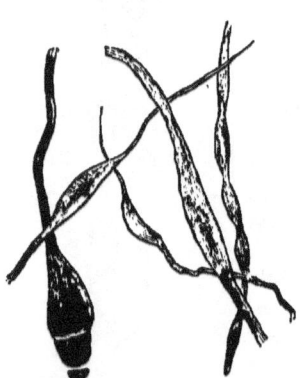

Fig. 50.—Several forms of hypertrophic varicosity of axis-cylinders from softened foci in the spinal cord. *Magn.* 200.

Regeneration of fibres in the central nervous system with restoration of function never occurs, in the higher animals at any rate.

While the segments of a cut peripheral nerve grow together again, a cut central nerve-tract is for ever put out of action.

Another form of nerve degeneration is associated with partial **hypertrophy of the axis-cylinder.** This is found in central fibres and is always a sign of irritation, as for instance in myelitic and encephalitic lesions; often, too, in neuro-retinitis the optic fibres of the retina, which are just as well considered as central fibres, suffer beaded enlargement.

In less severe conditions the axis-cylinder shows only slight swellings (varicose axis-cylinders, fig. 50). In advanced stages it may for a considerable distance be swollen out to six times its normal diameter, when it usually begins to exhibit transverse cleavage; fine granules of fat, usually arranged longitudinally, are often found in the enlarged pieces; their presence places beyond a doubt the degenerative character of the process. In fibres presenting this inflammatory change the separate swellings of the axis-cylinder may be so charged with fat granules as to present the appearance of fattily degenerated cells (*Unger*).

The medullary sheath is also in some cases of myelitis considerably thickened (*Leyden*). It is necessary, however, that one should carefully exclude cases in which the enlargement occurs *post-mortem*.

Calcified nerve-fibres which are not the processes of calcified nerve-cells are rare. *Förster* saw calcified fibres in the lumbar swelling of the spinal cord.

It was hoped that by the use of new reagents, such as saffranin (*Adamkiewicz*), other changes in nerve-fibres would be recognised.

2. NERVE-CELLS.

It is not easy to define a nerve-cell (or ganglion cell) from a histological point of view. It is possible to imagine that a bundle of primitive nerve-fibrillæ may at certain points of their course or even at their ends, suffer a thickening owing to a nucleus with its protoplasm insinuating itself among the fibrillæ. Thus would be produced theoretically the simplest and most generalised type of nerve-cell; it is practically impossible, however, to determine the existence of such conditions; so, too, it is very seldom possible to establish, under the microscope, another characteristic which beyond all doubt decides the nervous nature of a cell—namely, the continuity of one of its processes with a nerve-fibre. We are obliged in the majority of cases to seek for other evidence before we can decide whether or not we have to deal with a nerve-cell (see figs. 51, 52, 53).

The primitive form of a nerve-cell is the sphere; from this, by the prolongation of one of its axes, a spindle is produced. Never, however, is one diameter of the nerve-cell greatly diminished (as in tesselated epithelium) or greatly increased (as in muscle-fibre).

So it comes about that from every nerve-cell one process at least, but usually more than one, originates, which alters, but never completely obscures, the globular or spindle-shape of the cell.

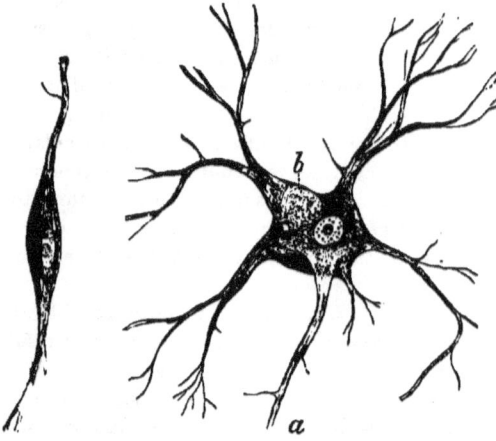

Fig. 51.— A cell from the anterior horn of the spinal cord of the pike. *Magn.* 150.

Fig. 52.—A cell from the anterior horn of the human spinal cord. *Magn.* 150.—*a*, Axis-cylinder process ; *b*, clump of pigment.

Fig. 53.—A pigmented cell from the substantia ferruginea. Human brain. *Magn.* 150.

All nerve-cells possess a granular protoplasm, which is for the most part prolonged some distance into the processes.

In the interior of the nerve-cell is a relatively large clear nucleus round or oval, or occasionally angular with rounded angles, and presenting besides a characteristic granulation, which occasionally constitutes a distinct network; an unusually large nucleolus is found within the nucleus, in this often a nucleolulus is to be seen. Even in very large cells in which the nucleus is completely covered with the protoplasm of the cell-body, the nucleolus still shines through owing to its high refraction. In the sympathetic it is common to find nerve-cells with two nuclei. A fibrillar striation is often clearly recognisable in the protoplasm of the cell, especially at the root of the cell-processes (fig. 51).

PIGMENTATION OF NERVE-CELLS.

It is exceedingly difficult to distinguish small nerve-cells from other cellular structures. No difficulty exists in the case of the larger ones, which are amongst the largest cells in the animal kingdom. They attain to a size of 0·1 mm., and even more in diameter; the electric cells in the spinal cord of malapterurus are as much as 0·21 mm. in diameter.

Other characters by which the nerve-cells may be recognised are the following :—

Many nerve-cells, especially the larger ones, contain a little heap of light yellow granules, regarded as a lightly-stained fat-like "pigment" substance. Usually the pigment is accumulated on one side of the cell near a process (figs. 52, 53, and 54). A dark-brown pigment is less common; it may almost fill out the cell-body, so that the nucleus remains as the only clear spot, and may extend for some distance into the processes. Such pigmented cells are grouped together in masses in two situations in the brain, the substantia nigra, and substantia ferruginea. They are scattered in other situations, as, for instance, the border of the vagus nucleus. Strongly pigmented cells are to be found outside the brain in the spinal ganglia and the ganglia of the sympathetic. Pigmented nerve-cells are rare in animals. No relation seems to obtain in Man between the general abundance of pigment and the pigmentation of the nerve-cells. The cells of the new-born child are free from pigment; pigmentation first appears later in life, and reaches its maximum in old age. Many large cells, as, for instance, Purkinje's cells in the cortex of the cerebellum and certain small cells, always remain devoid of pigment.

The chemical nature of the pigment is not yet understood. The clear pigment stains dark with osmic acid and Weigert's hæmatoxylin method. The dark pigment of the human brain becomes lighter when treated with concentrated sulphuric acid. A pigment is found in the nerve-cells of fresh-water molluscs, which turns green, blue, and finally indigo on treatment with concentrated sulphuric acid (*Buchholz*), and in acephalæ there exists a brownish-yellow pigment which turns, on the application of this reagent, a deep olive-green (*Rawitz*).

The pigment granules in the human nervous system appear, even under the highest magnification, always round or roundly angular. They can hardly be seen in perfectly fresh preparations.

Nerve-cells are further characterised by the behaviour of their nuclei towards hæmatoxylin. If one stains a section with alum-hæmatoxylin (p. 11), all nuclei, except the nuclei of nerve-cells, assume a deep blue colour; but even the largest nerve-cell nuclei present merely a blue-grey tone.

Nerve-cells are not invested by any proper cell-wall. In some places, such as the spinal and sympathetic ganglia, and in the accessory auditory nucleus, they are enclosed by capsules of epithelial cells, through which the processes, which are usually simple, come out (fig. 54). According to *Max Schultze*, the cells on the auditory nerve of the pike are invested with a myelin case.

Fig. 54.—Two cells from a human spinal ganglion. They have shrunk away from their capsules, on the inner surfaces of which the nuclei which line them are seen. Each point of the cell which remains connected with the capsule looks like a process. *Magn.* 200.

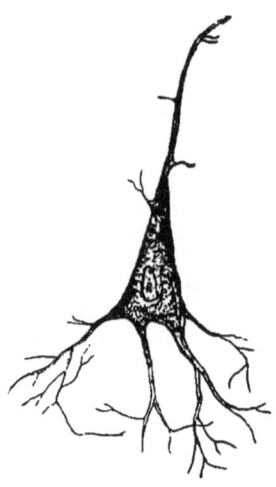

Fig. 55.—Pyramidal cell from the cortex of human cerebrum. *Magn.* 200.

The forms assumed by nerve-cells are observed partly in sections prepared by the several methods already described, partly in preparations of isolated cells. For separation, pieces of grey matter from the anterior horn of a spinal cord, as fresh as possible, are macerated in the following way:—They are placed in a weak straw-coloured solution of bichromate of potassium for two to four days, or by Ranvier's method in a mixture of one part absolute alcohol and two parts water. A little carmine or fuchsin may be added directly to the macerating fluid. The larger cells with their processes may then be easily isolated with the aid of a simple microscope. Fairly good permanent preparations may be made by spreading out the deposit containing the isolated cells on an object-glass, allowing it to dry, and covering with dammar varnish. Cells may also be easily isolated after the tissue has lain for fourteen days in a 0·1 per cent. solution of osmic acid. *Vignal* recommends arsenicated glycerin jelly for preserving teased preparations.

Much is to be learnt from such preparations, especially with regard to the arrangement of the cell-processes.

Anastomosis between two nerve-cells by means of a thicker process, as is often drawn and described, either does not exist at all or is only found as an abnormality.

The nerve-cell processes either divide dichotomously until they acquire the utmost tenuity, when one may look upon them as primitive fibrils which cannot (on account of their extreme delicacy) be followed farther (fig. 52); or else the process (as in the case of the apical process of the cortical cells, fig. 55) gives off little branches at right angles, and while it ever becomes thinner it presents slight enlargements at the points of origin of these side-branches. Probably one process at least of every [large] nerve-cell goes over into the axis-cylinder of a medullated nerve (in the case of many cells more than one process perhaps) before it breaks up into the finest primitive fibrils. This process may be distinguished as the chief or axis-cylinder process in contradistinction to the protoplasmic processes of *Deiters* or ramified processes of *Max Schultze*. The direct passage of a nerve-cell process into a nerve-fibre can only be seen under especially favourable conditions (*Koschewnikoff*). According to *Deiters* the axis-cylinder process is distinguished from the others by its hyaline appearance (fig. 52).

In the greater number of cells seen in preparations made according to the most suitable methods (silver or corrosive sublimate colouring of Golgi or Pal [methods which, as explained later on, really effect a precipitation around, not a colouring of, the process]) the axis-cylinder processes present appearances which cannot in any sense be regarded as specifically characteristic. It is better, therefore, to avoid hastily diagnosing such processes as do not, by their whole behaviour and relations, mark themselves out as axis-cylinder processes.

The number of processes given off by a nerve-cell varies, but seldom exceeds five or six. Apolar nerve-cells seem to be physiologically inconceivable. They are to be looked upon either as developing cells or as artificial products. *Rauber* regards such apolar cells as arrested developments; cells remaining in their original processless condition. It is difficult to understand the physiological value of unipolar cells, since, as a rule, their processes divide soon after they are given off from the cell. Many of them may, therefore, be looked upon as bipolar cells, the processes of which unite before they join the cells. *Ranvier* has proved that this is the case with the cells [of the spinal ganglia] which have a single T-shaped process. The finest ramifications of the processes join a nerve-network from which again the branches are probably re-associated into thicker fibres clothed with medullary sheaths.

The behaviour of the processes in the sympathetic of the frog is remarkable (*Beale, Arnold*); the cells of these ganglia give rise to a straight process usually attached to the cell by a conical base which runs for some distance without dividing, while a second process (the spiral fibre) surrounds the first with several spiral coils before it, too, is covered with a medullary sheath.

The higher we ascend in the animal kingdom and the more highly-developed the nervous system, the richer in processes do the nerve-cells become. This is easily proved by means of sections of homologous parts of the central nervous system. For example, one may compare the anterior-horn-cells of fishes, which are usually bipolar (fig. 51), with the stellate cells of mammalia (fig. 52). It stands to reason that the greater the number of conducting paths by which the nervous elements are bound together, the more complicated and varied will become the functions which they are capable of carrying out.

Nerve-cells which survive removal from the animal to which they belong are best obtained from invertebrates. According to *Freud*, the cells of the abdominal ganglion of the fresh-water crayfish may be examined in a living state in the blood of the animal.

The living cell is seen to consist of a substance arranged in a network and continued into the fibrillæ of the nerve-fibres as well as of a homogeneous basis substance. In the nuclei of these cells *Freud* has seen a variable number of bodies assuming a variety of forms (for the most part they are longer or shorter rods, twisted or forked threads, and so forth), which, as long as the cell lives, undergo obvious changes in shape and place.

E. Fleischl has observed movements of the whole nucleus in the fresh cells of the Gasserian ganglia of the frog under the influence of boracic acid.

Other important facts concerning the relations of living nerve-cells of the sympathetic and spinal ganglia to their processes have been discovered by the use of Ehrlich's method of introducing methyl-blue into the circulation of the living animal (*Ehrlich, Aronson, Smirnow*). In the sympathetic cells of the frog the straight process remains uncoloured, while the spiral fibre stains intensely blue; it is, therefore, possible to see that the coiled fibre breaks up into fine fibrils which surround the surface of the cell in the same way that cords enclose a balloon. The threads present button-like swellings. The sympathetic cells in the rabbit, and, indeed, both kinds of cell in the higher vertebrates, exhibit a similar structure. Since methyl-blue stains sensory fibres more easily than motor ones, the spiral fibre may, perhaps, be looked upon as the centripetal process of the cell (*Ehrlich*).

The difference in behaviour of neighbouring cells towards colouring

GRANULES.

reagents, and especially towards Weigert's hæmatoxylin ("chromophilous" and "chromophobic" cells), justifies the conclusion that they differ in function (*Flesch, Koneff, Benda*).

A quite distinct kind of nerve-cell is found in many situations, as, for example, the substantia gelatinosa [Rolandi], the retina, [the olfactory bulb], and the nuclear layer of the cerebellum. The statements made with regard to nerve-cells are not applicable to these so-called granule cells. They consist almost exclusively of granular nuclei of 5 to 8 μ in diameter, without any strongly refracting nucleolus. Their protoplasm is very scanty. Usually the processes and cell-protoplasm are not visible; in no cases can the very fine processes be followed far (fig. 56). The nuclei stain strongly with hæmatoxylin. This is a reason against regarding them, as is often done, as of equal value with nerve-cells. On the other hand, they do not completely agree in structure with other tissue elements, the connective-tissue cells, for example, nor is it possible to imagine what might be the use of such heaps of non-nervous elements in so many places in the nervous system.

Fig. 56.—Granules from the cortex of the cerebellum.

It is well to consider these "granules" (which are best exhibited in teased preparations of the nuclear layer of the cerebellum) as a peculiar tissue-adjunct of the nervous system.

[Some light is thrown upon the nature of these granule cells by a consideration of the situations in which they occur. In the retina they constitute a stratum many cells deep, the "inner nuclear layer." It is almost universally accepted that the non-medullated fibres attached to the bases of the rods and cones, or rather of their nuclei (the nuclei of the outer nuclear layer), pass through the outer molecular layer to the nuclei of the inner nuclear layer, with which they are connected before breaking up in the inner molecular layer. Each "granule" is, therefore, a minute fusiform bipolar cell intercalated in the course of a non-medullated fibre, between its epithelial terminus and the plexus into which it breaks up; from which plexus each large nerve-cell collects the products of many of these minute naked fibres, and associates them into a single medullated nerve for transmission to the central system. In the olfactory bulb, and for some distance along the olfactory tract, are found granules indistinguishable in appearance from those of the retina. Other conditions which obtain in the retina are also found in the olfactory bulb, and there is every reason for regarding the "granules" of the two organs as similar in function. Although methods of isolation have hitherto failed to show the continuity between the protoplasm surrounding the granule and naked nerve-filaments, it is safe to assume that the "granule" is the nucleus of a bipolar cell with non-medullated processes. The granule of the retina is the deposed epithelial cell which establishes a connection between the sensory epithelial cells and the central nervous plexus. According to the translator's theory, the large cells of the spinal ganglia have a similar origin and function (fig. 57), the astonishing difference in size in the two cases is not greater than

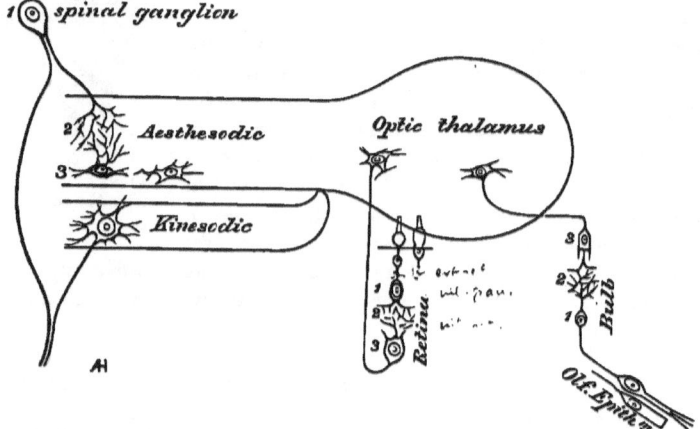

Fig. 57.—Diagram designed to show the homology of the "granules" of the olfactory bulb and retina and the cells of the spinal ganglia.—*1*, Bipolar cells; *2*, plexus of "molecular" substance; *3*, collecting or associating cells.

the difference between the fibres in the course of which the fusiform cells are respectively intercalated.

Looking at the vast number of "granules" which occur in the cortex, not only of the cerebellum but also of the cerebrum, and bearing in mind that the function of these great cortex-fields is to provide a plexus through which sensory impressions may flow over into motor tracts—a network of alternative routes connecting afferent and efferent fibres—it is tempting to suppose that the truly centralised portions of the nervous system are formed on the same plan as those older isolated clumps which, in the retina and olfactory bulb, still remain in the vicinity of sense-organs.

The layers of the cerebellum are arranged with great regularity. The granule layer invests the central medullary substance. A single sheet of Purkinje's cells is spread over the granule layer. The outer layer of the cortex is composed of molecular substance, containing but few cells, devoted to the support of the arborescent systems into which the apical processes of the Purkinje-cells break up. For reasons which will be detailed when the cerebellum is being described, the medullated fibres derived from the cells of Purkinje must be regarded as efferent. The afferent fibres which pour their impulses into this cortex-field break up within the nuclear layer into non-medullated fibres. Probably each non-medullated fibre bears a fusiform "granule" cell before it passes through the stratum of Purkinje's cells to join the plexus into which the processes of these cells break up.

The matter is treated of at some length in this place, because, if the translator's theory is correct, it is impossible to form the most fundamental conception of the constitution of cortex without allotting its proper position in this tissue to the "granule" cell.

Presumably the granules in the cortex cerebri are similar fusiform cells borne by the terminal twigs of afferent nerve-fibres. The central nervous system, according to this view, presents two kinds of tissue— central or primary grey matter, a plexus directly uniting sensory and motor fibres; peripheral or cortical grey matter, a second plexus in which are distributed the ramified processes of fibres which start in the lower or central grey matter, while it gives origin to descending fibres, the distal ends of which break up in the lower plexus. In either case the starting point of a fibre is marked by its trophic cell.]

Adamkiewicz has introduced a new morphological element into the description of peripheral nerves, the so-called nerve-corpuscles, which are seen in peripheral nerves on staining with saffranin after hardening in Müller's fluid. These are peculiar delicate fusiform cells which lie close against the fibres, and appear in cross-section as brown-red crescents. They stain with various alkaline aniline colours, and are altogether analogous with the so-called "plasma" cells,* which make their appearance as great coarsely granular cells, derived from connective-tissue cells, owing to a local exaltation of nutrition in this tissue. Hence, *Rosenheim* pronounces these nerve-corpuscles "plasma" cells only. They are absent in new-born animals, and only occur in numbers in old age.

Widely divergent views are still held with regard to the **histological meaning of nerve-cells.** It has even been denied that they are cells (*Arndt*), and, hence, the name "nerve-corpuscle" was proposed. We are not yet in a position to bring the varieties in shape, size, and pigmentation of the cells which we have already described into association with their physiological actions, still less can we account for the variations in the arrangement of their processes.

The variations in the pigmentation of the cells would seem to give us a clear indication of their functions, or at any rate of the special

---

* "Mastzellen"; allied to Waldeyer's plasma-cells.

features of their metabolism; but unfortunately, as yet, we do not understand this hint.

With regard to the size of cells we know that, as a rule, thick fibres belong to large cells and *vice versâ*. If we may also assume that the longest fibres are the thickest, the largest nerve-cells belong to the longest tracts. This statement is certainly not universally applicable, but it may have, at any rate, a limited value for the cells of any particular region, as, for instance, the pyramidal cells of the cortex. Just in the same way it is certainly not without significance that all the large cells of the cerebellar cortex have a uniform diameter.

An especial effort has been made to discover a difference between sensory and motor cells; or, as they would be better called, the cells standing in direct relation to sensory and motor paths. Evidences of such differences should be received with great reservation. Assuredly there must be many cells which are neither motor nor sensory in the proper sense of the words, but purely trophic or else associated with the higher psychic functions; and, lastly, there must be cells which, if one truly understood their functions, could not be ranged in any of these categories.

Passing over earlier attempts in this direction we will mention *Golgi's* views alone. He distinguishes two kinds of nerve-cells clearly characterised by their different behaviour under the corrosive sublimate method:—(1) Nerve-cells whose axis-cylinder processes, although they give off many lateral branches, pass, without losing their independence, into medullated fibres; [the doubt thrown upon the staining of nerve-elements by this method should be borne in mind]; (2) nerve-cells, whose axis-cylinder processes resolve themselves eventually by frequent division into a network of fibres. *Golgi* finds cells of the first category in the regions in which motor fibres take their rise; the second category comprises the cells characteristic of the centres in which sensory fibres end. He, therefore, distinguishes the former as motor, the latter as sensory cells. Hence, the curious result is reached that in the second class of cells the axis-cylinder process is not distinguished by the only character which marks it as an axis-cylinder process, namely, its direct passage into a nerve-fibre. This warns us to be careful in recognising axis-cylinder processes.

Our conception of the relations of the finest ramifications of the cell-process is in a very unstable condition. The most generally accepted view is that the cell-processes break up into fibres which finally anastomose with the processes of neighbouring cells in a close felt-work from which, on the other side, the axis-cylinders of nerve-fibres have their origin. Every nerve-cell, or almost every one, is, according to this view, in uninterrupted continuity with many others.

PATHOLOGY OF NERVE-CELLS.

Forel's view of the arrangement of processes is quite different. He thinks that the processes of neighbouring cells grasp one another, like the branches of contiguous trees, without continuity of substance, but he does not indicate the way in which he thinks the ends of these branches terminate. From a physiological standpoint, it is not necessary to exact a direct continuity of cell-processes. It is quite sufficient for the purposes of the physiological transference of impulses to imagine an interlocking of the filaments without continuity; something like the superposition of the spiral fibre upon the sympathetic cell as described by *Ehrlich*. Apart from genetic reasons the appearances presented in successful sublimate-preparations are rather in favour of the latter theory. Individual cells with their rich network are coloured, but no anastomosis between neighbouring cells is shown; [the metallic deposition does not surround the finest filaments of the felt-work].

The view first advanced by *Stricker* and *Unger* is based upon a most unusual expression of physiological principles. It may be formulated in the two following sentences :—1. There are all intermediate forms between connective-tissue cells and nerve-cells. 2. Nerve-cells and axis-cylinder processes give off-shoots which pass continuously into the connective-tissue network.

Pathological Changes in Nerve-Cells.—While nerve-fibres are subject to but few degenerative processes, the nerve-cells undergo in the living organism a great variety of pathological changes, always with the same physiological result, loss of function, and cell-death. Eventually the nerve-cells either disappear or some vestigial structure, depending for its character upon the nature of the change, alone remains to mark its situation.

Simple atrophy of the nerve-cell may occur (fig. 58); owing to

Fig. 58.—Simple atrophy of a nerve-cell from the oculomotor nucleus. Human. *Magn.* 150.

Fig. 59.—Commencing atrophy of a cell from the anterior horn of the spinal cord. Degeneration of the nucleus. *Magn.* 150.

Fig. 60.—Fatty-pigmentary degeneration of a pyramidal cell of the cortex cerebri. *Magn.* 150.

the shrivelling of the cell first in one and then in all dimensions, its processes are torn off at some distance from the cell. They often assume a corkscrew form. The nucleus becomes less distinct, and the last trace of the cell disappears, leaving occasionally an empty space. Sometimes the commencement of the atrophic process can be recognised in the nucleus which shrivels, loses its smooth outline, and often takes up an eccentric position near the periphery of the cell (fig. 59).

Fatty degeneration begins as an increase in the quantity of the pigment normally present (fig. 60). Since the light yellow pigment has certain properties in common with fat, and is apparently a body related to fat, we often speak of a **fatty-pigmentary degeneration.** The granular matter which accumulates in the cell in this form of degeneration is, however, much more like fat than pigment, and in the later stages of the disease the cell is simply filled with fat-granules. Since we cannot say how large a quantity of pigment a nerve-cell may normally contain, we cannot recognise the first stages of this degeneration. More and more fat accumulates in the cell until, in the later stages of the disease, it is simply a vesicle distended with fat which envelopes the nucleus. Lastly, the nucleus disappears, the fatty vesicle breaks to pieces, and the fragments are absorbed. This degeneration is found in the several forms of chronic atrophy, such as senile atrophy, general paralysis, and the atrophy of drunkards. When a fatty cell has lost its processes it resembles an ordinary fat-cell, and it may, if it retains its nucleus, remain as such.

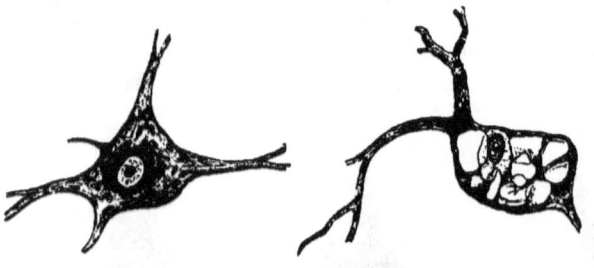
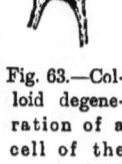

Fig. 61.—Granular degeneration of a cell of the anterior horn in myelitis. *Magn.* 150.

Fig. 62.—Cell of anterior horn with ten vacuoles in myelitis. *Magn.* 150.

Fig. 63.—Colloid degeneration of a cell of the anterior horn in myelitis. *Magn.* 150.

There is a peculiar **granular degeneration** which indicates an acute process (fig. 61). The cell body is, in this condition, dotted all over with granules which are larger and rounder or more distinctly oval than in the degeneration last described. They stain with carmine.

The distinctive features of the nerve-cell may for a long time remain unaffected.

It may also happen that, as the result of an irritative condition, the nucleus travels to the margin of the cell, or even lies for its greater part outside the cell-body. This is observed in dementia paralytica in the cells of the anterior cornua of the spinal cord (*J. Wagner*).

Vacuole formation is almost always, and indubitably in every case in which it reaches considerable dimensions, an indication of inflammatory processes, especially myelitis. The number of spaces in the protoplasm varies; there may be as many as ten, almost replacing the whole of the protoplasm, which persists only in the thin sepiments between the vacuoles, and at the roots of the cell-processes. Nuclei and processes retain in these cases their normal appearance. Attention must, however, be called to the fact that vacuoles may appear as the result of *post-mortem* changes. Near inflammatory lesions we sometimes meet with cells, the whole body of which is occupied by a great structureless hyaline drop of colloid. Such **colloid degeneration** (fig. 63) gives to the cells a peculiar globular form, such as is only possessed by a very few normal cells. The colloid-drop stains deeply with carmine.

As the opposite of pigmentation, a condition may be mentioned in which pigment normally present in nerve-cells is lost, **depigmentation.** At the same time the protoplasm loses its characteristic granulation, stains but weakly with carmine, and assumes, as seen in sections, merely a pale rose colour. This change is found in sclerosed portions of the brain, and hence is termed sclerosis of nerve-cells. The hyaline degeneration described by some writers is almost identical with this depigmentation. Later the cell changes in form and disappears; large masses of nerve-cells may in this way be missed from many regions of the brain. This form of atrophy is always the sign of a slow chronic process.

Hypertrophy of nerve-cells is not always easy to distinguish from *post-mortem* change. The appearance of the cell-substance is altered, usually becoming dim; hence the process is known as cloudy swelling or parenchymatous swelling; the nucleus is obscured. Only the higher degrees of this degeneration can be regarded as distinctly pathological. Varicose hypertrophy of single processes (the central processes of Purkinje's cells in the cerebellum for example) seems to be rare (*Hadlich*).

Calcified nerve-cells (fig. 64) have been found in the spinal cord as well as in the cortex of the cerebellum and cerebrum. In the spinal cord they are found in the anterior cornua as the result of infantile spinal paralysis and acute poliomyelitis; but they are more commonly found in the cortex of the cerebrum, arranged in groups, beneath superficial hæmorrhages (plaques jaunes), and after injuries in which the skull-case suffers directly and the brain indirectly, even when the brain appears to have escaped, so far as can be told at the

Fig. 64.—Calcified nerve-cells from the cortex cerebri beneath an apoplexy. *Magn.* 150.

Fig. 65.—A cortical cell (from the neighbourhood of a tumour) which has divided into a number of pieces. *Magn.* 150.

time. Hence *Friedländer* considered calcification as characteristic of acute changes in nerve-cells. Calcified nerve-cells are easily recognised even in the unstained preparation by their peculiar brilliance, as well as by the star-like arrangement of their processes. On the addition of sulphuric acid, they give off bubbles of carbonic acid gas, and crystals of sulphate of lime make their appearance. It will be inferred, without special explanation, that calcification is only a peculiar form of atrophy.

Lastly, a series of changes in nerve-cells of a much more active nature must be mentioned. All or almost all these, however, end in atrophy.

Nuclear division should be noticed. In inflammatory lesions changes in the shape of the nuclei are remarked. They are seen to be undergoing the kind of constriction which would end by their dividing. The process of nuclear division in inflammation was studied in detail by *Mondino*. He recognised caryo-kinesis in the large cells of the cortex of both cerebrum and cerebellum. **Cell division** has been seen to occur as the consequence of inflammatory processes, either primary or resulting from the irritation set up in the neighbourhood of a tumour, or by artificial means. *Robinson* induced division in the cells of the sympathetic; *Ceccherelli* in artificially produced encephalitis. A nerve-cell may divide into a great number of secondary cells which still, as a group, exhibit the original form of the parent cell (fig. 65, *Fleischl*).

B. NON-NERVOUS CONSTITUENTS.

1. VESSELS.

The structure of the blood-vessels of the interior of the brain is best studied by means of pieces not inconveniently small, which have been macerated for one to two days in a light-yellow solution of bichromate of potassium—it is well to put aside for examination a piece of cortex, of medullary centre, and of corpus striatum or optic thalamus—the vessels can without difficulty be detached with needles from the surrounding substance under water. Fairly large vessels with all their ramifications may thus be removed.

The vessel is now examined in a drop of distilled water or very dilute glycerin. It is undesirable to use pure glycerin or strong salt solution on account of the shrivelling of the coats of the vessels and consequent confusion which is induced. The vessel can be laid whole in picrocarmine or any other stain—in a watery solution of Bismarck-brown (1 in 300), for example (*Löwenfeld*)—and examined in water after a thorough washing. The nuclear structures of the vessel are clearly shown in this way. Such water preparations can be preserved for years unchanged by surrounding the cover-slip, after its margin has been allowed to dry, with dammar varnish. Preparations in weak glycerin last better than those in water only. For certain details of structure, normally or pathologically present, the vessels must be studied in section, after hardening.

Certain peculiarities of structure distinguish the blood-vessels of the interior of the central nervous system from all others. This is especially true with regard to the arrangement of the tunica adventitia. On account of these peculiarities it is worth while to describe separately the arteries, veins, and capillaries.

Genuine lymphatic vessels are not found in the brain or spinal cord. The lymph-paths can be shown to be clefts between the tissue-elements. Adventitial and perivascular lymphatic tracts surround the vessels, and the nerve-cells lie in lymph-spaces. Another system of lymph-spaces surrounds the connective-cells (*Rosenbach* and *Sehrwald*).

(*a.*) **Arteries.**—Four layers may be distinguished in the coats of all arteries of the brain-substance, except the smallest. From within outwards they are named—endothelium, membrana fenestrata, tunica muscularis, and tunica adventitia. It is highly probable that the space in which the blood-vessel lies is lined by a delicate limiting membrane which adheres closely to the brain-substance after the vessel is pulled out.

Endothelium (fig. 66, *a*) is a delicate membrane formed by the apposition of elongated cells, the outlines of which are brought out by the silver method of impregnation. The nuclei of endothelial cells are oval or hone-shaped, with their long axes arranged in the direction of the vessel. Lying on the nucleus or, in some cases, partly in its interior, is often seen a minute strongly-refracting granule of unknown meaning.

If in preparing the vessel the endothelium has been dragged upon, it is apt to happen that clefts in the delicate endothelial coat are produced, which have the appearance of the nuclei of fusiform cells with elongated processes.

Membrana fenestrata (fig. 66, *b*), which lies next to the endothelium, but does not adhere tightly to it, is a coarse elastic membrane

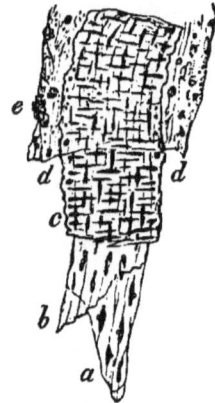

Fig. 66.—A middle-sized artery of the brain so torn as to expose each of its coats separately for a certain distance. *Magn.* 300.—*a*, Endothelium; *b*, membrana fenestrata; *c*, tunica muscularis; *d*, adventitia; *e*, pigment.

Fig. 67.—A small artery from the brain. Several clumps of pigment in the adventitia. *Magn.* 150.

Fig. 68.—A small vein from the brain-substance. *Magn.* 150.—*a*, A clump of fat granules on a small lateral branch; *b*, an indistinctly fusiform enlargement.

with a great tendency to arrange itself in longitudinal plaits. It contains neither nuclei nor other cellular elements. When strongly magnified it presents numerous clear points (? holes). This membrane it is which lends to the larger arteries, in which it should be specially studied, their look of longitudinal striation. Its existence is still to be proved in the smaller arteries, for it rapidly dwindles in importance with their diminishing calibre. It is not present in the smallest vessels or in capillaries.

Spindle-shaped plain muscle-fibres closely invest the membrana fenestrata, composing the **tunica muscularis** *seu* **media** (fig. 66, *c*). Without exception these muscle-fibres are disposed around the vessel with their axes, and hence, with their fusiform nuclei at right angles to it. The nuclei of the endothelium and of the muscle-fibres, therefore, cross one another at right angles (figs. 66 and 67). The muscle-wall appears on its outer side distinctly ribbed owing to the elevation of its fibres. While in the larger vessels the muscular coat is many-layered and constitutes the chief thickness of the vessel-wall, it consists in small arteries of but a single sheet. As the vessels diminish in calibre the muscle-cells change in shape, becoming progressively shorter and broader. Their nuclei change in the same sense. A single muscle-fibre is long enough to make more than one circuit round a small vessel.

In very large cerebral arteries longitudinal bundles of connective-tissue are sometimes to be observed on the outer side of the muscular coat. Usually the muscular coat stands free in a space bounded on the outside by the **adventitial coat,** or, shortly, **adventitia** (figs. 66, *d*, and 67). If it is isolated from the other coats it appears as a delicate sheet of connective-tissue sown with round or oval nuclei. On the periphery of these nuclei a distinct protoplasmic granulation is often visible. Many observers have shown, by treatment with silver nitrate, endothelial-cells both on the inside and the outside of the adventitia. Granules of pigment are usually found in the adventitial sheath. More rarely fat-granules also are present (fig. 66, *e*). We shall speak of these later on.

In sections of hardened brains, especially of animals, long and strong bundles of connective-tissue fibres are often seen traversing the nerve-substance to fix themselves with conical bases to the outside of blood-vessels. Not rarely such a bundle of fibrils can be followed in the opposite direction to a stellate connective-tissue cell (fig. 88). Since the adventitia in the hardened preparation lies close to the muscularis and even in the most carefully-isolated vessels no such processes are to be seen hanging on to the adventitia, it is necessary to suppose that the tube of brain-substance in which the adventitia lies is lined with a **limiting membrane.** Sometimes in sections which otherwise are hardly successful, one sees a great number of these prolongations very regularly arranged (fig. 69).

Between the adventitia and the muscularis a considerable space is seen in all isolated arteries, the adventitial lymph-space (Virchow-Robin space). The outside of the adventitia between it and the limiting membrane is also surrounded by a space, the perivascular space, or space of His. By lymph-space is meant in this sense any lymph-filled gap which may serve as the starting point of lymph-vessels. The lymph-spaces, especially the perivascular ones, serve for the rapid exchange of juices between the plasma and the several nervous elements. In very favourable injection-preparations from the new-born child one can convince oneself that tissue-spaces are injected which connect the perivascular-space with the space by which every cell is surrounded (the pericellular spaces). The pericellular spaces are visible in very thin sections, and occasionally their connections with perivascular spaces may be seen (fig. 70). Even though we allow that in the hardening process the pericellular and perivascular spaces are

Fig. 69.—An artery from the cortex cerebri in longitudinal section. *Magn.* 80. Numbers of fine fibres are seen streaming into the brain-substance.

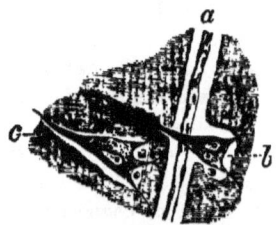

Fig. 70.—Section from the cornu Ammonis, showing perivascular and pericellular lymph-spaces. *Stained with carmine. Magn.* 150.—*a*, Capillary vessel in a perivascular lymph-space; *b*, pericellular lymph-space directly continuous with the former. Two leucocytes are seen in the pericellular space *c*, and one in the space *b*.

increased in size, owing to the shrinking of the tissues, the normal existence of such spaces is proved by the presence within them of lymphatic cells. It is probable that during the increase and decrease in size of the brain the lumina of the canals lined by the limiting membranes are subject to numerous fluctuations. Within these canals the complementary relations as to area of cross-section of the lumen of the vessel, and the adventitial and perivascular spaces, is perpetually varying. An increase in the lumen of the artery can only occur at the expense of one or both of the lymph-spaces.

(*b.*) **Veins.**—Only three coats can be recognised. The endothelium (fig. 68) is distinguished from that of the arteries by the rounder form of the cells and less regular arrangement of their nuclei.

The second layer which forms the proper vessel-wall consists, except for scattered plain muscle-fibres found especially in the larger veins, only of connective-tissue structures with numerous irregularly-scattered nuclei.

The adventitial lymphatic coat is a delicate membrane, essentially similar to that already described for the arteries. The points in which it differs will be noted later on. It may be assumed that the channels in the brain-substance occupied by the veins are lined by a limiting membrane.

(c.) **Capillaries** (fig. 71).—Capillaries may be regarded as the continuation across from arteries to veins of the endothelial coat of the vessels, for they consist of this coat only with a closely-adherent adventitial sheath. The endothelial coat has acquired, from the necessities of its independent position, a greater strength than it possesses in the arteries and veins.

Fig. 71.—Isolated capillaries from the cerebral cortex. *Magn.* 100.

(d.) **Fat and Pigment in the Adventitia of the Brain-Vessels.**—It has already been stated that pigment and fat-granules are to be regularly met with in the adventitia of the small vessels. A fuller explanation of the normal appearance is necessary.

Numerous cells filled with drops of fat are scattered throughout the brain of the new-born child; fat granule-cells. They are supposed to supply the material for myelin formation. Such fat granule-cells are found hanging on to the adventitial sheath of the vessels. A layer of fat is seen covering the adventitia in children in the first few years of life. After the fifth year some of the fat-granules are always found to have assumed a distinct yellow colour, being changed into pigment. In the brains of adults the presence of cells containing yellow or yellow-brown granules may be looked for with confidence in the adventitia of the arteries. The cells present every gradation of granules; in some they are small and irregular, in others numerous and large. Many reagents, concentrated sulphuric acid for instance, fail to affect this pigment. Osmic acid gives it, especially when the pigment is light in colour, a shade of grey.

It is quite otherwise with the adventitia of the veins. Pigment is only present in small quantities. Fat surrounds the veins in every brain examined. It may be scattered irregularly over the adventitia in the form of small drops. Fat granule-cells are, on the other hand, often met with, looking under a low power like dark spots on the vessels. Fat-granules and fat granule-cells are sometimes scattered over the adventitia (fig. 68); sometimes they form a ring around the vessel, giving to it a fallacious appearance of fusiform enlargement.

It must be granted that the fat in the adventitia is a remnant of the embryonal period; later, it undergoes, especially in the region where metabolism is most active, a chemical change into pigment, probably by oxidation. The pigment is not, therefore, a degeneration-product of blood-pigment which has transuded from the vessels. In all its chemical characters it differs from blood-pigment, and is to be, equally with the fat, looked upon as a normal appearance. The oxidation of fat into pigment occurs to a less extent around the veins, owing to the small amount of oxygen which the blood in these contains.

(ε.) **Pathological changes in the small Brain-Vessels.**—In considering changes in the vessels it is necessary in the first place to determine which coat of the vessel is diseased, for the process is quite different in the several coats. Disease of the adventitia contrasts, for example, with disease of the muscularis.

It is worthy of remark, too, that one often meets with alterations of the vessel-walls of the brain which have given rise to no symptoms during life, but which yet are distinctly pathological in their nature and affect the nutrition of the brain.

Granular pigment, which might easily be mistaken for normal blood-pigment, is found in the brain as a residuum of hæmorrhages, and, perhaps, also as the consequence of prolonged hyperæmia. **Hæmatoidin** (the name given to this pigment) differs from normal blood-pigment in certain unmistakable respects. In colour it is a browner red, and it shows a tendency to crystalise in rhombic prisms, which may be found in the fat granule-cells, sometimes several together (fig. 72). A similar pigment derived from the blood is, as a rule, found in the surrounding brain-substance. If these characters are not sufficient to differentiate the pigment, its chemical behaviour affords an unmistakable test. On addition of concentrated sulphuric (or other mineral) acid the pigment passes through a series of colour changes, becoming green, blue, violet, and, finally, dissolving. This reaction can easily be obtained with any piece as large as a millet seed from the contents of an old hæmorrhagic cyst. Even after the brain has been hardened in chromic acid, a very small piece placed on a slide and allowed to dry affords this play of colours on addition of the acid. The blue-green spots are visible to the naked eye.

A peculiar pigmentation is sometimes met with in the adventitia of the vessels which enter at the base of the brain. Elongated cells, with thick knotty processes, are quite filled with dark-brown pigment. The nucleus alone appears light as if it were a space punched out of the cell (fig. 73). Beside these, many round **pigment cells** are seen scattered singly or united into chains. Such cells occur normally in the pia mater, especially on the ventral surface of the medulla oblongata. No pathological importance, therefore, attaches to them. They have simply been drawn up along the course of the vessels from the meninges into the substance of the brain.

Fig. 72.—Cells with hæmatoidin crystals from the walls of an old apoplectic clot. *Magn.* 200.

Fig. 73.—A moderate-sized artery from the corpora striata, with numerous pigment-cells in its adventitia. *Magn.* 80.

Fig. 74.—Capillary vessel from a case of melanæmia. *Magn.* 200.

A fourth kind of pigment is to be mentioned, having its origin in the blood, and, therefore, not belonging to the adventitia, although occasionally fixing itself in it. This pigment, **melanin**, occurs in the brains of people who have suffered from violent epilepsy (fig. 74). It appears as very fine black granules, only rarely occurring on the outside of the vessels, and gives to the brain-substance a striking grey colour. This pigment also offers great resistance to chemical reagents. It is easily seen that the blood-corpuscles carry these pigment-granules, for large granules are found in emboli which have led eventually to rupture of the vessel.

Collections of fat in the adventitia may also attain to pathological importance, as, for instance, in cases of softening of the brain and spinal cord. It is not possible to place a limit upon the amount to which accumulation of fat-granules in the adventitia may attain in health. In disease the quantity becomes so great that the vessels appear to the naked eye as thick yellowish-white columns.

Fatty degeneration of the muscular coat presents essentially different features (fig. 75). In the earliest stage of this condition bright shining fat-drops appear between the muscular fibres of the arteries. Later on the fat-granules fill the muscle-fibres themselves, which become dull in appearance, their nuclei lose their distinctness, and the muscle-coat constitutes simply a dim yellowish tube invested by the adventitia. Although such arteries with fattily degenerated media must have lost considerably in resistance and elasticity, they may be found in quite healthy brains, even of young people. It must be unconditionally accepted, however, that such a condition gives a tendency to rupture of the vessels, and consequent hæmorrhage.

Calcification of the vessels is not rare. It occurs under different forms. Simple calcification of the media (fig. 76) is to be met with in healthy persons, even children. Calcification may occur in patches, or the whole artery may be converted into a tube of chalk invested by its adventitia. Examined macroscopically, the vessels look like white threads. They grate when pressed upon with a needle. Angular fragments of these calcified tubes are seen in sections. To make quite sure of their nature, a drop of sulphuric acid is allowed to run in under the coverslip, when bubbles of carbonic-acid gas escape.

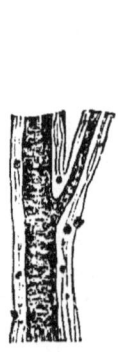 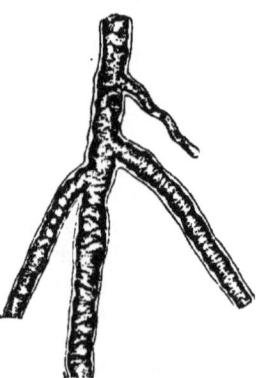 

Fig. 75.—Fatty degeneration of the muscular coat of a cerebral artery. *Magn.* 150.

Fig. 76.—Calcification of the muscular coat of a vessel of the brain. *Magn.* 150.

Fig. 77.—Calcification of an artery of the brain affecting the adventitia as well as its other coats. *Magn.* 150.

The second form of calcification of vessels has a more important pathological significance. It begins in the adventitia, but soon extends beyond the limits of this coat, bulging into the brain-substance as rounded, knotty, chalky structures (fig. 77). The more advanced phases of this process are only found as concomitants of other diseases of the brain.

Calcification of the capillary network is sometimes met with in circumscribed regions; in the nuclear layer of the cerebellum for instance.

HYPERTROPHY OF VESSEL-WALLS. 143

Connective-tissue overgrowths originating in the media affect especially the veins. At first the lumen of the vessel is not altered, while its circumference is increased; therefore the endothelium as well as the adventitia remain intact. **Fusiform hypertrophy** of the vessels of the brain is thus produced, especially at the spots where the brain-vessels give off numerous fine branches, almost at a right angle, as, for example, the vessels going into the brain from the great veins at its base or the finest branches of the meningeal arteries which are destined for the surface of the cortex (*Neelsen*). As the process

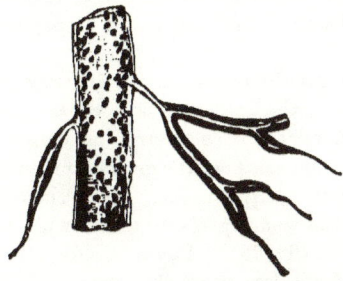

Fig. 78.—A vein from the brain showing fusiform hypertrophy of its lateral venules resulting in their obliteration. *Magn.* 150.

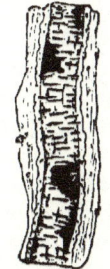

Fig. 80.—Atheromatous degeneration of the tunica intima of an artery of the brain. The dark patches of atheroma are seen to reach no farther than the muscular coat, which appears as a bright crenulated seam sharply marked off, both from the lumen and the surface of the tube, differing in this respect from fig. 66. *Magn.* 150.

Fig. 79.—Pseudo-hypertrophy of the muscular coat of an artery of the brain. *Magn.* 150.

advances the lumen of the vessel is diminished, even to obliteration. The adventitia is drawn into the process. The part of the vessel in front of the overgrowth, now put out of function, dwindles to a connective-tissue thread (fig. 78). This kind of obliteration of the vessels is commonest in the veins, and especially in the veins of old atrophied brains. After the fiftieth year it can also be found in almost every case in the arteries. Even in children's brains it is not invariably absent.

Very extensive connective-tissue hypertrophy of the vessel-wall, often accompanied, however, by considerable enlargement of the lumen of the vessel, is met with in places where the nervous tissue is sclerosed. It occurs also as the result of irritation, the inflammation which surrounds tumours for example.

In the brains of animals affected with rabies, *Golgi* found frequent caryomitosis of the nuclei of the muscle-fibres of the vessel-wall; in a small number of cases he could, in addition, see such division-figures also in the connective-tissue epithelium and nerve-cells throughout the whole nervous system.

A form of degeneration of the muscular coat, especially important in connection with hæmorrhages, is the condition known as **pseudo-hypertrophy** (granular degeneration). In groups of muscle-fibres fine roundish granules make their appearance; the granulation increases, the fibres fuse together, and so an opaque, usually wedge-shaped mass is formed, the rounded base of the wedge projecting somewhat beyond the general periphery of the media (fig. 79). A granular disintegration of the muscularis supervenes for a longer or shorter extent of the vessel-wall (*Löwenfeld*).

Fig. 81.—Beaded enlargement of a large artery of the brain. *Magn.* 50.

Fig. 82.—Miliary aneurysm of very small vessels. The dilatations are partly filled with blood. *Magn.* 50.

Fatty atheroma of the intima is recognised, as a rule, by the dark granular patches on the inside of the vessel, the subjacent muscularis remaining unchanged (fig. 80). If a teased vessel mounted in glycerin is under examination, it is often possible to loosen some of these

ANEURYSM OF BRAIN-VESSELS.

atheromatous patches by pressing on the cover-slip; the detached piece floats about freely within the tube until it gets wedged, as a rule at the point of junction of two branches. This gives us an idea of the process by which an embolus is formed in a small artery. When more strongly magnified it shows an amorphous mass containing shining fat-granules. In many cases of apoplexy such atheromatous degeneration is visible in the arteries, and hence it is possibly not an unusual condition, especially in old brains.

Colloid degeneration is a not uncommon condition of the vessels of the spinal cord; their walls are converted into a shining hyaline mass, taking a deep colour with carmine. More than one condition distinguished by special chemical reactions is included under the name of colloid degeneration.

Aneurysm of the brain-vessels, especially localised aneurysm, occurs in various forms. Paralytic dilatations of the small arteries are to be seen in the brains of chronic lunatics (fig. 81); a striking irregularity in

Fig. 83.—Ampullar dilatation of the adventitial lymph-space surrounding an artery of the brain. *Magn.* 50.

Fig. 84.—Packing of an adventitial lymph-space with leucocytes. *Magn.* 100.

the calibre of the vessels characterises them. The muscularis exhibits a series of bead-like dilatations separated by constrictions. They are caused by deficient innervation of the vessel-wall—a condition of partial paralysis. A higher form of the condition constitutes **miliary aneurysm**, which, however, may also arise in other ways. Larger and smaller aneurysmal dilatations are especially common in the neighbourhood of apoplectic lesions; but it would be a mistake to suppose that all hæmorrhages are due to the rupture of such dilated vessels (fig. 82).

Miliary aneurysms are generally found in the small arteries and capillaries. They are more numerous the smaller they are. For the most part they are globular or fusiform, and situate on the side of the vessel, with which they are sometimes connected by a stalk. They are very rare in the vessels of the spinal cord (*Hebold*).

The **adventitia** may also be **hypertrophied** in certain situations; sacciform dilatations of the vessels are thus formed in otherwise healthy brains. When they reach a certain size they make conspicuous spots in the brain-tissue, which are to be looked upon as **lymph-cysts** (fig. 83). When these dilatations of the adventitia are large and freely scattered throughout the brain-substance a sieve-like appearance is produced, the *état criblé* as it has been termed. Spaces exactly similar in appearance are produced when by shrinking of the nervous tissue the perivascular spaces are dragged upon and enlarged.

Cystic formations are commonest in the grey; sieve-like degenerations in the white substance.

The **contents of the adventitial lymph-spaces** deserve especial consideration.

When the cover-slip is made to press upon and so spread out the adventitia its spaces are seen to contain certain formed elements, amongst which lymph-corpuscles are the chief. Besides these, drops of fat, pigment-granules, vesicular cells of very large size (altered leucocytes, perhaps), and even blood-corpuscles are also seen within this space. The presence of many red blood-corpuscles suggests aneurysma dissecans, a rupture of the inner coat rather than pure diapedesis.

The quantity of leucocytes within the adventitial lymph-space may be so great that the muscular coat appears to have a distinct covering of them. This condition (which has been mistaken for nuclear proliferation, but is, in fact, due to increased emigration of white blood-corpuscles) is met with in inflamed and hyperæmic conditions of the brain and also in progressive paralysis (fig. 84).

The elements already described are normally present in the lymph-space; but distinctly pathological products also are to be found within it. In purulent meningitis these spaces are filled for some distance down into the brain-substance with pus-corpuscles.

Of great importance is the appearance of **neoplastic** elements (sarcoma- and carcinoma-cells) in the lymph-spaces, especially in the neighbourhood of tumours. It shows that in the brain the lymph-paths are the most important tracts along which such growths extend.

In syphilis accumulations of peculiar cells are found in the adventitial lymph-spaces. The cells are large, transparent, nucleated, and not unlike the cells of embryonal tissue. Further, there are to be found in the same situations (and also within the perivascular spaces in cases of long-standing infantile paralysis (*Leyden*) and in myelitic lesions) accumulations of cells like the cells of endothelium. Endothelial cells may also be heaped up into papillose excrescences of the adventitia (*Arndt*).

In various inflammatory conditions in the spinal cord, and also in rabies, a peculiar structureless colloid mass is found to be discharged around the arteries, especially the larger ones. It stains more or less strongly with carmine; when this mass, which originates in the blood, saturates, as it may do, the arterial wall itself, it gives to it when stained a peculiar brilliance. Similar colloid effusions are sometimes to be seen around the arteries in otherwise normal cords.

Lastly, attention must be called to the fact that the **contents of the vessels** also deserve consideration. Often the blood within them remains almost unchanged. In other cases it is coagulated in a special manner, a central cord of fibrin surrounded by a network of threads filling up the vessel. The endothelial coat may be loosened and lying in the lumen. The coagulation takes, at other times, a different form without our being able to trace the influence upon it of local causes; peculiar globular masses of coagulum are seen lying either separately or associated in groups; but, still, they are only products of the blood-plasma.

Special attention must be called to the constituents of the emboli which block the vessels. At times (as in leuchæmia) they are formed by roundish masses of white blood-corpuscles; or, again (after fractures), the emboli consist of drops of fat; or they may be formed (in epilepsy) of the peculiar pigment already mentioned, or, in still other cases, of atheromatous patches detached from the vessel-wall. Pieces of inflammatory lymph from the heart or great vessels are only exceptionally carried as far as the small intra-cerebral vessels. The elements of neoplasms which have found their way into the blood, and even bacteria, occur in the emboli of brain-vessels. One must be cautious, however, in the last case to exclude the products of *post-mortem* putrefaction.

Not seldom such diseases of the vessel-walls as aneurysm or calcification are limited to certain layers of the cortex of the cerebrum and cerebellum. So, too, with regard to the contents of the vessels, special forms of coagulum may distinguish particular strata of the cortex. This fact makes it appear probable that the several layers of the cortex are, up to a certain point at any rate, independent of one another in respect to the nutrition and innervation of their vascular networks.

2. EPITHELIUM.

The attempt has been frequently made in recent years to range with the nervous elements the epithelial cells which clothe the cavities of the central nervous system. Although we cannot, in the present state of our knowledge, refuse to accept such a view, an unconditional inclusion of the epithelial cells amongst the nervous elements is at any rate premature. We shall return to this question.

The adult human central nervous system is most unfavourable for the study of ventricular epithelium. Not only is it difficult to obtain from the human subject specimens of epithelium in a sufficiently fresh condition for observation before changes have set in, but it appears that the epithelium in question is better developed in lower animals than it is in Man, and also that in Man it undergoes, locally at any rate, changes in character after childhood.

The ventricles of the brain, as well as the central canal of the spinal cord, are in animals lined with ciliated epithelium. The epithelial cells (fig. 85) are renewed from below, and their bases are continued into processes (usually one, rarely two, for each cell) which may be followed far down into the nervous substance. In the frog they continue the direction of the principal axis of the cell. In the spinal cord of Proteus anguinus *Klaussner* followed these processes, some as far as the posterior nerve-roots, others into the anterior commissure.

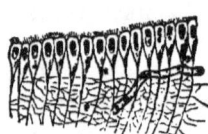

Fig. 85.—Ventricular epithelium of the frog. *Magn.* 200.

Cilia on the free surface of the epithelium are seen in lower animals, especially in the fresh condition, but also after hardening. They project into the ventricle or the central canal, as the case may be. Each cell contains a large oval nucleolated nucleus.

The epithelial cells are not everywhere equally well preserved in the human nervous system. They are least altered in appearance in the central canal of the spinal cord, the floor of the fourth ventricle and the aqueduct of Sylvius.

The cells are less perfect in other places, and the constant presence of cilia is so difficult to prove that their existence is doubted by some people. Together with the connective-tissue [neurogleia] layer, on which they rest, they constitute the so-called **ependyma.** Hence they are often termed ependyma cells.

It appears even more probable that, as already stated, the epithelium of the whole medullary canal stands in the closest relation to the nervous elements, both cells and fibres. Most, if not all, nerve-cells in the central organs are formed from epithelial cells, although the passage in all its details, from a cylindrical epithelial cell to a nerve-cell, is not yet clear. In ammocœtes, according to *Hermes*, certain ventricular epithelial cells grow with great rapidity, developing into nerve-cells which subsequently sink into the underlying substance of the brain. [See also the account on p. 29 of the development of neuroblasts as described by *His*.] A direct passage of the processes of epithelial cells into bundles of nerve-fibres has often been described; with gold these processes stain like nerve-fibres (*Freud*).

3. SUPPORTING-TISSUE.

(*a*.) **Connective-Tissue.**—Throughout the nervous system there exists a tissue which not only presents all the characters exhibited by the connective-tissue of other organs, but even passes over into direct continuity with indubitable connective-tissue. Certainly it is a connective-tissue in which the intercellular substance is reduced to a minimum. In most places it takes the form of a close network of fine fibres which can be followed to connective-tissue cells.

It is not surprising that very divergent views obtain as to the histological meaning of this tissue, found as it is in an organ which opposes the greatest difficulties in the way of its examination.

For the study of the connective-tissue cells of the central nervous system it is best to take little pieces of the fresh brain or spinal cord, and macerate them for two or three days in a straw-coloured solution of bichromate of potassium or in 0·1 per cent. osmic acid. The tissue can be subsequently coloured at pleasure. It will always yield a considerable number of connective-tissue cells, which vary in appearance according to the region from which the preparation has been taken.

Fig. 86 shows a cell from a radial septum of the human spinal cord. Numerous fibres of the utmost tenuity stream out from a granular nucleus which often is not very distinct. The processes which may attain to a length of 0·5 mm., are usually disposed in two groups which run in diametrically opposite directions. A proper cell-protoplasm is usually wanting; the cell body is only represented by flat appendages of the nucleus, usually faintly granular, which soon resolve themselves into cell-processes, which are further distinguished by their characteristic stiffness. In some cells they hardly divide at all, in others, division of the processes is very frequent. The processes of some of the cells radiate in all directions, as, for example, in the cell shown in fig. 87, from the ependyma of the lateral ventricle (spider-cells).

It is connective-tissue cells of this latter kind which, when prepared by Boll's method, are pointed out as Deiters' cells.

A somewhat different appearance from the cells just described is presented by most, although not all, of the connective-tissue cells of

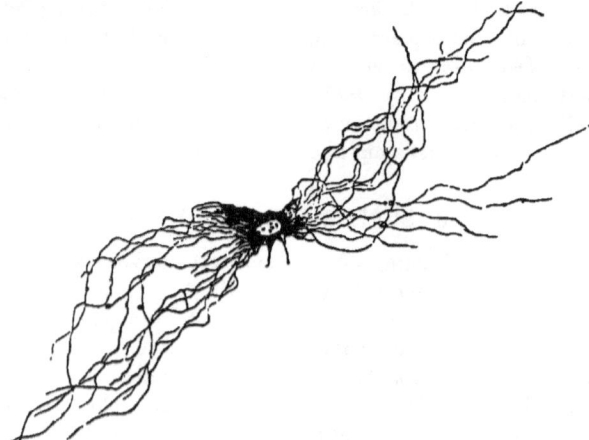

Fig. 86.—Isolated connective-tissue cell from the human spinal cord. *Magn.* 800.

the white substance of the cerebrum, cerebellum, and pons. In these situations are found angular cells, often arranged in rows, which

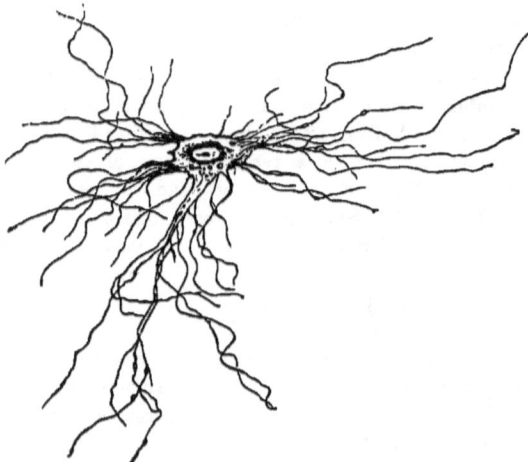

Fig. 87.—Isolated connective-tissue cell from the ependyma of the lateral ventricle. *Magn.* 800.

seem to exhibit, on superficial examination, a resemblance to epithelial cells. Especially in osmic acid preparations, one can convince oneself that these cell platelets are nothing more than connective-tissue cells from which the finest processes stream off like tufts of hair.

In many of these supporting cells of the central nervous system no nucleus is visible; it seems to be converted into chitin. In others the protoplasm is so scanty that the processes appear to start from the nucleus itself.

The disposition of the cell-processes varies according to the locality. Thin sections yield the best results, but the fine network is very difficult to pick out, with most staining methods, from the other histological elements. After staining with alum-hæmatoxylin, which gives to the nuclei an intense blue colour, their arrangement is easily grasped, but, as far as the processes are concerned, more is usually seen in carmine preparations, and this not infrequently in sections which seem from other points of view hardly successful. Golgi's sublimate method sometimes affords pictures of the most surprising distinctness (fig. 89).

The whole central nervous system is permeated and supported with a fine scaffolding of which cells constitute the nodal points. This is the stroma in which the nervous elements and their vessels are disposed. Sometimes disease so destroys these latter structures that

Fig. 88.—Section of the white matter of the brain. *Magn.* 100.—*a*, Small blood-vessel, the wall of which is connected with a spider-cell.

Fig. 89.—Longitudinal section of the spinal cord. *Magn.* 80.—*a*, White substance; *b*, grey substance. The section is stained after Golgi's corrosive-sublimate method and shows at *c* three pin-shaped crystals of the sublimate.

only the connective-tissue skeleton remains, as if in a corrosion-preparation.

Manifold local peculiarities are presented in the finer details of arrangement of this supporting-tissue. A thin layer of closely-felted connective-tissue forms, almost exclusively, the outer layer of the cortex of the cerebrum. In sections it appears as a dark border when moderately magnified. In the radial septa of the spinal cord the processes of the connective-tissue cells are arranged in close bundles. Together with the vessels they form the main constituents of these septa. In the white substance of the spinal cord are seen here and there the nuclei of spider-cells, the processes from which stream between the nerve-fibres, enclosing each individually in a network (figs. 37 and 89). In the white substance of the cerebrum and cerebellum still other peculiarities are added to the characters already assigned to connective-tissue cells. In these regions more distinctly than elsewhere processes of spider-cells are frequently seen to run with a straight course to the limiting membrane which invests the blood-vessels. To this they attach themselves with conical bases (figs. 69 and 89). The cells which have already been mentioned as disposed in rows also give off processes which take part in the formation of the network between the fibres. Attention must be called to the fact that in the brain the connective-tissue network is distinguished by its extreme delicacy and tenuity. In the spinal cord it is much coarser.

An account of the arrangement of its connective-tissue will be given in connection with each organ. The connective-tissue supporting the epithelium of the ventricles seems to be peculiar in the following respect:—The mass of fibres present in the ependyma suggests the conclusion that these fibres are not only the direct processes of its connective-tissue cells, but also the products of its intercellular substance.

We cannot here explain the extent to which one is justified in supposing that the connective-tissue cells and their processes share in the lymph-supply within the brain; but in this connection the funnel-like endings of the fibrils on the side of the perivascular lymph-space may be pointed out.

Pathological Changes in the Connective-Tissue of the Central Nervous System.—Amongst the pathological changes which affect the connective-tissue, the first to be mentioned should be that form of overgrowth which gives rise to **sclerosis.** Despite the number of special observations, opinion is still divided with regard to this certainly variable process.

Above all things a great importance is always assigned to the intercellular substance which we have seen plays for the rest but a subordinate part in the nervous system. The number of connective-tissue cells as recognised by their conspicuous nuclei, is, as a rule, not only not increased in sclerosed spots, but (in chronic processes especially) the amount of intervening substance is so far out of proportion to the number of the cells, that it is often difficult to find well-preserved specimens of the latter.

On the other hand, an increase in the number of cell-elements in sclerosis is not rare. It is produced by the division of pre-existing cells or, as more often happens, by new cell-formation. The material for this increase is afforded by the lymph-cells, which, as the result of irritation, wander in increased numbers out of the vessels into the nerve-substance, become fixed, give off processes, and, lastly, are metamorphosed into connective-tissue cells. Especially in the earlier stages of this process, in dementia paralytica for example, numbers of round cells are found in the brain-substance, the origin of which from the blood-vessels is proved, not only by their filling up the perivascular spaces, but also by their grouping within the nervous substance in the neighbourhood of blood-vessels (fig. 84). It is obvious that the nerve-elements, although they take no part in this process, must suffer from the overgrowth of the connective-tissue to such an extent that eventually they come to grief.

The granulation of the ependyma of the ventricle so commonly observed depends upon an overgrowth of the subepithelial connective-tissue which, breaking through the epithelium, appears uncovered in the ventricular cavity.

Very little is known as to the actual nature of the pathological changes which occur in the connective-tissue of the central nervous system.

In many cases of sclerosis the number of fibrils which originate in a single cell is greatly increased. Nuclei are seen from which innumerable delicate fibres, usually short, spread out in every direction (fig. 90).

Vincenti describes connective-tissue cells into which red blood-corpuscles have immigrated from the coats of the vessels *viâ* the processes.

Under the influence of continuous pressure, due, for instance, to the proximity of a tumour or a hæmorrhage, the connective-tissue cells swell and assume a turgid, glassy look; their nuclei disappear; the refraction of the processes is changed, so that, like the cells, they become more distinct; sometimes the connective-tissue cells come to resemble nerve-cells; in the neighbourhood of a hæmorrhage they are apt to imbibe a little blood-pigment. These are the changes which characterise inflammatory swelling of the connective-tissue. Proliferation of nuclei also occurs; some cells occasionally taking on the form of irregular plaques containing as many as twelve to fifteen nuclei (fibroplastic bodies of *Hayem*).

(*b.*) **Neurogleia.**—The last constituent of the nervous system to be described in this place is a substance which appears peculiarly suitable for filling out the spaces which are not occupied by the other elements; taking its place in the constitution of the general mass without offering any resistance to the free circulation of nutrient juices. This is the function of the neuroglia which, in the form of an excessively fine granular mass, constitutes the matrix of the grey substance; this said, its description is exhausted.

The neuroglia must be regarded as a peculiar kind of intercellular substance, for the cells from which it takes origin are not found in the adult organism. The free nuclei often pointed out as belonging to the neuroglia are probably nothing more than leucocytes escaped from blood-vessels. The fine granules of the neuroglia apply themselves to the processes of both nerve-cells and connective-tissue, and sometimes are seen attached to isolated elements. They were called by *Boll*, interfibrillar granules (*cf.* figs. 86 and 87).

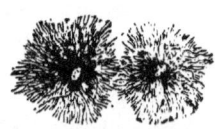

Fig. 90.—Connective-tissue cells with numerous short processes from the corpora quadrigemina which were affected with sclerosis in a case of dementia paralytica. *Magn.* 250.

Their chemical nature, as well as their morphological constitution, precludes us from regarding them as part of the connective-tissue. They must be looked upon as elements peculiar to the nervous system. The total amount of the neuroglia is very small. Most of what was hitherto regarded as ground-substance has been proved to be a network of nerve-fibres, medullated and non-medullated.

[It is very difficult, if not imposssible, to trace the genesis of any such ground-substance in the animal kingdom. The lower the animal in the scale of vertebrates, the thicker are the ultimate processes of the nerve-cells, and the more distinctly does the grey matter present the form of a network of protoplasmic strands. In the spinal cord of a fish, the cod for example, it seems to be fairly certain that the grey matter consists of a plexus of nerve-strands with cells for the nodal points of the plexus, of supporting-tissue (cells with delicate non-ramifying processes), and no other elements except those belonging to the lymph, the plasma of which is often coagulated in the process of hardening. As the animal scale is ascended the network of nerve-processes becomes denser, its strands finer, and, consequently, the grey matter appears more and more homogeneous. It is, however, by no means necessary on this account to formulate the existence of an intercellular nervous substance. No one has as yet been able to demonstrate in the mammalian brain or cord the union of the branching systems of contiguous nerve-cells, a union which seems, nevertheless, to be a physiological necessity, and it may be that histologists of the present day stand with regard to the nervous system, in the same position in which Harvey stood to the vascular system when he postulated the existence of communicating channels between the arteries and veins forty years before capillary vessels were discovered by Malpighi. It is, perhaps, an exaggeration to speak of the nerve-substance of which the finer cell-processes are made as semi-fluid; but, undoubtedly, it presents a strong tendency to lose its fibrillar form after death, and it is possible that the intercellular substance here referred to under the name of neurogleia consists partly of cell-processes, which the microscope fails to resolve, and partly of the coalesced substance of such processes.

It would be beyond the scope of a text-book to attempt any account of the bibliography of neurogleia, or to give a critical digest of the large number of divergent views as to its nature held by histologists; but it may be pointed out that it is not necessary to believe in the existence of a non-cellular ground-substance, and that from some points of view it is as well to apply the term neurogleia to the whole of the supporting-tissue of the central nervous system, on account of its development from epiblast, restricting the expression "connective-tissue" to supporting-tissue of mesoblastic origin. This is the meaning which most English writers attach to the term.

The influence of even a very weak electric current in maintaining the molecules of a fluid in fibrillar arrangement, restraining them from running together into a drop, may perhaps afford an analogy which will help us to understand how it is that, so long as life lasts and nervous impulses traverse the grey matter, its protoplasm retains a filamentous form, whereas, the same tissue examined after death, no matter how highly it is magnified, seems to consist of arborescent cell-systems ending in granular ground-substance. It is premature as yet to attempt to picture to oneself the structural connection between cell-process and cell-process along which the nerve impulse selects its path. The constitution of the ultimate ramifications of the cell-branches, and consequently the resistance which they interpose in the course of the nerve current, may depend upon nutritive conditions which vary incessantly.]

4. OTHER TISSUE-ELEMENTS WHICH OCCUR IN THE CENTRAL NERVOUS SYSTEM.

Besides those elements already described, which actually take part in building up the central nervous system, there are others which always, or almost always, put in an appearance when the nervous system is diseased, and afford by their nature and arrangement important evidence as to the character of pathological processes.

(1) **Fat granule-cells**, already repeatedly noticed, are perfectly round cells with, usually, distinct nuclei, and filled out with brilliant round drops of fat (fig. 91). As a rule, they are lymphoid cells, which have either stuffed themselves with fat for the purpose of transferring it to developing medullated nerves; or else have taken into their substance, with a view to eventually carrying it away, the fat set free in the degeneration of such medullated fibres. When degeneration is taking place at any spot, drops of myelin are shown by Weigert's method of colouring to be present in quantities in neighbouring fat granule-cells. Fat granule-cells are also formed from nerve-cells and connective-tissue cells by fatty degeneration, and even plain muscular fibres are supposed when fattily degenerated to have a similar fate (*Huguenin*).

Fig. 91.—Two fat granule-cells from the spinal cord. The area in which they occurred was secondarily degenerated. *Magn.* 250.

The presence of fat granule-cells is most easily detected by squeezing a little piece of the fresh tissue under the cover-slip. They are always to be looked for when degenerations are occurring in the spinal cord as well as in embolic lesions of the brain. Under a low power they appear as distinct dark spots; an estimate as to their quantity may be made in this way. When they are very numerous, lying in heaps, yellowish white flakes and stripes, dark in transmitted light, are visible, even with the naked eye.

(2) **Amyloid bodies** are seen under the microscope as clear strongly refracting round or oval bodies characterised by their brilliancy when slightly magnified. Dilute tincture of iodine stains them, when fresh, a light blue colour, the subsequent addition of a little sulphuric acid turns them a deep violet.

One may always count upon finding large numbers of amyloid bodies if one examines the tractus olfactorius of a person not too young. Their presence indicates a retrogressive atrophic process. Hence, they are found in sclerosis as well as in tabes dorsalis, and on the surface of the brain in very old people (*Kostjurin*). They are also frequently to be seen in the ependyma of the ventricle, especially on the mesial wall of the optic thalamus, and in the inferior horn when the cornu Ammonis is sclerosed in epilepsy.

Sections give the best information as to their relative position. They stain more or less darkly with carmine, but they come out most beautifully when very slightly stained with alum-hæmatoxylin, when they assume a fine blue colour, and are, in the majority of cases, distinguished from the nuclei of connective-tissue cells by their greater size. They are especially common in regions in which much connective-tissue is present, as, for example, in the outer layer of the spinal cord, the apex of the posterior horn, the thicker septa, &c.

No satisfactory explanation of the origin of amyloid bodies has yet been given. The substance of which they are composed is not really related to starch, but to albumen. They do not appear to be formed from cells (*Rindfleisch*). They always put in an appearance where nerve-substance is slowly disintegrating, and may possibly represent the last phase in the series of chemical changes associated with the atrophy of the nerve-fibres.

(3) A detailed description of the **neoplasms** which have their seat in the central nervous system would lead us too far.

(4) **Leber's corpuscles** are strongly refracting transparent globular bodies, about as large as the nuclei of the largest nerve-cells, which appear to be formed in the inside of the non-medullated nerves to which they are found attached. They were described under this name by *Vincenti*. They are supposed to be essentially different from amyloid bodies in chemical constitution, and are found especially where nerve-substance is affected by a tumour.

(5) Various **bacteria,** as, for example, the bacilli of typhus and splenic fevers (*Curschmann*) migrate into the central nervous system, where their presence may be revealed either by the microscope or by cultural experiments. By agglomeration in the finer blood-vessels they may even give rise to embolism (*cf.* p. 147).

SECTION IV.—MINUTE STRUCTURE OF THE SPINAL CORD.

STRUCTURAL FEATURES COMMON TO ALL THE CENTRAL ORGANS.

The gross anatomical features in the structure of the central nervous system, as seen with the naked eye, have received due attention in the second section. We must now prepare ourselves to appreciate the physiological meaning of the several organs by carefully studying their anatomical connections; this is the task of minute anatomy, a most difficult task, and still far from receiving its solution.

Certain general considerations which will aid us in the solution of other problems, as well as in dealing with the spinal cord and the brain, must first be set forth; the subsequent explanation of details will be thereby facilitated. The more our knowledge of the minute structure of the central organs widens the more possible does it become to present to the reader an exact and detailed "general anatomy of the central nervous system." It is a thankworthy task to deduce from the overwhelmingly numerous details which research in this field daily brings to light the resulting laws and general rules which first introduce order into the choas of more or less misunderstood anatomical combinations.

First, having regard to our limited space, we will speak of the particular facts which belong to this part of the subject.

We may start with the two kinds of nerve-elements of which the nervous system is built up, the cells and fibres.

The nerve-cells are the real nerve-centres. To the fibres belongs only the task of conducting the stimuli transferred to them. Many other functions of the nervous system besides simple conduction belong to the cells. The cells are the stations; the fibres the railroads which connect the stations together.

The nerve-cells are not scattered about irregularly, nor do they occur singly in the nervous system; but they cover, for the most part, extensive areas, in which they are collected in groups. In such places the character of the ground-tissue is also changed, the medullated nerves are mixed with many which are not medullated, the blood-vessels are very numerous and divide in a characteristic manner; so that, even to the naked eye, such regions, rich in cells, present appearances by which they can be recognised. While those parts of the nervous system which consist almost entirely of medullated fibres exhibit an almost pure white colour (the white matter or medulla), the regions rich in cells are distinguished by different tints of red-grey or yellow-grey (the grey substance). The intensity of this colouring is not the same in all brains. Various circumstances combine to give the grey matter a darker or less dark tint. The capillary network is narrower in the grey substance than elsewhere, and hence the condition of the brain with regard to its blood-supply has a great influence upon its colour. Small, much-convoluted brains in which the pigmented elements are probably pressed closely together (as, for example, in cases of premature synostosis of the skull), contrast with the less fissured brains, showing often a strikingly strong colour in such parts of the grey matter as the cerebral cortex, nucleus caudatus, putamen, and cortex of the cerebellum. The brain of the negro is not darker than that of the white races. The brain is darkest in pathological conditions resulting from violent epileptic fits.

The formation in those parts of the brain which are rich in nerve-cells but yet, owing to the preponderance of white fibres, do not exhibit the distinctive features of grey substance, is known as substantia reticularis (formatio reticularis).

All anatomico-physiological investigations must make the grey masses their starting points, and must next take up the question of the manner in which these are brought into connection with one another by the white tracts. This method of procedure cannot at all times be logically followed out, since there are many cases in which we have but a very superficial knowledge of the connections of the grey masses; while, on the other hand, many groups of fibres cannot with certainty be traced to their destinations.

In comparing the grey masses of the central system with one another, the conclusion that they are not all of equal morphological value is forced upon us. We are not yet in a position to assign to each its place, but must be content with a classification which is in the main correct and which we believe will never be altogether superseded, although as regards detail it may, by subsequent observations, be extended and completed.

The following kinds of grey substance must be distinguished:—

(1) The cortex of the cerebrum which everywhere covers the surface of the secondary fore-brain.

(2) The cortex of the cerebellum.

(3) The region in which the peripheral nerves originate, namely, the grey masses in the spinal cord, and the corresponding portions of the brain from which the cranial nerves take origin. With

these grey masses may be included the inner lining of the third ventricle which is their direct continuation, although it does not give origin to any peripheral nerves. These grey structures are summarised by *Meynert* in the expression "central cavity-grey" (centrales Höhlengrau). They are the proper primary central grey masses of the nervous system to which the other organs arranged in the classes 1, 2, and 4 are but adjuncts.

(4) The central ganglia. Theoretically this group is well defined since it includes all structures which do not find a place in classes 1 to 3, so heterogeneous a group does it make, however, that it is best to regard it as a temporary association of elements which cannot be otherwise classified.

Just as in the grey masses so also in the white we can recognise distinctions, but these after all depend upon the arrangement of the grey masses. Every fibre may be regarded as a conducting path either connecting two nerve-cells or a nerve-cell and a peripheral end-organ, be it motor or sensory. As each terminal apparatus is independent and endowed with a well-defined function we may regard it as of equal value, physiologically, with a nerve-cell. Indeed an end-organ may be looked upon as the terminal station of the road which extends outwards from the nerve-network.

Topographically nerve-fibres may be separated into two large groups—

(1) Fibræ homodesmoticæ; or fibres uniting together two homologous points of similar grey masses, as, for example, the two anterior cornua of the spinal cord or two spots in the cortex of the cerebrum.

(2) Fibræ heterodesmoticæ; which bring into connection two grey masses of unequal value, or else unite a central region with an end-organ. It is possible further to subdivide this group as, for example, into fibres connecting the cerebral cortex with the central ganglia, fibres connecting the periphery with the nuclei of origin, &c.

The physiological importance of these distinctions needs no explanation.

The term **tract** is used to signify the connecting road between two central grey masses, or between a grey mass and an end-organ. We have almost always, however, to do with a more or less complicated combination of roads, bringing stations of several different distances apart into functional relation with one another. Hence, for example, we may speak of a cortico-muscular tract, meaning thereby the whole group of nerve-fibres (perhaps interrupted at more than one point by the intercalation of grey masses), along which an impulse starting in the cortex must travel, if it is to induce a movement in a certain muscle. In the same way we speak in common life of the Berlin and Vienna Railway, although we know quite well that Dresden and Prague lie between the two termini.

Following this comparison a little longer. The route just mentioned is by no means the only connection between Berlin and Vienna; not only can we take the line through Breslau and Oderberg, avoiding Dresden and Prague, but various alternative routes from Berlin to Dresden, or from Prague to Vienna, are offered to us; further, we are in a position to go direct from Dresden to Vienna without touching at Prague, and so forth. When, for example, it happens that owing to a landslip the line between Dresden and Prague is impassable, the connection between Berlin and Vienna is not thereby interrupted. The richer the network of rails the more numerous are the connections, the "tracts" between the two chief termini.

Now let us transfer these observations to nerve-tracts. The most highly specialised nervous system, capable of performing the greatest variety of functions, is the one in which the paths connecting its grey masses are most numerous. Already (p. 126) we have called attention to the fact that homologous nerve-cells have more processes, and the processes ramify more freely the higher we ascend in the animal scale. So, too, in higher animals the number of medullated fibres originating in the nerve-plexus is increased; especially is this true of the fibræ homodesmoticæ which connect together homologous portions of the grey substance. The corpus callosum is a striking example of this law. In birds it is almost wanting, in animals lower than birds it is very small; it only attains to its highest development in Man.

From this follows a result not difficult to prove by gross anatomy. The higher the animal the greater is the quantity of white substance, relatively to the grey, found in its central nervous system.

Since the cells of the grey substance are the true apparatus of the higher cerebral functions, it might be inferred, *a priori*, that the higher the intellectual development of the animal the greater would be the relative amount of grey substance; the perfect action of the brain depends, however, upon the most intimate association possible of all its centres with one another.

Danilewsky has shown this changing relation of the white substance to the grey by a chemical method.

Another lesson may be learnt from the railroad illustration. Supposing the line between Dresden and Prague is interrupted, I can still, if I choose, adopt a method, somewhat slower, perhaps, of travelling from the one place to the other—I can drive. So, too, in the central nervous system, when one track is interrupted other collateral routes are still at our command. It would be quite wrong to conclude that, because a function is still performed after certain fibres are destroyed, these fibres have nothing to do normally with the conduction of the said impulses.

It follows that one must be very careful in assigning an object to nerve-routes, especially when they exceed an internode (the distance between two nerve stations) in length.

Just as we are certain that no such things as apolar nerve-cells exist, so also is it impossible to conceive of the existence of nerve-cells which do not stand in connection with other organs. Probably they are all connected with organs lying on two sides at least. The simplest scheme of a primitive nervous system (fig. 92) supposes the presence of a single cell, c, receiving, on the one side, a centripetal, sensory, nerve-fibre, ps; and giving origin, on the other, to a centrifugal, motor, fibre, pm. This conception we can enlarge into a cell-group instead of the single cell, c; and bundles of fibres instead of the single fibres, $ps.c$ and $pm.c$.

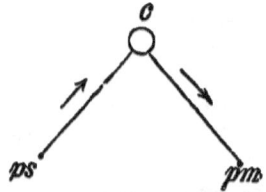
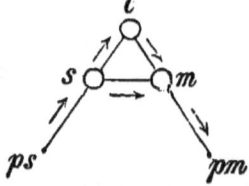

Fig. 92.—c, Central nerve-cell; ps, sensory periphery; pm, motor periphery.

Fig. 93.—c, Central nerve-cell; ps, sensory periphery; pm, motor pheriphery; s, intercalated sensory cell; m, intercalated motor cell.

The first step in complication consists in the interruption of both centripetal and centrifugal fibres by cells, and these two cells, s and m, are united by a fibre, sm (fig. 93).

Alternative routes from s to m are now presented. The impulse can travel directly from s to m, or, on the other hand, can pass through c.

If it is considered how much more complicated than this is the arrangement of the nervous system, in even lowly organised animals, we grasp the fact that the connections between the several parts of the system are almost inconceivably complex. It may almost be asserted that every part of the nervous system is brought into functional connection with every other part; there are variations in the intimacy of the connections which are in no way dependent upon the situation of the particular regions connected together. No isolated regions, no islands performing their function in complete independence of the other parts are to be found in the nervous system. Returning to our illustration—just as fast direct trains run between certain stations, while others are connected only by slow trains which often stop; so in the nervous system we find, as the result of physiological experiment, that the oftener a track is broken by nerve-cells the longer does it take for an impulse to traverse it.

It is possible in physiological experiments to reduce certain regions of the nervous system to an isolated condition with independent functions; for example, after the spinal cord is cut across, the caudal part still lives and performs its functions, although its connection with the rest of the central nervous system is completely severed.

The conclusion is inevitable that the several parts of the larger more extended grey masses may be united together in one of two ways. In the first place a direct connection between their different points may be established by means of long bundles of fibres; for example, in the case of the cerebral cortex the frontal and parietal lobes are directly united by such bundles, which are usually termed "association" fibres; while a second possibility of communication presents itself through the medium of the close network which everywhere enters into the constitution of the grey matter. Every point of the cortex may in this way be connected at pleasure with every other point in the same hemisphere. This functional connection between the several parts of the grey masses has not yet been sufficiently recognised.

It has already been mentioned that we have to distinguish two kinds of nerve-fibres from one another—the homodesmotic and heterodesmotic fibres uniting co-ordinate and sub-ordinate grey masses respectively. A further distinction between the fibres depends upon whether they unite together centres lying on the same side only, or, by crossing over in the middle line, bring into connection stations which lie in opposite halves of the body. Either class of fibres may cross in the middle line—homodesmotic fibres when crossed constituting a **commissure** (fig. 94, cc'), heterodesmotic fibres a **decussation** (cg' and $c'g$).

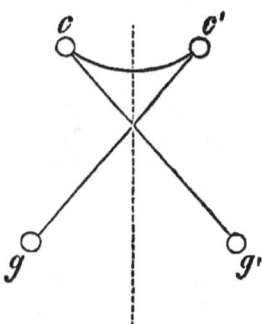

Fig. 94.—c and c', Cells of the cortex; g and g', nerve-cells of a different category; cc', commissural fibres; cg' and $c'g$, decussating fibres.

Here we may remark that the central nervous system appears to be symmetrical in structure as far as the main lines on which it is laid down are concerned. Excluding purely teratological or pathological differences between the two sides, however, certain striking, although inconstant, deviations from symmetry are exhibited, especially between the two sides of the great brain. Deviations from symmetry are commoner, and more conspicuous in highly developed brains than in those lower in the scale.

Certain peculiarities in the manner of arrangement of peripheral fibres—fibres, that is to say, which terminate in motor or sensory end-organs—are of special importance in helping us to comprehend the arrangement of nerve nuclei.

By **nucleus, or nucleus of origin** of a peripheral nerve is meant the group of nerve-cells in which the nerve begins or ends within the central system.

One or more nuclei (which form part of the grey masses classed on p. 159 as group 3) belong to each peripheral nerve. It has never

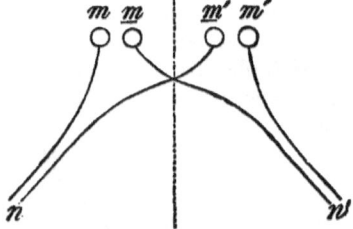

Fig. 95.—n and n', Motor nerve-roots; $m\,\underline{m}$, $\underline{m'}$ and m', cells of the nuclei of origin of the two sides; m and m' for the uncrossed, $\underline{m}$ and $\underline{m'}$ for the crossed fibres.

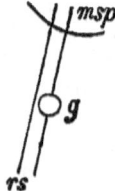

Fig. 96.—rs, Posterior sensory nerve-root, of which some fibres go directly into the spinal cord, msp; while other fibres only reach it after interruption in a cell of the root ganglion, g.

been proved in a single case that a nerve-fibre runs directly from the periphery to the cortex cerebri; the same assertion may be made with the highest degree of probability in the case of the cerebellar cortex.

The expression **nerve-root** is used in two quite different senses, likely to give rise to misapprehension; each use of the term is, however, so thoroughly naturalised that it would be very difficult to displace it.

By "nerve-root" is meant in gross anatomy the bundles of fibres by which the nerve appears to come out from the brain or cord, which bundles seem, therefore, to be the commencement of the nerve, the *peripheral root*. On the other hand, the fibres by which, within the central system, the nerve is continued to its nucleus are known as the *central root*. These fibres it is which lead the nerve back to its real origin. The trigeminus arises from two peripheral roots, but has at least six roots which run within the cerebro-spinal axis.

For all motor-nerves it may be asserted that they have two sets of root-fibres—(1) the fibres which join the nucleus or nuclei on the same side (fig. 95, $n\,m$, $n'\,m'$); (2) the fibres which cross to the opposite side of the body ($n\,\underline{m}'$, $n'\,\underline{m}$). A part of the nerve always enters a decussation, therefore, and the fibres which take this course are the more numerous, in proportion as the muscles which they innervate are unaccustomed to function independently of those on the opposite side of the body. The groups of muscles which usually act bilaterally, the muscles of the pharynx for example, are more fully supplied with fibres from the opposite side of the body than those which, like the muscles of the fingers, act independently.

A similar typical origin cannot be proved for sensory fibres. We know of cases, the olfactorius for example, in which the nerve ends, to all appearance, in a nucleus on its own side only.

[This case, however, requires special treatment, for it appears that the grey matter belonging to the nose and the eye has never been centralised to the same extent as obtains in the remaining sensory nerves. Nerve-plexus, which in the primary centres of other sensory nerves forms part of the central grey tube, lies, in the case of the nose and the eye, still in the vicinity of the sense-organ, taking part in the formation of the olfactory bulb and retina respectively. In these two cases, which stand amongst sensory nerves in a group by themselves, the fibres originating in the local clump of grey matter decussate abundantly on their road to the brain.]

A special difficulty presents itself in the case of the sensory nerves, in homologising the spinal ganglia which are borne by the posterior roots of the spinal nerves with the corresponding groups of cells in connection with the cranial nerves. This difficulty has been further increased by the discovery of *Freud* [and *Max Joseph*], that of the fibres which enter the distal pole of the spinal ganglion, only a part are connected with its cells (fig. 96, g). It is necessary, therefore, to distinguish in the fibres which connect the central nervous system with the spinal ganglia two groups—(1) those which coming without interruption from the periphery are peripheral nerves in the strict sense of the word; and (2) those fibres which having been broken in the cells of the ganglion are, in this part of their course, really central fibres.

A corollary to what has been already said is, that no two parts of the periphery of the body are directly united together by nerve-fibres without the intervention of cells.

Every nucleus of a peripheral nerve is connected with other parts of the central nervous system. The paths by which these various connections are maintained may be classified as follows :—

1. Connections with the corresponding nucleus of the opposite side;
2. ,, with other nerve nuclei;
3. ,, with various secondary ganglionic centres;
4. ,, with the cortex cerebelli;
5. ,, with the cortex cerebri, directly or indirectly.

Commissural fibres between corresponding nuclei of the two sides probably always exist; they are only demonstrated with absolute certainty for some nerve-centres; for example, the oculomotor nuclei (*Nussbaum*), the hypoglossal nuclei (*Koch*), &c. *Flechsig* believes that it is possible to prove the existence of commissural fibres between the nuclei of the three first sensory nerves, and infers that such fibres exist in all cases.

Examples of the connections with dissimilar nuclei (class 2) are very numerous, some crossed, others not crossed. The "posterior longitudinal bundle" (p. 223) probably consists of fibres connecting together nuclei, which lie one behind another. A directly connected sensory and motor nucleus, together with their attached nerves, represent the simplest reflex arc.

Connections with secondary ganglia (3), as the thalamus, globus pallidus, corpora quadrigemina, corpora geniculata, olive, and so forth, are proved in many cases, and may well obtain in all without exception.

How the cortex cerebelli (4) is connected with the nuclei of origin of sensory nerves is by no means always clear, but in the case of the spinal nerves and a part of the auditory nerve, we think we know the route by which a connection is established.

An undeniable direct connection between the cortex cerebelli and a motor nerve-nucleus has yet to be found.

Especially interesting, both from an anatomical and a physiological point of view, are the so-called central connections of the nerve-nuclei (5), the tracts, that is to say, by which they are connected with the cortex cerebri. In all cases the connection is supposed to be a partly crossed one, but it is extremely difficult to ascertain the relative numbers of the crossed and uncrossed fibres. Probably, the proportion, even for the same nerve, is not the same in any two subjects. For many of the nerves belonging to the hind-brain we have to seek these crossing fibres in the fibræ arcuatæ (including the striæ acusticæ); the fibres cross one another in the raphe at a very acute angle.

We may hope to come to a better anatomical and physiological understanding of the central nervous system when we have learned to classify the, at present, almost incomprehensible mass of details, with such order and method as will doubtless be made possible by further observations.

When we are able to speak of structures in groups which now seem to exact individual treatment, a clearer general view of the system will be obtained. The optic tract, for example, is, in a certain sense at least, homologous with sensory conducting paths in the posterior columns or other parts of the spinal cord. When this homology is further traced we shall come to a clearer understanding of the meaning of many structures which are in connection with the optic tract.

It is undesirable to press homologies too far, although a scheme embodying the main features of the central conducting paths is of inestimable value. We recognise at once departures from what we consider the normal arrangement and seeking to account for these deviations, we discover the faults in our scheme and the real explanation of these differences.

The relations exhibited in a simple form by the spinal cord yield information which we may use in studying the more complicated medulla oblongata and brain.

Motor-spinal and motor-cranial nerves may be supposed to present similar central connections; a parallelism may likewise be traced in the case of the sensory nerves in the two regions respectively. This consideration makes it possible for us to throw light into many dark places. It affords us many hints as to the objects we should have in view in anatomical investigations, and as to the connections of fibres which seem to be physiologically necessary, or which we should expect from analogy to be present.

The methods of investigation which it may be hoped will lead us to a clearer conception of the arrangement of the nerve-bundles—their schematisation out of the apparent chaos—have already been detailed.

The difficulty in elaborating a simple and comprehensive scheme of the construction of the nervous system depends upon the great variety of elements which must be ranged in their places; if all varieties are allowed for, a perplexing and highly complicated scheme is the result.

Many difficulties have to be overcome before a scheme satisfying all our requirements will be produced. This is, however, the place to sketch in outline some of the generalisations on this subject.

Luys takes his starting point from the central ganglia of the great brain (nucleus caudatus, nucleus lenticularis, and optic thalamus); they form the proper central point, towards which all nerves converge from the two sides. There are two principal systems of converging fibres—(1) "fibres convergentes inférieures," including all the fibres which travel from the periphery to the central ganglia, without regard to the direction in which they conduct; and (2) "fibres convergentes supérieures," including all the cortical fibres, for these in a similar way seek the central ganglia. All fibre-roads of the first category are broken on their way to the ganglia by still other grey masses. All cross the middle line, although the fibre-systems of the two sides of the body remain distinct from one another. The other kind of fibres, "fibres convergentes supérieures," pass without crossing and without interruption from the cortex to the ganglia. They are united, however, with those of the opposite side by a special commissural system.

Meynert, in formulating his scheme, commences with the cortex of the great brain, as being the organ devoted to conscious processes. All routes which serve as media of communication between the cortex of the cerebrum and the outer world are grouped together in a chief system. Through the fibres of this system sense-pictures are projected on the perceptive cortex, and, further, not only are movements of one's own body the source of sensations of movement which are represented in the brain in the same way as phenomena of the outer world, but the cortex also, by means of the motor-tracts, reflects outwards again the states of stimulation, information with regard to which is transferred to it by means of sensory nerves. The whole of these conducting paths Meynert, therefore, terms a "projection system."

The cells of the cortex are connected with the corresponding cells of the opposite side by "commissural-systems" and with cells of distant parts of the same hemisphere by "association-systems." The medullated fibres which connect the lobes of the great brain with the cortex of the cerebellum fall into a special class.

The "projection system" is divided into segments by the intercalation of two kinds of grey matter. The first segment consists, for the most part, of fibres radiating from the central ganglia to the cortex—the corona radiata. The second segment extends from the basal ganglia to the grey matter surrounding the central cavities in the peduncular system. The third segment of the projection system is made up of the peripheral nerves which have their origin in the grey matter bordering the cavities from the aqueduct of Sylvius down to the end of the spinal cord.

[The *translator* in a "plan" of the central nervous system, published in 1885, argued that it consists fundamentally of but two parts, or tissues, the older plexus of grey matter accumulated in the vicinity of the central canal (the "central grey tube") and the more recent and more plastic mantle of grey matter which covers the surface of the cephalic vesicles (the "peripheral grey tube"). These two tubes are connected by cell-processes, which, protected by medullary sheaths, make up the mass of white matter which intervenes between the central and peripheral tubes. Each intrinsic fibre grows out from a cell at its proximal end (having regard to the direction in which impulses travel), and terminates distally, not, as a rule, in a single distributive cell (as in the case of the visceral fibres), but by ramifying into a number of fine processes.

The central grey tube is in direct relation with both anterior and posterior peripheral roots. Its plexus is divided metamerically into "centres" for nerves. It includes the optic thalami.

The peripheral grey tube comprises two chief fields of cortex, the cerebellar and cerebral mantles, connected by afferent and efferent fibres with the several metameric clumps of the lower or central grey tube, and is, therefore, itself divided into areas connected *indirectly* only with the several peripheral nerves.]

The above is a superficial survey of the schemata of *Meynert, Luys* [and *Hill*]; it is hardly possible to give an account of other views as to its plan of construction. *Aeby* takes his stand upon the segmental constitution of the spinal cord, each metamer of which belongs to an anterior and posterior root. In the brain a similar segmentation may be traced, but it affects the stem region only, not the cortex. *Aeby* analyses the relations of the grey masses and the tracts of fibres on this basis.

Flechsig has designed a "plan of the human brain," but it is impossible to reduce his views to an abstract; it may just be mentioned in this place that he summarises the conducting paths in the four following chief systems:—(1) the (relatively) direct connection of the cortex with motor and sensory nerves; (2) system of the optic thalamus; (3) system of the pons; (4) system of the tegment, to which belong the fibres of the corpora restiformia, as well as certain columns of the spinal cord

1. TOPOGRAPHY OF THE SPINAL CORD.

The internal structure of the spinal cord is best studied by means of sections of the hardened organ cut transversely. The comprehension of the anatomical relations of its several parts is, however, aided by studying sections cut in other planes. For this purpose pieces of the spinal cord from the cervical or lumbar enlargement, of from 1 to 1·5 cm. long, are cut into sagittal sections (antero-posterior sections parallel to its long axis). Frontal sections (transverse, but parallel to the long axis) are also prepared. Oblique sections are useful. Longitudinal sections intended for staining by Weigert's method are best mounted in celloidin (p. 9). For the preparation of faultless sections it is desirable to let the cord lie for from three to six weeks in chromate of potassium and subsequently in alcohol, as this treatment ensures its cutting well. Freshly-prepared chromic cords of animals (horse, ox, &c.) give beautiful preparations, even without hardening in alcohol.

The structure of the cord changes from region to region; therefore, it is desirable to study a large number of sections taken at intervals from the whole length of the cord, in order that one may be enabled to recognise approximately the level from which a given section was taken. The lowest magnification suffices to arrange these sections in their order. We shall, therefore, first describe the appearances of sections as seen under a No. 2 objective (of *Hartnack* or *Reichert*).

It is best to make two parallel series of sections in this, as in many other investigations, treating each series *secundum artem* with a special method of double staining. The one series is coloured with carmine or picrocarmine and then alum-hæmatoxylin; the other with Pal's modification of Weigert's method and then with picrocarmine. Other staining methods may be used.

In the ordinary method of taking the brain out of the skull-case, the spinal cord is usually cut across at the level of the second or third cervical nerve.

We will commence our description with a section through this region (fig. 97, half of the cord only represented). In the first place we notice that a complete section is divided into two almost symmetrical halves. On the ventral side the fissura longitudinalis ventralis, *Fsla*, sinks into the substance of the cord. After reaching in depth about a third of the antero-posterior diameter, it splits into two short lateral divisions. From the sulcus longitudinalis dorsalis, *Fslp*, a connective-tissue septum (septum medianum dorsale, *Smd*) dips inwards about half as far again as the ventral fissure, which it almost meets. Only a narrow bridge of nerve-substance which unites the two halves of the grey matter together, the commissura medullæ spinalis, *Cm*, intervenes between the two fissures.

The white investing sheath and the central grey substance are clearly differentiated in the spinal cord. In each half of the cord the grey matter is surrounded by white substance on all sides with the exception of two spots :—

1. The grey commissure, *Cg*, which surrounds the central canal, *Cc*, and unites the grey masses of the two sides.
2. The place where the posterior roots come out on the dorsal surface of the cord in the sulcus lateralis dorsalis, *Sld*.

The grey matter on either side of the section we are now considering appears as an elongated area with its long axis placed almost sagittally; in the dorsal half it bends a little sideways. The grey masses of the two hemispheres, taken together, make an H, the cross bar of which is formed by the grey commissure. The larger part of the grey matter lies on the ventral side of the cord, and is known as the anterior horn, *Cra* (cornu anterius), the more slender portion is directed backwards as the posterior horn, *Crp* (cornu posterius).

Considered in their continuity the anterior and posterior horns, extending as they do as veritable columns throughout the whole length of the spinal cord, may well be termed, as is often done, the anterior and posterior columns.

The short lateral bulging of the grey matter opposite the commissure is pointed out as the lateral horn, *Til* (middle horn, tractus intermedio-lateralis). The re-entrant angle between the posterior and lateral horns is filled in with trabeculæ of grey substance, between which room is left for the passage of columns of fibres; the network of grey strands constitutes the processus reticularis, *Pr* (by many persons, *Goll* for instance, termed the lateral horn).

The anterior horn is round, while the posterior horn is fusiform in shape. The much drawn-out point of the spindle (apex cornu posterioris, *Ap*) is continued to the sulcus lateralis dorsalis. The posterior horn is connected with the rest of the grey matter by the basis cornu posterioris; dorsal to the base, it is constricted into a neck [cervix cornu posterioris] while the real body of the spindle is known as the head [caput cornu posterioris].

Two kinds of grey matter are usually distinguishable from one another when the preparation is faintly stained with carmine—substantia spongiosa and substantia gelatinosa.

The latter, which stains darkly, is limited to two regions—(1) immediately around the central canal (substantia gelatinosa centralis); (2) in a part of the posterior horn where it forms the cap on the top of the caput cornu posterioris (substantia gelatinosa Rolandi, *Sg*). Dorsally it extends into the apex, while it is concave ventrally. The spongy substance is far more considerable in amount.

The anterior roots, *Ra*, are seen springing from the anterior horn; they take their exit in several (3-8) thin bundles of medullated nerves which traverse the white substance in a horizontal plane while distinctly curving outwards in their course. Even with the least magnification it is possible to see in the anterior horn the large nerve-cells which give origin to the fibres of the anterior roots. A group of smaller cells, more closely pressed together, is seen in the lateral horn. Large cells are scattered in the processus reticularis. From the latter region in some sections at this level (although not in the particular one which is now being described) distinct bundles of nerves are seen coursing towards the periphery in arches directed dorso-laterally. These are the root-fibres of the accessorius Willisii.

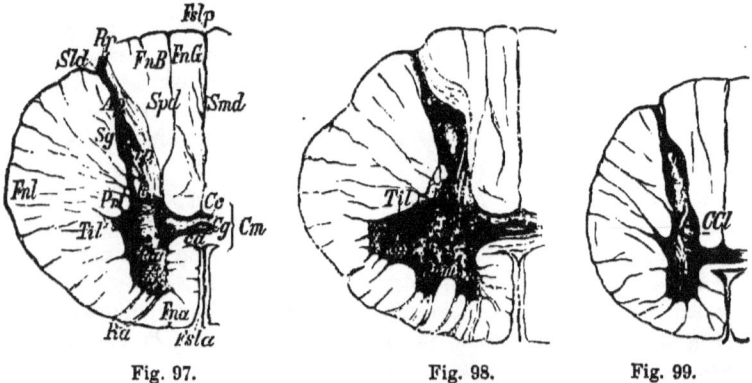

Fig. 97.   Fig. 98.   Fig. 99.

Figs. 97-103.—Transverse sections through the human spinal cord. Carmine staining. *Magn.* 5.

Fig. 97.—Section at the level of the third cervical nerves.—*Fsla*, Fissura longitudinalis anterior; *Fslp*, fissura longitudinalis posterior; *Fna*, anterior column; *Fnl*, lateral column; *FnB*, Burdach's column; *FnG*, Goll's column; *Smd*, septum medianum dorsale; *Spd*, septum paramedianum dorsale; *Sld*, sulcus lateralis dorsalis; *Rp*, radix posterior; *Ra*, radix anterior; *Cra*, anterior horn; *Crp*, posterior horn; *Til*, tractus intermedio-lateralis; *Pr*, processus reticularis; *Sg*, substantia gelatinosa Rolandi; *Ap*, apex; *k*, respiratory bundle of Krause; *Cm*, commissura medullæ spinalis; *Cg*, commissura grisea; *ca*, commissura alba; *Cc*, central canal.

Fig. 98.—Transverse section at the level of the sixth cervical nerves.—*Prm*, Processus cervicalis medius cornu anterioris; *Til*, lateral horn.

Fig. 99.—Transverse section at the level of the third dorsal nerves.—*CCl*, Clarke's vesicular column.

POSTERIOR ROOTS.

The dorsal or posterior nerve-roots, *Rp*, are seen entering the posterior horn at the sulcus lateralis dorsalis, *Sld*. Part of their fibres can be followed to the mesial side of the posterior horn, another part streams directly into the substantia gelatinosa Rolandi. The fibres of the first group describe, in their course through the posterior column, more or less open arches, and seem to sink into the grey substance of the posterior horn, in which they can be followed ventrally for a considerable distance.

The white substance of the cord is usually divided into several columns.

(1) The posterior column, which extends on either side of the septum medianum dorsale as far as the posterior horn. A constant septum of connective-tissue (septum paramedianum dorsale, *Spd*), starts at the surface and passes inwards with an inclination towards the median septum, and often gives off a branch directed outwards

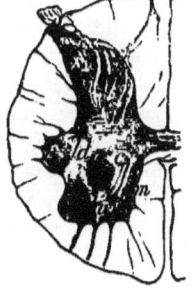

Fig. 100.   Fig. 101.   Fig. 102.   Fig. 103.

Fig. 100.—Transverse section at the level of the twelfth dorsal nerves.—*CCl*, Clarke's column.

Fig. 101.—Transverse section at the level of the fifth lumbar nerves.—*m*, Medial. *lv*, latero-ventral, *ld*, latero-dorsal, *c*, central groups of cells of anterior horn.

Fig. 102.—Transverse section at the level of the third sacral nerves.—*m*, Medial, *ld*, latero-dorsal group.

Fig. 103.—Transverse section through the inferior part of the conus medullaris at origin of nervus coccygeus.

towards the posterior horn. It splits the posterior column into two well-defined subdivisions, of which the mesial or smaller is termed Goll's column or the funiculus gracilis, *FnG*, the larger one is Burdach's column or funiculus cuneatus, *FnB* (ground bundle of the posterior horn).

(2) The lateral column, *Fnl*, usually considered as extending from

the outer margin of the posterior horn to the most laterally situate bundle of the anterior root.

(3) The anterior column, *Fna*, surrounding the ventral and mesial surfaces of the anterior horn.

It has been recognised for a long time that the division between the anterior and lateral columns is an artificial one, and, hence, they are often united under the name of antero-lateral column.

In addition to the three white columns already mentioned, the white commissure, *ca* (commissura alba), still remains to be described. It lies on the ventral side of the grey commissure, and forms a narrow bridge across from one of the anterior columns to the other.

Lastly, a small, but often very conspicuous, bundle of nerve-fibres is cut transversely in this section (the respiratory bundle of *Krause*, *k*), which lies in the basis cornu posterioris on the mesial side of the processus reticularis.

If we now muster our sections in order from the third cervical to the end of the spinal cord we meet with changing conditions, subject, however, to not inconsiderable individual variations.

At the fourth cervical the picture presented by a section is almost the same as at the third, save that, on close inspection, a commencing enlargement of the anterior horn is perceived. At the fifth cervical the enlargement becomes more pronounced. The total cross-section of the cord is by this time obviously increased, especially in its transverse diameter. It has assumed an elliptical form, the amount of eccentricity of the ellipse being very variable. Root-bundles of the accessorius are no longer distinguishable.

The cervical enlargement reaches its maximum at the level of the sixth cervical nerve (fig. 98). Here the anterior and lateral horns are fused together into a considerable mass, forming in cross-section an equilateral triangle. A little grey protuberance (processus cervicalis medius cornu anterioris, *Prm*) projects from the middle of the ventral side of the anterior horn, giving it a triangular shape. A group of large nerve-cells is seen in each corner of the triangle. Although the anterior and lateral horns are fused together it is usually possible to recognise, near the dorsal border (formerly the lateral border) of the anterior horn, the closely aggregated cells of the former lateral horn, *Til*.

The posterior horn has also increased in size, but not so obviously as the anterior, without, however, losing its elongated form. Still it must be pointed out that the increase in size of the posterior horn is almost restricted to its mesial side, so that now the apex rises with a step from the posterior column, a peculiarity which (despite many changes in form) it retains throughout the rest of the spinal cord. The processus reticularis loses in development, and so, too, does the so-called respiratory bundle.

The cervical enlargement is still at a maximum at the level of the seventh cervical nerve, beyond the eighth it rapidly decreases. The processus cervicalis medius sinks away and the ventral side of the anterior horn is bounded instead by a slightly concave line.

At the level of the first dorsal nerve the lateral horn grows rapidly smaller, at the same time retiring mesially. It projects like a beak from the lateral border of the grey matter (fig. 99). The characteristic group of cells which constitute the lateral horn is extended towards the posterior horn. Thus it comes about that the total cross-section of the grey matter again assumes the form of the letter H which we remarked in the upper cervical region; the two sections are, however, easily distinguishable, for the one from the dorsal region has the characteristic features which we have just remarked—its grey matter is narrower and more slender, the respiratory bundle is absent, the processus reticularis poorly developed, the posterior horn, directed a little outwards, rises by a step on its mesial side. At the level of the seventh or eighth cervical a group of cells, not found at higher levels, makes its appearance in the basis cornu posterioris, near its mesial border, *CCl*. The fibres of the posterior roots arch around this column, the general ground-substance of which is somewhat lighter in colour than the rest. In it are contained at first only scattered cells of large size. It has been named Clarke's column (columna vesicularis, dorsal nucleus of *Stilling*). Only in the dorsal region do the cells of this column constitute a well-defined group which causes a mesial bulging of the posterior horn. In many preparations of the upper dorsal cord these cells are altogether absent.

Except for the slow increase in size in Clarke's column from above downwards, it is impossible to distinguish the several sections of the dorsal cord from one another. Clarke's column is best developed at the eleventh or twelfth dorsal, and at this level the total amount of grey substance begins again to increase; this is the commencement of the lumbar swelling (fig. 100). Here the posterior horn again inclines outwards, recalling the disposition in the cervical cord; the great size of Clarke's column, however, allows of no mistake.

In the region of the cord, from which the lumbar nerves come off, the cross-section of the grey matter increases both in the anterior and the posterior horns. Nevertheless, the total size of the cord in the lumbar region can never equal that of the cervical enlargement, for the constantly diminishing amount of white matter observed in descending the cord tells in the total cross-section. The difference in relative amount of grey and white substance is obvious (fig. 101). The section is nearly cylindrical in this region.

In comparison with its shape in the cervical region the anterior horn is noticeably more rounded. So, too, is the posterior horn, the main mass of which, owing to the shortening and thickening of the apex, approaches nearer to the dorsal surface. At the fourth, and still more at the fifth, lumbar nerve where the grey matter is most abundant the lateral horn again acquires a certain amount of independence after having been involved in the upper lumbar region in the rounded enlargement of the anterior horn. Here, too, the large nerve-cells are more distinctly collected in groups than anywhere else. Their arrangement, however, is not quite constant, and hence they have been variously described. Between the second and third lumbar nerves Clarke's column again completely disappears. In the anterior horn the cells are arranged as follows :—

1. A mesial group (m) not very well defined, to which the whole of the mesial border of the anterior horn belongs.
2. A latero-ventral group (Lv).
3. A latero-dorsal group (Ld) which represents the lateral horn.
4. A middle group (c) almost in the centre of the anterior horn.

The general appearance of the cross-section is somewhat changed, owing to the anterior longitudinal fissure cutting more deeply into the substance of the cord, whereby the anterior commissure is carried almost to the middle of its sagittal axis. The septum paramedianum is often wanting below the lower dorsal nerves. If it is present it inclines inwards towards the posterior septum, and Goll's column, which is ill-defined, appears as a narrow plano-convex band lying against the dorsal septum.

From the point of exit of the lowest fibres of the anterior root of the fifth lumbar nerve, the spinal cord rapidly diminishes in size until it terminates in the filum terminale. The white sheath diminishes much more quickly than the central grey masses which rapidly obtain the preponderance.

The form of the grey horns is not much altered, but they become plumper; the posterior horn especially appears more uniformly rounded. The grey commissure becomes broader and approaches nearer to the posterior surface, at the level of the lowest sacral nerves where the cord scarcely attains 3 mm. in diameter, but very little room is left for the posterior columns.

Of the several groups of nerve-cells described above, only the latero-dorsal group (the representatives of the lateral horn), Ld, and the mesial group, m, remain by the time the third sacral nerve is reached.

At the level of the fourth sacral nerve no distinct groups, but merely scattered cells, are seen.

Even at the end of the conus medullaris (fig. 103), the region from which the nervus coccygeus springs, the typical formation of the spinal cord is evident. The filum terminale is nothing but a tube of epithelium with a thin covering of grey substance, the last remnants of the central grey matter of the cord.

As already noticed, the relation which the grey and the white matter bear to one another in amount changes considerably from the cervical region to the conus medullaris. Although there are individual differences, it is as well to tabulate the average measurements made by *Stilling* in a man twenty-five years old.

At the level of the attachment of the lowest Root-fibres of the following Nerves.	Cross-Section in Square Millimeters.			Proportion of White to Grey.
	Whole Cross-Section.	The White Substance.	The Grey Substance.	
Cervical III,	84·15	71·40	12·73	5·6
,, IV,	85·55	72·82	12·73	5·7
,, VI,	91·55	74·23	17·32	4·3
,, VIII,	78·12	62·92	15·20	4·1
Dorsal I,	65·39	53·73	11·66	4·6
,, IV,	57·67	50·26	7·42	6·8
,, IX,	42·07	33·94	8·13	4·2
,, XII,	52·32	41·71	10·61	3·9
Lumbar II,	57·62	41·01	16·61	2·5
,, V,	62·57	39·24	23·33	1·7
Sacral I,	51·96	28·63	23·33	1·2
,, III,	22·27	9·45	12·73	0·74
,, V,	9·54	4·94	4·60	1·07*
Coccygeal,	4·94	2·47	2·47	1·00*

It must be remarked that in *Stilling's* last five tables of measurements compiled from observations upon other subjects, the white substance from the third sacral nerve downwards, almost without exception, falls below the grey in amount, the slight excess of white matter in the conus medullaris (*) of this case must therefore be regarded as unusual.

2. HISTOLOGY OF SPINAL CORD.

The sections of the cord which have already served for the study of the more conspicuous variations in structure at different levels may now be used for work with higher powers.

Beginning with the **white sheath** we find that it appears at first sight to consist almost exclusively of longitudinal fibres. In cross-sections, stained in carmine, they "look like little suns" (fig. 104). The diameter of the fibres varies; in man from 1 to $25\,\mu$; in the horse they may be as large as $50\,\mu$; in the spinal cords of some fish, particular fibres are even larger. Everywhere thick and thin fibres

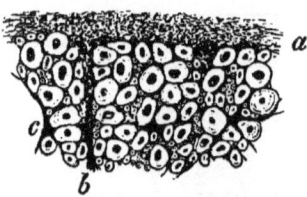

Fig. 104.—Cross-section of the anterior column of the spinal cord. *Stained with carmine. Magn. 150.*—*a*, Peripheral grey investment; *b*, a small septum. In the white substance besides the nerve-fibres cut across, some of which are coarse and others fine, three distinct stellate connective-tissue cells are seen; one of these is indicated by the letter *c*.

are mixed together; but certain local peculiarities in grouping are to be noticed. Many thick fibres occupy the peripheral region of the anterior and lateral columns; in the angle between anterior and posterior horns (the central part of the lateral column) thin fibres preponderate. In the posterior column, Burdach's column contains not a few coarse fibres, while Goll's column is entirely made up of fairly fine fibres. The difference between these two subdivisions of the posterior column is especially pronounced in the cervical cord. The larger the fibres which it contains the lighter does any particular region appear when stained with carmine; the smaller its fibres the darker its colour, especially when looked at with the naked eye or a simple lens. Goll's column, for instance, is in the cervical cord conspicuously darker than its neighbour.

The periphery of the medullary substance is separated from the pia mater by a thin layer of grey matter, 5 to $40\,\mu$, or exceptionally as much as $100\,\mu$ thick. This is the cortical layer of the cord, *a*, (fig. 104). It consists of fibrous connective-tissue containing a great deal of finely granular neurogleia [non-cellular matrix]. From the pia mater septa, some thicker, some thinner, pass through the white substance, carrying numerous vessels with them. The septa consist of connective-tissue with more or less neurogleia derived from the cortical layer. They split the white matter into columns, divided again, by lateral septal plates, into fasciculi. The septum medianum dorsale is the largest of these septa, the septum paramedianum is also large. Many large connective-tissue cells are interposed between the fibres, c (fig. 104). Their processes are, as a rule, disposed in the direction of the nerve-fibres (fig. 89). Hence in a carmine-stained transverse section many dark dots are seen which may be either slender, naked axis-cylinders or processes of connective-tissue cells.

Besides the longitudinal fibres, numerous transverse bundles, as well as fibres, which run obliquely, are also to be found in the medullary substance of the spinal cord. They are:—

(1) The anterior root-bundles which in the cervical and lumbar regions are made up almost exclusively of thick fibres, but contain in the dorsal region many thin fibres also (*Siemerling*).

[The thick fibres 18 to 20 μ in diameter are the motor fibres of skeletal muscles, the thin fibres of some 2 or 3 μ in diameter are destined for the fibres of the visceral system and the blood-vessels. With a view to writing the anatomy of the sympathetic system, *Gaskell* has made transverse sections of all nerve-roots attached to the cerebro-spinal axis in the dog, and, as *Schwalbe* had done for Man, determined the situations in which the small nerve-fibres occur. The ramus visceralis consists of small medullated fibres which enter into the formation of both anterior and posterior roots. The account is not yet complete, but it appears that their outflow occurs in the facial and glosso-pharyngeal nerves; for the thoracic region, and to a certain extent also for the stomach, in the vagus nerve; some of the fibres (accelerator) for the heart, and fibres for the abdominal viscera, leave the spinal cord from the second dorsal to the second lumbar nerves, whilst the dilator fibres for the pelvic viscera accompany the roots of the second and third sacral.

No small fibres are found in the roots of the cervical or lower lumbar and first sacral nerves. The series of visceral roots is interrupted in these two regions.

According to the manner of their peripheral distribution, the visceral fibres may be divided into two groups—(A) those which enter distributive cells in the ganglia which lie nearest to the vertebral column, the "lateral ganglia" or ganglia of the "sympathetic chain," as it is usually called in human anatomy; (B) those which retain their medullary sheaths as far as the ganglia which lie in the course of the larger blood-vessels, the "collateral ganglia," and only by the intervention of their cells are broken up into bunches of non-medullated fibres. These two sets of fibres are differently distributed throughout the three regions of the cerebro-spinal axis to which visceral roots are attached. Class B, or fibres which pass by the lateral ganglia without losing their myelin-sheaths, are found in all three regions. Class A is further divisible into the motor fibres for the alimentary canal which leave the cerebro-spinal axis in the vagus nerve, the motor fibres for the walls of the blood-vessels which are entirely restricted to the thoracic outflow, and some others. It appears that the two classes, A and B, are as widely distinct in function as in anatomical disposition. Class A includes motor, accelerator, constrictor fibres, fibres for circular muscles, all katabolic or exhaustive in action. Class B comprises the inhibitory, retarding, dilator nerves, nerves supplying longitudinal muscle-fibres, or in other words, all the visceral fibres possessing an anabolic or restorative function.]

(2) The posterior roots containing thick and thin fibres.

(3) The white commissure which lies at the bottom of the anterior longitudinal fissure, and attains to a thickness of nearly half a millimeter. In most mammals the white commissure does not form a single bundle (or as it may better be described, having regard to its continuity, a compact membrane), as it does in Man, but it is made up of many little bundles which cross the anterior fissure at different levels, and pierce the anterior columns instead of lying completely beneath them.

(4) Finally, fibres stream off from the grey matter on all sides and run outwards through the white columns, sometimes for a considerable distance, before bending over and assuming a longitudinal course.

In the **central grey mass**, as already mentioned, two substances are to be distinguished :—A. **Substantia spongiosa**, the ground-substance of which is made up of neurogleia and connective-tissue cells. As shown by staining in hæmatoxylin, the connective-tissue cells are somewhat more thickly distributed than in the white matter. Their processes, as shown by preparations impregnated with corrosive sublimate, are scattered in all directions, but chiefly follow the long axis of the cord.

In successful carmine preparations (especially if alcohol has not been used to harden them), but still better, in sections stained after Weigert's method, the spongy substance is seen to be traversed in all directions by medullated nerves of the most varied calibre (fig. 105). They interlace in all directions, and appear in transverse sections of the cord, cut obliquely and transversely as well as exposed in length. Division of fibres [within the grey matter] is not to be detected. In many places, however, they follow defined directions as may be pointed out in the following regions :—

The anterior root-fibres are seen to diverge shortly before plunging into the real grey matter of the anterior horn. Within the grey matter they spread out in a broad brush, both brain- and caudal-

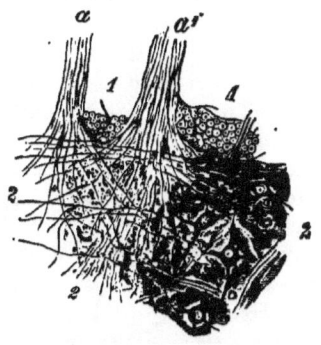

Fig. 105.—Junction of the anterior root-bundles with the anterior horn. Lumbar region. *Magn.* 30.—*1*, Anterior white column; *2*, anterior horn; *a*, *a′*, two root-bundles. On the right side four cells belonging to a nerve-cell group are to be seen.

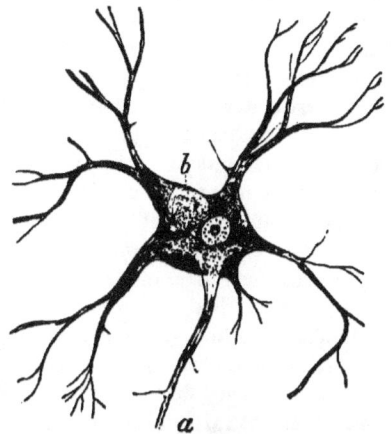

Fig. 106.—A nerve-cell from the anterior horn of the human spinal cord.—*a*, Axis-cylinder process; *b*, clump of pigment-granules.

wards as well as sagittally, as found by comparing longitudinal with transverse sections. The outermost fibres on either side diverge so far that they form a fibrous layer between the grey and white substance.

In the anterior horn bundles of fibres are seen to converge towards the white commissure.

The arched fibres of the posterior roots retain their independence far into the substance of the posterior horn. The ground-substance of Clarke's column, as well as of the grey matter enclosed within the concavity of the substantia Rolandi, is conspicuous by its clearness. It is made up chiefly of a vast number of delicate medullated longitudinal fibres. A bundle of the finest fibres in the posterior root (the border zone of Lissauer) is found in the apex cornu posterioris.

Besides the plexus of medullated fibres the grey substance contains a second network of non-medullated fibres. It is difficult, as a rule, to bring this second plexus into view, only occasional fibres being visible. Their cross-sections appear as fine dark dots.

There are several kinds of *nerve-cell* in the spinal cord, those of the anterior horn being the most conspicuous (fig. 106). These are frequently termed motor cells, since it is generally understood that they give origin to the fibres of the anterior root. These large cells are not limited to the anterior horn, however, for they form groups in the lateral horn where this is best developed, as in the cervical and lumbar enlargements, and they also extend into the processus reticularis. They have a number of processes (from five to eight), giving them in cross-section a stellate form. In size they vary from 35 to 100 μ. Since their processes taper off from the cell-substance it is not possible to define sharply the limits between processes and body, and, hence, various estimates of size are given. Each cell possesses a large round nucleus, as much as 18 μ in diameter, with distinct nuclear bodies and nucleolus. [In stained sections, the large clear vesicular spherical nucleus, with its usually single darkly-stained nucleolus, enables one to recognise any nerve-cell of the more conspicuous type with certainty.] They always contain a clump of yellow pigment. The size of the cells is supposed by *Pierret* to bear a direct relation to the length of the fibres to which they give origin (p. 130). Hence, they are largest in the lumbar swelling—somewhat smaller in the cervical swelling, and smallest in the dorsal region. Separate processes of these cells can be sometimes followed for a great distance, usually into an anterior root-bundle, but not rarely also into a fibre-bundle which enters the lateral column. It has already been mentioned that, especially in the lumbar enlargement, the cells are collected into groups isolated by grey matter which stains more deeply (fig. 105). The darker colouring is due to the smaller number of medullated, and the greater quantity of non-medullated fibres, as well as to the greater richness in vessels. Each cell is more or less distinctly surrounded by a pericellular space.

The **cells of Clarke's column** are somewhat smaller, 30 to 60 μ [in transverse diameter], less well provided with processes, and richer in pigment (fig. 107). They are slightly more elongated in their longitudinal than they are in the transverse axis. One or two processes leave the sides of the cell, and almost constantly a single process is attached to either pole. They join the cell more abruptly than the processes of the cells of the anterior and lateral horns, and hence the cell presents a rounder form. Longitudinal sections show us that the pigment hardly ever lies to the side of the nucleus, but is almost always accumulated at one of the poles. Their nuclei are large and conspicuous like those of the cells of the anterior horn. Their processes can often be followed a long distance in longitudinal sections without being seen to divide.

Another kind of cell, although presenting every transitional form from those of the anterior horn, is scattered through the spongy substance. They are smaller in size (even as little as 15 μ in diameter), and have fewer processes; appearing, therefore, triangular or spindle-shaped. Attention should be especially directed to certain standpoints with regard to them.

(1) In the centre of the grey matter lying between the lateral horn and the grey commissure spindle-shaped cells are disposed (fig. 108, g) with a process directed dorso-laterally towards the arched fibres of the posterior root, of the fibres of the mesial portion of which they may be looked upon as the probable source.

[Other connections for the spindle-shaped cells at the base of the posterior horn have been suggested. *Gaskell* thinks it possible that they give origin to the motor-fibres of the muscles of the alimentary canal, the inhibitory fibres coming, as he may almost be said to have proved, from the cells of Clarke's column. The *translator* has on several occasions pointed out that every nerve-fibre is a process of a nerve-cell; that it grows out from the cell in the direction in which it afterwards conducts impulses; that when the fibre has a considerable calibre and traject, a large cell is needed for its nutrition. If these positions can be maintained, it follows that anatomists have been in error in assigning all the conspicuous cells of the spinal cord to extrinsic nerves; the fibres connecting the spinal cord with the brain must start from cells sufficiently large (an unknown relation) to provide for their nutrition. The assumption that the fusiform cells at the base of the posterior horn are connected directly with the fibres of the posterior roots is contrary to the observed course of their axis-cylinder processes (*Gerlach*), and still more opposed to the experimental evidence derived from cutting the spinal nerves, which shows that sensory fibres have their nutrient cells in the spinal ganglia, or in some cases still nearer the periphery, but *not* in the spinal cord. The *translator* thinks it probable that, amongst others, the scattered cells beneath the substantia gelatinosa Rolandi bear the same relation to the ascending intrinsic fibres as the pyramidal cells of the cerebral cortex bear to the fibres of the pyramidal tract.]*

---

* For the sake of clearness the above statements are made without qualification, but certain facts with which they seem at the moment irreconcilable are not overlooked. With regard to the assumption that fibres grow in the direction in which they subsequently conduct impulses—in the case of the motor roots of the peripheral nerves this is obviously true, and much might be said *a priori* in favour of the probability of its being the universal rule. The fibres starting in the olfactory bulb, the retina and the nerve-cells of the lamina spiralis of the cochlea grow inwards towards the cerebro-spinal axis; in other sensory nerves the law, if it be a law, clearly does not apply. The posterior roots of the spinal nerves grow from the cells of the spinal ganglia in opposite directions, one process towards the centre, the other towards the periphery. There may, however, be a morphological reason for this which will reduce it to the level of a further adaptation rather than an

(2) At the apex of the lateral horn throughout the whole dorsal cord, and the neighbouring parts of the cervical and lumbar regions, a column of small closely-packed cells, most of them fusiform, is found sharply marked off from the larger cells which the lateral horn also contains.

(3) Amongst the scattered cells of the substantia spongiosa, those which lie in the middle of the posterior horn deserve attention; probably they are, as represented in fig. 108 (*h*) in connection with the lateral root-bundles of the posterior root.

B. Scarcely anything is yet known as to the histology of the **substantia gelatinosa.** It consists of a special ground-substance which stains intensely with carmine, and is more like the cortex of the cord than anything else. Through the apex cornu posterioris the two substances are in connection. It is a question whether the substantia centralis and the substantia Rolandi are of the same nature. Neither of them is as yet well understood.

The substantia gelatinosa shows in the neighbourhood of the posterior horn a peculiar striation parallel in direction with the fibres of the posterior root, but only partly to be referred to them. It contains also fairly numerous cell-elements, of which certain can be pointed out as connective-tissue cells. Larger nerve-cells are also seen in it, on its edge for the most part. In the lower sacral cord, at a level from which all other large cell-elements have disappeared, certain large scattered vesicular cells situate in the substantia gelatinosa are very conspicuous. Many of the cells of the substantia gelatinosa are distinguished by their delicacy; when the tissue is stained in the usual way they are recognised by their light colour; they may well

---

exception. If, as there are reasons for thinking, the original sense-organs were situate in the locality of the present spinal ganglia, whether in their present situation or farther afield matters little, and if these organs were, at this time, the only end-stations of sensory nerves, it follows that the extension of sensibility to the surface generally was due to a circumferential growth of the sensory nerves. After the loss of the sense-organs associated with the spinal ganglia, the circumferential growth of sensory nerves still in all probability bears testimony to their mode of origin.

The statement is also made that some of the fibres of the posterior roots have their nutrient cells farther afield than the ganglia. This assertion is made on the strength of *Joseph's* observation, that after cutting the posterior roots on the distal side of the spinal ganglia, not all the fibres passing through the segment containing the ganglion live (*cf.* p. 21). *Joseph* has not, however, demonstrated the presence of living fibres among the dead ones of the peripheral stump, any more than he has, after severance of the root on the proximal side of the ganglion, detected the presence of living fibres between the cord and the point of section. The situation of the nutrient cells of the fibres which do not retain their vitality, although left in apparent connection with the ganglion, is not, therefore, determinable as yet.

be regarded as nervous elements (*H. Virchow*), although their histological meaning is held to be doubtful by some anatomists (*Lustig*).

[The complete history of the substantia gelatinosa has yet to be written. This substance is not distinguishable in the earliest stages of the development of the central nervous system, whereas in what might be called the middle period, after the appearance of the fibre-columns which ensheath the grey matter, and especially about the fifth month of intra-uterine life, the cap of gelatinous substance which covers the posterior horn is extremely large and conspicuous owing to its being packed with small nuclei which stain darkly in carmine.

According to *Corning*, the substantia gelatinosa first appears as a local thickening of the inner layer of the grey matter in its dorsal portion. This thickening extends outwards over the posterior horn to the point of entrance of the posterior roots. Finally, it divides into two portions, or rather the greatly developed cap of the posterior horn breaks away from the original formation which lies in the dorsal wall of the central canal. In new-born animals the same authority tells us it consists of two kinds of cells—the first with large clear nuclei, and oval or fusiform cell bodies; the second with darkly staining nuclei and indistinguishable cell bodies. The contrast between the cells of these two classes, the one clearly nervous and the other belonging to the neurogleia, is still more marked fifteen days after birth. The nerve-cells do not lose their indifferent character until after the cells of the anterior horn and the grey matter of the posterior horn are differentiated.

Fig. 107.—A nerve-cell from Clarke's column as seen in a longitudinal section of the spinal cord of the horse. *Magn.*150. *The arrow points towards the brain.*

If we bear in mind the origin of the sensory ganglia from rudiments laid down beyond the borders of the medullary plate, but caught in between the lips of the plate as it closes into a tube (fig. 3), and consider that the posterior roots grow into the cerebro-spinal axis from the sensory ganglia, we shall see that it is not impossible that the substantia gelatinosa is also formed from this ganglionic rudiment. If this be the case, it belongs not to the medullary plate, but to the ingrowing sensory root, being, in fact, the brush of filaments into which the short posterior root breaks up on entering the cord, together with their small-cell connections and supporting neurogleia.

His concluded from his observations that the substantia gelatinosa is formed from immigrant cells, but looked upon these cells as mesoblastic or connective-tissue.

The substance of Rolando is supplied by the arteries of the posterior roots, whereas all the rest of the grey matter receives its blood from arteries which dip down into the anterior sulcus. This is a far-reaching distinction, for the arteries of the roots are strictly segmental, whereas the sulcal arteries arise from the common anterior spinal.]

The substantia gelatinosa centralis surrounds the central canal and spreads out a little, especially in the cervical and dorsal regions into the grey commissure. It, too, consists of a ground-substance (doubtless of the nature of neurogleia), scattered connective-tissue cells, and more or fewer angular cells which may well be derived from the epithelium.

A description of the **central canal** is best inserted here. Its epithelial lining has already been referred to (p. 148). The cross-section of the central canal varies; in the upper cervical region it is irregular, but usually almost square, and sometimes very wide. It begins to be reduced to a narrow cleft about the level of the fifth or sixth cervical nerves. The cleft is placed frontally coinciding in direction with the grey commissure, but often gives off from its centre a dorsal branch. The frontal extension predominates throughout the dorsal region, although this diameter of the canal becomes less and less, until at last it is almost circular in cross-section. In the lumbar cord the frontal gives way to a sagittal extension; a disposition which is still more marked in the sacral cord and in the conus medullaris, where the canal consists of an open ventral and a narrower dorsal portion. Near the filum terminale the canal dilates into an irregular cavity (ventriculus terminalis, sinus rhomboidalis inferior) [probably not homologous with the large open sinus rhomboidalis of the bird's lumbar cord]. The canal appears to end blindly in the upper part of the filum terminale. In birds the central canal opens out widely in the region of the whole lumbar swelling. This opening is called the sinus rhomboidalis.

The central canal presents important individual variations in form. Only rarely in the human adult is it found to be completely pervious as in the child, and in all animals. In most cases it is partially deformed. The caudal part of the cord from the sacral region downwards presents, as a rule, a continuous canal; so does usually the lumbar cord and the cervical cord from the fifth nerve upwards.

The blocking of the central canal is due to the overgrowth of the epithelium which lines it, as well as usually of the epithelial cells scattered about in the substantia gelatinosa centralis, and the sub-epithelial connective-tissue. When the overgrowth affects portions of the margin of the canal and not the whole of it, it may give rise to septa dividing the canal into several (as many as five it is asserted) parallel canals. This is the cause of the condition described as double or treble central canals.

The central canal lies in the **commissure** midway between the anterior fissure and the septum posterius. The commissure is divided into a dorsal grey and a ventral white portion.

The *white commissure* (commissura alba, less suitably called anterior commissure) will not be described in this place, but it may with propriety be pointed out here that, although the name commissure is retained as a classical term, the structure is probably a decussation, and contains but few commissural fibres properly so-called.

The *grey commissure* (commissura grisea *seu* posterior) contains the central canal, which lies a little to the ventral side of its centre, especially in the lower part of the cord.

The portion lying between the white commissure and the substantia gelatinosa centralis may be termed the ventral (or anterior) grey commissure; the portion lying dorsally to the substantia gelatinosa centralis is then the dorsal (posterior) grey commissure. In the lower lumbar cord the posterior grey commissure, which as far as this point is very thin, not more than 30 to 100 μ through, begins to increase in thickness reaching in the lower sacral cord a sagittal diameter of as much as 1 mm. Sometimes the grey commissure is also strongly developed in the upper cervical cord. Dorsally the grey commissure is drawn out into a ridge which is directly continuous with the septum medianum posterius. Both grey commissures possess the same ground-substance as the substantia gelatinosa centralis, and both contain medullated nerve-fibres, which cross the middle line, although they are more numerous in the posterior than they are in the anterior commissure. Longitudinal fibres also appear in both.

3. COURSE OF FIBRES IN SPINAL CORD.

Sections from the normal adult human organ are insufficient to reveal the course of fibres in the spinal cord. But little light would have been shed upon the relations of the various fibre-systems, which are by no means simple, without the help of pathological and developmental observations.

Root-fibres, which are the direct continuations within the cord of fibres which form the roots of the spinal nerves, are to be looked upon

188   RELATION OF POSTERIOR ROOTS TO CORD.

as constituents of peripheral nerves; like them they are early covered with myelin, sooner indeed than any portion of the white columns of the cord. The dorsal roots ought not perhaps to be treated unconditionally as peripheral nerves since some of them have already suffered an interruption to their course in the cells of the spinal ganglia, but yet they belong to them both histologically and histogenetically (see pp. 28 and 165, and fig. 96).

The **anterior roots** (fig. 108, 1 to 6) pierce the anterior columns of the cord obliquely from below upwards, the more obliquely the nearer

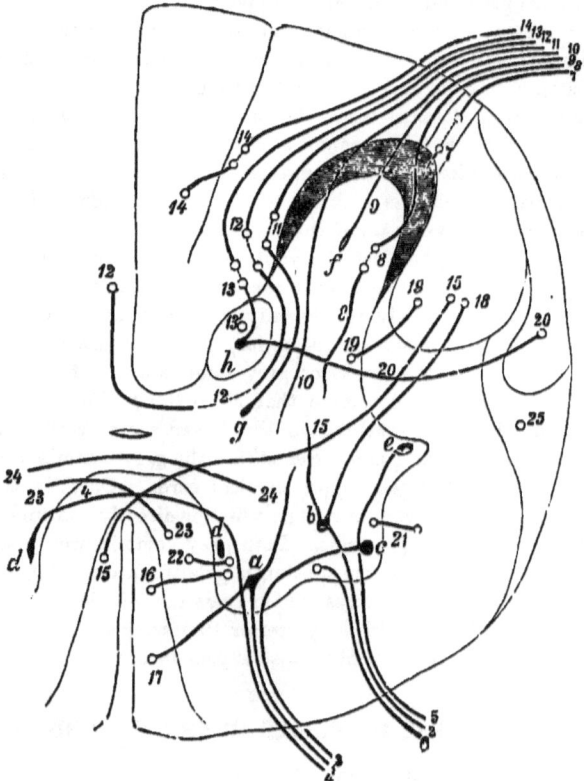

Fig. 108.—Diagram showing the course of fibres in the spinal cord. *Longitudinal fibres are represented by white circles. The nerve-cells are black. Further explanation is given in the text.*

the section approaches the end of the cord, and pass towards the anterior horn with a curve, concave outwards. The fibres which are almost exclusively large ones (in the cervical and lumbar cord at any rate) spread out in a brush as they approach the grey matter (*cf.* fig. 105). The greater number, if not all, the fibres of the anterior root end in the motor cells of the anterior (*a, b, c, d*) and lateral (*e*) horns.

It must be noticed, however, that within the grey matter the fibres spread out in all directions, and the most mesial fibres (3) of the root may come into relation with the most laterally situate group of cells (*c*).

Further, many fibres (6) go to cells situate farther forwards, others to cells farther down the cord. A part of the root-fibres (4) goes across to the other side in the anterior commissure to end in the more fusiform cells (*d*) which lie along the mesial border of the anterior horn; hence, some of the fibres of the anterior commissure are medullated very early. This completes the general scheme of the origin of the motor nerves.

The exact termination of the fibres which pass through the anterior horn without interruption to join the longitudinal fibres of the lateral column, or in other cases of the anterior column, is not known for certain. According to *Birge's* calculations, the correctness of which is doubted, however, by *Gad* [for it is obviously very difficult to recognise the same cell in several sections, or to make certain that the cell counted to one section is not the same cell which has already made its appearance in the section above], the number of cells in the anterior horn (in the frog at least) corresponds exactly to the number of issuing motor-fibres, so that one may consider each fibre as having its origin in a cell near its point of exit.

Already we have seen (p. 41) that the experimental method has thrown light upon the relation of the several groups of nerve-cells to groups of muscles. The attempt has also been made to determine whether certain defined groups of cells of the anterior horn innervate definite muscles. We are dependent in this matter upon pathological observations, and, as yet, the results are inconsiderable. It is possible that the middle group of cells at the level of the fourth and fifth lumbar nerves belongs to the calf-muscles (*Kahler* and *Pick*), and that the lateral group of cells in the lower cervical cord supplies the thenar muscles (*Prevost* and *David*).

The origin of the **posterior roots** in which thin fibres everywhere accompany thick ones (*Siemerling*) has already been traced in outline. The finest most laterally situate root-fibres (7) assume, soon after their entrance into the cord, a longitudinal direction forming the area (the boundary-zone of *Lissauer*) which corresponds approximately to the apex. This region stains strongly with carmine owing to the fineness of the fibres and the abundance of delicate intervening connective-tissue. Soon the fibres abandon the longitudinal direction and enter the gelatinous substance in the plane of the section, so that the total cross-section of the boundary zone does not increase from below upwards.

The thicker root-fibres are divisible into a lateral and a mesial portion.

The lateral smaller portion (8 to 10) enters the apex and passes into the gelatinous substance, where it divides into many little bundles which, "like the degrees of longitude streaming from one of the poles on a globe," divide the gelatinous substance into segments.

Arrived in the posterior horn, many of these fibres continue farther ventralwards through the substantia spongiosa (10), others assume a horizontal direction and run, as is supposed, both brain- and caudal-wards (8). Since the substantia spongiosa contains but a small proportion of the fibres which must have entered it at various levels, it follows that the majority of fibres leave it again. Some of the fibres may end in the cells which it contains (f), but nothing certain is known as to the further course of by far the greater number.

The mesially-situate bundles of the posterior roots (11 to 14) bend medianwards in wider or closer curves as soon as they enter Burdach's column, and, turning longitudinally, cannot be farther followed in transverse sections of the cord. From almost the same region, however, in which these arching fibres are lost other arched fibres originate and sweep into the grey substance of the posterior horn. There can be no room for doubt that these fibres are the continuation of those which have entered Burdach's column lower down and run some distance in it longitudinally. They can be easily followed for some distance farther into the substance of the posterior horn. A portion of these fibres (11) are distinctly connected with the fusiform-cells (g), near the grey commissure. More mesially-lying fibres (13) enter Clarke's column (in regions in which this column is present) and doubtless come into relation with its cells (h). Nothing more can be said with regard to the fine fibres (13^1) which originate in Clarke's column than that they stand in close relation with the fibres of the posterior roots. We may remind the reader that the cells of Clarke's column usually give off two longitudinal processes (fig. 107). The fibres in the posterior grey commissure are partly the continuations of the posterior roots (12). Since, however, it has been shown on fundamental grounds that a partial crossing of posterior root-fibres cannot be allowed, it is important to decide in the first instance whether these fibres of the posterior commissure are the direct continuation of root-fibres, or whether, as is much more probable (chiefly on pathological grounds), the root-fibres have already been interrupted in cells.

This latter view is especially supported by the fact that in tabes dorsalis, after the posterior roots are largely degenerated, the posterior grey commissure is still rich in fibres. Certain it is that what we have already said is far from having exhausted the course of the posterior root-fibres. It may be maintained, with a great show of probability, that a large number of these fibres (8, 10) break up in the fine network of the grey substance, being finally connected by its means with nerve-cells, perhaps even with the large cells of the anterior horn. It is also asserted that some of the posterior root-fibres (14) traverse Burdach's column, and, after a longer or shorter course, assuming a longitudinal direction, run brainwards in Goll's column. It must, however, be allowed, as the result of experiments on the guinea-pig, that the posterior root-fibres do not run so directly into Goll's column (*Rossolymo*). This appears more probable when one remembers that Goll's column consists of fibres which only acquire their myelin sheaths at a much later date than the fibres of the posterior roots.

There is no fact in the fine anatomy of the brain clearer or less controvertible than the origin of the posterior root-fibres in nerve-cells, as discovered by *Kutschin* and *Freud* in the spinal cord of Petromyzon. *Klaussner* has shown somewhat the same thing in Proteus, and if we add to this the results of observations on the human spinal cord, it follows that the older view of the connection of posterior roots, not with nerve-cells, but with the nerve-plexus, is best allowed to drop.* The fusiform cells marked in the diagram, *g*, deserve

---

* It may possibly be no disadvantage to the reader that Professor Obersteiner and his translator look at this matter from different points of view. The principal reasons for thinking that sensory nerve-fibres do *not* run, without dividing, into nerve-cells, which is tantamount to saying do not join cells at all, or not at any rate cells of the larger kind, are the following :—(1) In mammalian cords no such connection can be demonstrated; whereas nothing is more obvious than the junction of motor fibres and large cells in the anterior and lateral horns, a similar connection of the posterior root-fibres cannot be exhibited. (2) After section the stumps of anterior root-fibres attached to the spinal cord retain their vitality, while the stumps of posterior root-fibres die; it is, of course, possible that a nerve-fibre may join two cells together, and yet depend for its nutrition solely upon one of the cells; but such evidence as is at present available seems to show that such is not the case, but that each fibre is connected with a single nerve-cell, from which it draws its supply of nutriment; or rather it obtains from the cell an influence which enables it to benefit by the food-stuffs in the lymph-bath by which it is surrounded. (3) Morphological evidence, which cannot be summarised in this place, leads the translator to believe that nerve-cells first appeared in the vicinity of sense-organs, that the plexus so formed was withdrawn into a sheltered and centralised cerebro-spinal axis, but that the nerve-cell, the processes of which join the epithelial sensory cells, always remains in the vicinity

especial attention in this connection. The cells of the spinal ganglia in which many nerve-fibres have their origin are, perhaps, nothing more than extended portions of the grey substance of the spinal cord (*Hensen, Schenk*); [or possibly, as thought by the *translator*, portions of the nervous system which still occupy their primitive situation in the vicinity of the spots in which sense-organs used to exist; portions of the system, that is to say, which have not been absorbed into the cerebro-spinal axis]. It may well be supposed that the diverse anatomical relations of the fibres of the posterior root answer to differences in function.

But little, however, can be alleged in this connection; the fibres which run brainwards in Goll's column may conduct sensations from the muscles; while the cells of Clarke's column are, perhaps, the halting stations on the roads along which visceral sensations travel.

The **white matter** of the cord is divided into several parts, the boundaries of which have been determined by studying its pathology and development. It must be premised that the number of longitudinal bundles which it is possible to distinguish from one another is constantly being increased by continued observation, and that the differentiation of separate groups of fibres promises to become more and more detailed. First, we must confine ourselves to the universally acknowledged facts, and then consider the divisions established by *Flechsig* as the result of his application of the developmental method.

A distinction between the long and short tracts in the spinal cord is first to be insisted on. Short tracts unite places in the grey matter which lie near together, while the long tracts consist of fibres which can be followed into the medulla or even farther. In sections this

of the sense-organ, or in the neighbourhood of its original location. Thus in the olfactory organ the non-centralised portion of the primitive nerve-plexus is found in the olfactory bulb; in the eye in the retina; in the case of other vanished segmental sense-organs in the spinal ganglia. From these cells, processes or nerves grow outwards towards the periphery, inwards towards the spinal cord. In the olfactory bulb and retina these proximally running processes (of the granules or small spindle-cells) undoubtedly break up and join a plexus; if the homology holds, the same arrangement may be taken for granted in the cord, the substantia Rolandi being the homologue of the gelatinous substance of the bulb and the inner molecular layer of the retina.

With regard to the connection in Petromyzon of posterior root-fibres and cells of the posterior horn, observed by *Freud*, it is very doubtful whether these cells are the homologues of the cells at the base of the posterior horn in higher vertebrates, whereas it is certain that the constitution of the posterior root in Petromyzon is different to the constitution of the posterior root in mammals (*D'Arcy Thompson, Ransom*).

difference is not demonstrable, but it is brought out in pathological or experimental lesions of the cord. The short tracts show degeneration in the neighbourhood of the injury only, while degenerated bundles in the long tracts can be followed brainwards and caudalwards through the whole cord. This distinction between short and long routes must not be looked upon as absolutely differential, nor its physiological importance exaggerated, even as a criterion of secondary degeneration. Nor must it be regarded as holding true in all cases.

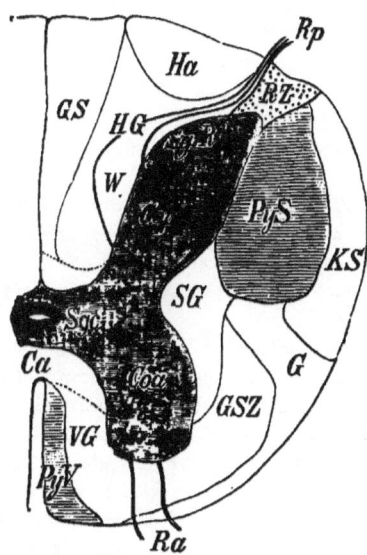

Fig. 109.—Diagram showing the subdivisions of the white columns of the cord.— *PyV*, Anterior pyramidal tract ; *VG*, anterior ground-bundle ; *Ca*, anterior commissure ; *Ra*, anterior nerve-roots ; *GSZ*, mixed lateral zone ; *SG*, lateral ground-bundle ; *G*, Gowers' bundle ; *KS*, direct cerebellar tract ; *RZ*, border zone ; *Rp*, posterior nerve-roots ; *HG*, ground-bundle of the posterior column consisting of the root zone, *W*, and the postero-external tract, *Ha ;* *GS*, Goll's column ; *Coa*, anterior horn ; *Cop*, posterior horn ; *SgR*, substantia gelatinosa Rolandi ; *Sgc*, substantia gelatinosa centralis.

In the anterior columns of each side we have to distinguish four long tracts. The several divisions in the white matter are not, however, the same at all heights, so, first as a paradigm, we will examine a somewhat schematised transverse section of the cervical cord (fig. 109).

(1) The anterior pyramidal tract, *PyV* (Türck's bundle), lying on

either side the anterior fissure, forms the mesial tract of the anterior column, and very often extends outwards along the free ventral surface of this column.

(2) The lateral pyramidal tract, PyS, which in the dorsal region occupies a large part of the lateral column.

(3) The lateral cerebellar tract [direct cerebellar tract], KS, a small column lying between the margin of the lateral column and the lateral pyramidal tract.

(4) Gowers' bundle, G, lying partly as a small marginal column on the ventral side of the direct cerebellar tract, partly in the midst of the lateral column on the ventral side of the lateral pyramidal tract.

As short tracts there may be distinguished in the antero-lateral columns of the cervical cord—

(1) The anterior ground bundle, VG, the part of the anterior column, that is to say, which remains after allowing for the anterior pyramidal tract.

(2) The tract, SG, bounding on the outer side the grey matter of the anterior horn, and filling up the space between the posterior horn and the lateral pyramidal tract, lateral ground-bundle.

(3) The mixed zone, GSZ, in the lateral column which comprehends all that now remains of it.

Only one long tract situate mesially, Goll's column, GS, is with certainty distinguished in the posterior column. To the short tracts belongs in part the ground-bundle of the posterior column, HG (the lateral or Burdach's column). This can be divided into two parts, one of which, the root-zone, W (zone radiculaire, bandelette externe), comprises the region, $H\ddot{a}$, through which the posterior root-fibres bend; the other, which lies peripherally, contains no root-fibres.

Between the posterior and lateral columns is placed a little tract, the border zone of Lissauer, RZ, which corresponds to the caput cornu posterioris.

The order in which the separate constituents of the white substance of the cord become medullated is as follows :—

(1) The anterior and posterior root-fibres.
(2) The ground-bundle of the anterior column.
(3) The ground-bundle of the posterior column.
(4) The mixed lateral zone.
(5) The lateral ground-bundle and Gowers' column.
(6) The mesial posterior column.
(7) The lateral cerebellar tract.
(8) The lateral and anterior pyramidal tracts (which in man are first medullated at the time of birth).

The **pyramidal tracts** are continued into the anterior pyramids of the medulla oblongata which they join, the one directly, the anterior pyramidal tract, PyV, and the other, the lateral pyramidal tract, PyS, after crossing. Different views are held as to the manner in which the direct pyramidal tract is reinforced on its road to the brain. The most probable view is that fibres leaving the crossed tract, PyS (fig. 108, 15), turn ventrally and mesially, and arrive at the other side after passing through the grey substance of the anterior horn and the white commissure. On the other hand, it is possible that fibres coming out of the mesial border of the anterior horn reach the anterior pyramidal tract in the most direct way, and perhaps these fibres (17) arise from the anterior horn cells (2) or nerve-plexus (16). Or again, it is not excluded that reinforcing fibres from the anterior horn cells join the opposite anterior pyramidal tract by way of the anterior commissure. After all the size of the anterior pyramidal tract is subject to great variations in different individuals. It is possible, however, to formulate the law that the more the anterior pyramidal tract of one side is developed the smaller will be the lateral pyramidal tract of the opposite side. The better the anterior pyramidal tract is developed the farther it can be followed caudally. Exceptionally it can be recognised in the lumbar swelling, although in most cases it disappears in the dorsal cord.

The form and situation of the lateral pyramidal tract changes at different heights. Where it is best formed it extends ventrally from the posterior horn, almost to a line carried transversely through the posterior commissure. From the cervical to the sacral cord it exhibits a constant diminution in its cross-section; by the time the lumbar swelling is reached it is reduced to a small layer on the outer side of the caput cornu posterioris. The increase in the lateral pyramidal tract from below upwards (which is especially marked in the swellings) is chiefly due to fibres (18, 19) which stream into the lateral column from the lateral border of the grey horns. These bundles afford a direct or indirect connection with the nerve-cells of the grey substance.

Sometimes, after hardening in bichromate of potassium, but without further staining the lateral pyramidal tract, especially in the cords of animals, is distinguished by a slight difference in colour.

The lateral pyramidal tract rests against the grey substance of the posterior horn, and especially against that part which lies nearest to the periphery. It is cut off from the periphery in most places by the lateral cerebellar tract, but reaches it near the apex cornu posterioris from the eleventh or twelfth dorsal nerves downwards, and for a short space in the neighbourhood of the third cervical nerve (*Gowers*).

Descending degeneration in the cord, whether the lesion is situate in the brain or in the cord itself, affects only the crossed and direct pyramidal tracts (PyS and PyV). The degeneration-areas correspond almost exactly with the developmental areas (fig. 110). In one-sided lesions of the brain which give rise to descending degenerations, it is the direct pyramidal tract (PyV) of the same side, and the crossed pyramidal tract (PyS) of the opposite side which are affected. On close examination, however, it is frequently possible to discover a degeneration, although less marked, in the lateral pyramidal tract of the same side, and, perhaps, even in the anterior pyramidal tract of the opposite side. Descending degenerations are diffuse in the dog (*Schiefferdecker*). *Marchi* and *Algeri* found scattered degenerations in every part of the cord after injury to the cortex cerebri of the dog; descending degeneration of Burdach's columns was especially conspicuous when the portion of the cortex which they destroyed lay behind the motor zone of the opposite side.

[We do not as yet understand in all respects the course taken by degenerations secondary to destruction of the cortex. An attempt has been made by *Sherrington* to determine the exact number of degenerated fibres found at all levels of the cord, and the situation of these fibres in the cord; and although these observations throw much additional light upon the question of the fibre-dependencies of the cortex, the results are in some points still inexplicable. It is found, for example, that a strictly localised destruction of the arm-area in the monkey (or even of the face-area?) gives rise to the appearance of degenerated fibres throughout the whole cord; while, on the other hand, lesion in the leg-area induces degeneration which in great part stops short in the cervical enlargement. Every unilateral lesion induces degeneration in both lateral pyramidal tracts, the smaller degeneration (on the same side as the lesion) being most pronounced in the cervical and lumbar enlargements immediately above which regions it is often absent. It follows, therefore, that the degeneration on the same side as the lesion is a "recrossed" degeneration. Not only may the number of degenerated fibres not decrease regularly from above downwards, but the lower regions may even contain more degenerated fibres than the upper ones. It is obvious, therefore, that fibres branch as they descend in the white columns, and as degenerated fibres can often be traced for a long distance in pairs it is inferred that the fibres divide into two. These pairs, termed "geminal" fibres, are much less frequent in the recrossed than they are in the crossed pyramidal tracts.

Sherrington suggests that the pyramidal tracts conduct cortical visceral as well as cortical somatic fibres, and that the degeneration in the lumbar region following lesion in the arm-area may be explained in this way. The conclusion of *François-Franck*, that the disturbances of the vascular and visceral systems which follow stimulation of the cortex in *curarised* animals are true instances of cerebral visceral epilepsy lends support to the view that visceral and somatic fibres lie side by side in the same tract. This is the only suggestion of which we are aware which attempts to explain the apparent discrepancy between the anatomical tract and its functional import; but the obvious morphological improbability of the same area of the

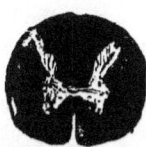

Fig. 110.—Descending degeneration after a one-sided lesion of the brain. Section through the upper part of the spinal cord. The anterior pyramidal tract of one side, and the crossed pyramidal tract of the opposite side are degenerated. The healthy lateral pyramidal tract shows a somewhat lighter staining. *Magn.* 2.

Fig. 111.—Ascending degeneration in the cervical swelling. Goll's column, *GS*, degenerated on both sides; and, to a less degree, the lateral pyramidal tract, *KS*, and Gowers' tract, *G*. *Magn.* 2.

cortex representing muscles of one segment of the body, and vessels or viscera of a different segment, stands in the way of its acceptance. Since so much of our knowledge of the course of fibres is derived from a study of degenerations, it is very important that we should know the exact causal relation between destroyed grey matter and degenerated white tracts.]

The **lateral cerebellar tract**, *KS*, is not found below the level of origin of the upper lumbar nerves. Ascending from this level through the lower dorsal cord it grows rapidly in cross-section; as it reaches the upper dorsal and lower cervical nerves its increase becomes slower. Its greatest increase in size occurs in the region in which Clarke's column of cells is best developed. The coarse fibres composing it can be followed without crossing up into the substance of the cerebellum. With lesions of the spinal cord situate above the first lumbar nerves (or after destruction of the posterior roots at a corresponding level, *Edinger*), the lateral cerebellar tract, *KS*, degenerates upwards towards the brain (fig. 111); no such degeneration follows injuries to the spinal cord below this level. *Pick* has observed that the cells of Clarke's column (fig. 108, *h*) give off processes, which as nerve-fibres traverse the lateral column horizontally (horizontal cerebellar bundle, 20), and end in the lateral cerebellar tract. Hence it follows that the lateral cerebellar tract receives its fibres from Clarke's column, perhaps from it only.

Gowers' bundle (first described by *W. R. Gowers*, ascending antero-lateral tract, lateral system of the lateral column of *Bechterew*) begins as low down as the lumbar cord, and shows a continuous increase of fibres (25) from thence upwards. It likewise degenerates in an ascending direction, in many cases at any rate, and may well be looked upon as a direct sensory route from the spinal cord to the brain.

The remaining portions of the anterior and lateral columns are made up of short tracts about which very little can as yet be asserted. The **anterior ground-bundle** and the **anterior mixed lateral column-zone** (mixed lateral zone) seem to have a similar physiological importance. From the border of the grey substance numerous fibres everywhere bend into these columns, which appear to be made up chiefly of such fibres (21, 22). It is quite possible that fibres from the anterior horn pass *viâ* the white commissure to the anterior ground-bundle (*VG*) of the opposite size (23).

The **mesial posterior tract** (Goll's column) like the other long tracts increases in section constantly from below upwards. Hardly recognisable in the sacral cord, Goll's column consists in the lumbar cord of a small convex tract lying against the septum posterius, but hardly reaching either the posterior commissure or the dorsal periphery of the cord. Farther brainwards the wedge shape predominates, but its ventral angle is never a sharp one, for, especially in the cervical cord, the column tends rather to spread out as it approaches the grey commissure, which it never quite reaches. It would seem that this crescentic portion (fig. 109) of the column which lies nearest to the grey commissure is especially important; a sharp delimitation of Goll's column is only possible above the middle of the dorsal cord. Like the lateral cerebellar tract, it degenerates upwards (fig. 111). Goll's column receives its fibres from the posterior roots in an indirect way not yet determined. That it stands in relation with the posterior roots of the same side is indubitable; a similar connection through the posterior commissure (12) with the posterior roots of the opposite side is not certain, although probable. The fibres travelling brainwards in the posterior column always incline towards the middle line, the fibres which have joined last occupying its lateral part. Thus it comes about that in the upper cervical region Goll's column only contains the fibres from the lower extremities, while most, perhaps all, of the fibres received from the upper extremities lie in Burdach's column (*UG*). The fibres connected with the sciatic nerve are found at the level of the cervical swelling in the most dorsal portion of Goll's column, close up against the periphery; and it is especially noteworthy that after the posterior nerve-roots have been cut across, degeneration of the fibres of the posterior column is invariably limited to the same side, whereas both clinical observations and experiment prove that the nerves from the skin cross over very soon after joining the cord. It is probable that they are chiefly meant to convey impressions of muscular sense (*J. Wagner*). The fact that in whales, in which the extremities are not developed, the posterior columns are exceedingly small harmonises with this view.

The **ground-bundle of the posterior column** is made up of the mesial portions of the posterior roots (fig. 108, 11 to 14). In the ground-bundle they are seen during two portions of their course, one horizontal, the other longitudinal. The alterations in the course of the fibres of which it is made up gives a net-like structure to the cord in this region. When cut transversely this part of the white columns, HG, degenerates upwards, but only for a short distance, and in quickly decreasing amount; at the level of one or two nerve-roots higher than the section the degeneration has disappeared. It is chiefly the root-fibres which degenerate. Beside these root-fibres other longitudinal fibres must be present, but their connections are not yet known. Root-fibres are absent from the peripheral portion of that field which lies on the mesial side of the point of entrance of the posterior roots.

It remains to recapitulate the fibres which enter into the formation of the **white commissure**; the following bundles may be looked for in it (*cf.* fig. 108):—

1. Anterior root-fibres coming from the fusiform cells (d) on the mesial side of the opposite anterior horn, 4.

2. Fibres passing from the anterior pyramidal tract (PyV) of one side to reach the lateral pyramidal tract (PyS) of the opposite side, 15.

3. Fibres passing from the anterior horn into the anterior ground-bundle (VG) of the opposite side, 23.

4. True commissural fibres connecting the two anterior horns probably occupy the ventral part of the grey commissure, 24.

The other fibres described as making up the anterior commissure are so uncertain that it is not worth while to mention them.

Very commonly the spinal cord is spoken of nowadays as made up of **segments** (metamers). Each metamer consisting of the portion of the cord to which an anterior and a posterior root belong. Each segment ought to be regarded as a "spinal unit" for a definite region of the body. Such a view as to the constitution of the body, based chiefly upon comparative anatomy, does not find a place in the scheme adopted in this book, for in higher animals we find this primitive division of the spinal cord largely obscured by the longitudinal extension of the hinder roots, as well as by the arrangement in long reflex-arcs (recognised by *Gad*), and in other ways.

[There are severa points of fundamental importance upon which we need information with regard to the columns of white fibres which surround the grey matter of the cord, but of which we are unfortunately unable as yet to give an account. The development of the fibres which compose these columns is still wrapped in mystery. The rudiments of the white sheath of the cord appear very soon after the grey matter is first recognisable as such. The white matter exhibits in its earliest condition a distinct radial striation, an appearance which is exaggerated by the arrangement of the nuclei of its embryonic myelin-cells and neurogleia-cells in radiating rows. Into this tissue penetrate the fibre-processes of the neuroblasts, most of them being undoubtedly on their road to form peripheral nerves. Whether or not any of the neuroblast processes turn upwards or downwards in the cord cannot be stated at present. However probable it may be, therefore, that the longest ascending fibres are processes of cells in the cord while the longest descending fibres are the processes of cells in the cortex of the brain, short fibres in like manner having their trophic cells on the side from which they carry impulses, no conclusions on these points can be based as yet upon histogenetic data.]

4. VESSELS OF THE SPINAL CORD.

The spinal cord is partly supplied with blood by arteries which come from the vertebral arteries, partly by branches coming from the intercostal, lumbar, and sacral arteries, which enter the spinal canal through the intervertebral foramina and reach the cord along the anterior and posterior roots.

Just before the junction of the two vertebrals, to form the basilar artery, a somewhat slender branch arises from each of them (or not infrequently from one only) inclines across the ventral surface of the medulla oblongata towards the artery of the opposite side, and reaches the anterior fissure usually on the cerebral side of the upper cervical cord. Here the two vertebro-spinal arteries unite into the unpaired arteria spinalis anterior, which can now be followed caudalwards as far as the conus medullaris. It corresponds to the anterior fissure. Occasionally this union between the two vertebro-spinal arteries takes place lower down the cord, at the level of the fourth, fifth, or even sixth spinal nerves, or, on the other hand, the vessels separate and reunite repeatedly.

The affluents which discharge into the arteria spinalis anterior are carried to the ventral side of the spinal cord along the anterior roots. Their number although variable is always small; sometimes there are only three. The most posterior is always the largest. This arteria spinalis magna is to be found, according to *Adamkiewicz*, between the eighth dorsal and third lumbar vertebræ on either side.

From the anterior spinal artery, *Spa* (fig. 112), frequent strong branches pass at right angles into the anterior fissure, the arteriæ sulci (*s*). Other small branches (arteriæ radicinæ) course laterally along the anterior roots to take part in the formation of a plexus on the surface of the lateral column.

The relation of arteries to the dorsal surface of the spinal cord is somewhat different. Here, too, an artery (arteria vertebro-spinalis posterior, or arteria spinalis posterior) arises from each vertebral artery, but in this case it lies to the outer side of the posterior spinal roots, and does not join with the artery of the opposite side. The arteries are not, however, independent of one another, for they form a chain of anastomosis both on the lateral and the mesial side of the posterior roots; and these anastomotic chains are not only united together by numerous cross-branches, but they receive minute affluents from without along the course of almost every posterior root. Arterial twigs also pass from these vessels mesially towards the posterior longitudinal fissure, whilst others join the before-mentioned lateral plexus.

At the conus terminalis a lateral branch comes off from the anterior spinal artery on each side (rami cruciantes of *Adamkiewicz*), which anastomoses with the arteries on the dorsal surface. Striking, too, is the zig-zag course of the arteries in the region of the conus medullaris.

The different branches and twigs which are spread out in the pia mater on the surface of the cord are characterised by the numerous anastomoses, some finer, some coarser, which they form.

Amongst the numerous veins on the surface of the cord the unpaired vena spinalis anterior deserves especial mention. It runs parallel with the artery of the same name.

Passing on to the **blood-vessels in the interior of the spinal cord** (fig. 112) attention is to be called to the great wealth of vessels in the grey substance as compared with the white.

All the arteries in the cord-substance can be arranged in two systems; (1) those in the zone of the arteriæ sulci; (2) those in the zone of the vaso-corona (*Adamkiewicz*).

The arteriæ sulci advance from their origin in the arteria spinalis anterior to the bottom of the anterior fissure, where they turn on the ventral side of the white commissure, either to the right or left (bifurcation is rare according to *Kadyi*), as the arteriæ sulco-commissurales, *sc*. These go into the grey substance of the anterior horn, where they break up into a close capillary network, which occupies the greater part of the cross-section of the grey matter. The portion of the white substance which borders on the grey also receives, according to *Kadyi*, branches from these arteries. One branch of especial size (*cl*) goes to Clarke's column of which it is the exclusive supply. Soon after its entrance into the grey substance each sulco-commissural artery gives off brainwards a considerable anastomotic branch, and a like branch in the caudal direction; in this way an uninterrupted anastomotic chain is formed through the whole length of the cord. Formerly it was thought that the spaces at the sides of the central canal, which are for the reception of these arteries and their accompanying veins, were only destined for longitudinal veins (the central veins).

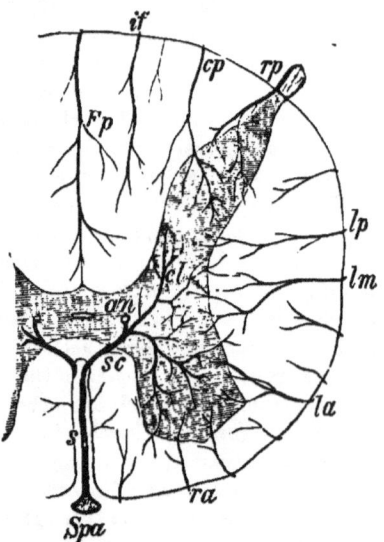

Fig. 112.—Semidiagrammatic representation of the arteries in the interior of the spinal cord.—*Spa*, Arteria spinalis anterior; *s*, a. sulci; *sc*, a. sulco-commissuralis; *an*, anastomotic branch of the same; *cl*, a. columnæ vesicularis; *Fp*, a. fissuræ posterioris; *ra*, aa. radicum anteriorum; *sp*, a. radicis posterioris; *cp*, a. cornu posterioris; *if*, a. interfunicularis; *la*, *lm*, *lp*, aa. laterales anterior, media, et posterior.

Under the title of vaso-corona may be included all the arterial branches which stream in a radiate manner into the substance of the spinal cord from the periphery. Of these, the finer are destined for the white substance only, while the coarser reach to the grey matter. The peripheral portions of the grey matter, like the white columns, receive their branches from both systems in an irregular manner.

This debatable ground comprehends about a third part of the cross-section of the cord (*Kadyi*).

The arteria fissuræ posterioris, the largest of the branches belonging to the vaso-corona, runs in the septum posterius nearly up to the grey commissure, supplying numerous branches to both sides.

Usually a large artery, *if* (arteria interfunicularis), courses in the septum paramedianum between Goll's and Burdach's columns; on the whole the more important branches are to be found in the connective-tissue septa. Arteries also accompany the anterior and posterior roots to the grey substance, *ra*, *rp* (arteriæ radicum ant. et post.). An artery on the mesial side of the posterior root traverses Burdach's column and loses itself in the caput cornu posterioris, *cp* (arteria cornu posterioris). Two fairly constant arteries extend from the pia mater into the lateral column and the adjoining grey matter, the arteriæ laterales anterior (*la*) et media (*lm*), the latter corresponding to about the middle of the lateral column. An arteria lateralis posterior (*lp*) is less constant. The veins follow the course of the arteries; the venæ sulci, however, are not of sufficient calibre to accommodate all the blood admitted by the arteriæ sulci, the rest goes into the veins of the vaso-corona, and so to the posterior part of the periphery of the cord (*Kadyi*).

5. PATHOLOGICAL CHANGES IN THE SPINAL CORD.

The diseases of the spinal cord, which are interesting from the standpoint of pathological anatomy, can only, owing to the abundance of material, receive a cursory notice in this place.

Pathological changes in the cord may either originate within itself or may extend into it from other organs—*e.g.*, brain, nerve-roots, bones, meninges. Hence primary and secondary lesions are distinguished, each of which again may be either localised or diffuse. It is characteristic of localised diseases of the spinal cord that they are confined to well-marked regions of the white or grey matter, morphological regions, the limits of which are not at first overstept, and hence reciprocally, much information as to the definition of these regions is obtained by a study of their diseases. Diffuse diseases do not respect these boundaries.

Acute and chronic processes are first to be distinguished.

Acute Diseases of the Spinal Cord.—Anæmia and hyperæmia occupy the chief place; after these may be mentioned hæmorrhage into the spine-substance (apoplexia spinalis, hæmatomyelia). Hæmorrhage occurs spontaneously (in which case it affects the grey matter, and immediately adjoining white substance more especially) as the result of injuries or diminished atmospheric pressure, or it may be secondary to myelitis. If the rest of the cord be healthy the effused blood shows a tendency to spread through the cord in the direction of its long axis. Acute diffuse myelitis has either a traumatic origin (as in concussion of the spinal cord), or is toxic, due to the poison of tetanus, &c. Owing to its greater richness in blood-vessels, the grey matter is more affected than the white, except in cases such as the rare tubercular infiltration, in which the process extends inwards from the meninges. Softening of the cord extending a long way in a longitudinal direction may result from such a myelitis.

Localised acute myelitis attacks the anterior horns. The large nerve-cells are thereby destroyed, as, for example, in the spinal paralysis of children and in acute poliomyelitis of grown-up people.

Chronic Diseases of the Spinal Cord.—(*a.*) PRIMARY LOCALISED DISEASE may affect—

(1) The large cells of the anterior horn : chronic poliomyelitis.
(2) The lateral pyramidal tracts : primary lateral sclerosis.
(3) The ground-bundle of the posterior columns: posterior sclerosis (tabes dorsalis).
(4) The longitudinal fibres which traverse Clarke's column (also in tabes dorsalis).

Two different regions may be diseased at the same time—*e.g.*, the lateral pyramidal tract and the cells of the anterior horn in amyotrophic lateral sclerosis. All varieties of chronic primary localised lesions are, as a rule, bilaterally symmetrical. Tabes dorsalis is characterised by degeneration of the root-zone in the posterior ground-bundle, associated with degeneration of the small fibres in the interior of Clarke's column, as well as in Lissauer's border zone. The regions affected are those into which the posterior root-fibres can without doubt be followed. Later on Burdach's and Goll's columns are attacked, so that, in all cases of tabes dorsalis in which the upper extremity suffers, the whole of the posterior columns, from the calamus scriptorius to the conus medullaris, with the exception perhaps of some fibres in the grey commissure, are found to be sclerosed. In very advanced cases, even the posterior grey commissure is sclerosed. In cases in which the upper extremity is not affected, the degeneration is entirely limited to Goll's column, and does not extend so far ventrally as the grey commissure. It must, however, be pointed out that possibly tabes dorsalis is not a primary disease of the spinal cord; the starting point of the process is generally considered to be the posterior roots, if not the peripheral nerves.

Primary disease of the whole of Goll's column is described by *Pierret;* *Vierordt* also gives a case of descending primary disease of this column.

A special form of combined localised degeneration (found in hereditary ataxia) is that in which the mesial division of the posterior column degenerates in company with the lateral pyramidal tract, only extending usually as far forwards as the crossing of the pyramids.

The strict localisation of the degeneration is in many of these cases fallacious; rather we have to do with a primary meningitis (pseudo-localised or pseudo-systemic degeneration), the injury to the spine being secondary to that of its sheath (fig. 113). Sometimes it happens that inflammation of the meninges is set up by the degeneration of the posterior columns, and this again extends to neighbouring parts of the periphery of the cord as a "margin-degeneration," which later on affects the long fibre-tracts (*Borgherini*).

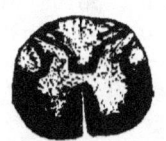

Fig. 113.—So-called combined systemic disease of the spinal cord. Section from lumbar region. *Magn.* 2. *Pal's staining.* Both posterior columns, with the exception of the part bordering on the grey matter, as well as the lateral pyramidal tracts, and to a certain extent the whole periphery of the cord, appear diseased.

(*b.*) SECONDARY LOCALISED DEGENERATIONS.—We have already several times pointed out the directions in which secondary degenerations travel in the cord; they descend in the anterior [direct] pyramidal tract, PyV, and the lateral [crossed] pyramidal tracts, PyS; they ascend in Goll's column, GS, the lateral [direct] cerebellar tract, KS, and in Gowers' column, G, as well as in the fine longitudinal fibres of Clarke's column.

Secondary degeneration of the cord follows destruction of the pyramidal tracts in the brain in eleven days (*Kahler* and *Pick*).

How far tabes dorsalis and other localised diseases are to be classed amongst secondary degenerations has just been discussed.

The micromyelia of microcephali finds a place here; the diminutive size of the column depending upon the want of development of the brain (*Steinlechner*). Secondary degenerations of the posterior columns have also been described.

(*c.*) PRIMARY DIFFUSE DISEASES.—(1) Transverse, diffuse, chronic myelitis which spreads out more or less over the whole transverse section of the cord (fig. 114).

(2) Central myelitis, myélite periépendymaire, syringo-myelia, fig. 115, three expressions which, though not strictly speaking identical, are very often used for the same anatomical result. In syringo-myelia,

properly so-called, one finds in the interior of the cord a tubular cavity, usually large enough to put the little finger into, of variable length, but seldom extending farther caudalwards than the lumbar region. It always lies to the dorsal side of the central canal, although the canal not infrequently opens into it. A central glioma is often the cause of the hollowing out. It is a remarkable fact that syringo-myelia of the cord is most common in those regions in which blocking of the central canal by overgrowth of the ependyma usually occurs. Hydromyelia, a dilatation of the central canal itself, a condition analogous to chronic hydrocephalus, may well be distinguished from syringo-myelia.

(3) Disseminated sclerosis (insular sclerosis, sclerose en plaques), fig. 116.

 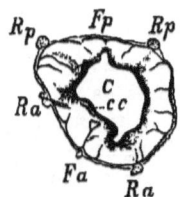 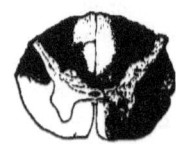

Fig. 114.—Chronic transverse diffuse myelitis. Upper lumbar cord. *Staining with alum-hæmatoxylin.*

Fig. 115.—Syringo-myelia. *Carmine staining. Magn. 2.*—*C*, central cavity; *cc*, central canal; *Fa*, fissura longitudinalis anterior; *Ra*, and *Rp*, anterior and posterior roots.

Fig. 116.—Disseminated sclerosis in the cervical cord. *Pal's staining. Magn.* 2.

The degenerated spots are of variable size and may occur in any part of the cross-section of the cord. They spread over from the white substance to the grey, and *vice versâ.* Sclerosed spots are found at all levels in the cord, but lie oftener in the lumbar cord than in the regions above. Many attempts have been made to refer disseminated sclerosis back to a primary meningitis. More likely the disease has its origin in connection with the intraspinal vessels. Generally it attacks both brain and cord at the same time. The sclerosed patches in the brain are usually large and numerous; they are found chiefly in the region of the pons and in the white substance of the hemisphere. Often the ependyma of the lateral ventricle is the starting point of large lesions.

(4) Tumours of the spinal substance may also be reckoned amongst diffuse diseases of the spinal cord. With the exception of gliomas they commence for the most part in the membranes and belong, therefore, to the next category.

(5) Lastly, diffuse diseases of the blood-vessels must be mentioned, such as the formation of numerous miliary aneurisms, a very rare disease, but observed by *Koehler* and *Spitzka* throughout the whole length of the cord.

(*d.*) SECONDARY DIFFUSE DISEASES.—These affections depend, as a rule, upon pressure from outside (compression-myelitis), due to such causes as inflammatory thickening of the dura mater spinalis in the cervical region (pachymeningitis cervicalis hypertrophica); or the pressure may be due to the growth of tumours within the vertebral canal, either proceeding directly from the membranes as gummata, myxomata, and sarcomata, with a tendency to cavernous formation, or solitary tubercles, lipomatous overgrowth of the peridural fat or echinococci (generally outside the dura); but by far the commonest causes of these secondary diseases in the spinal cord are troubles affecting the vertebræ—caries, for instance, or more rarely tumours.

Meningitis spinalis may, owing to the accumulation of exuded lymph, produce general compression of the cord; or a much more interesting direct concentric extension of the inflammatory process to the spinal cord, myelitis annularis. Various diseases of associated tracts of the cord may, as already mentioned, be attributed to primary meningitis.

It is impossible, owing to the uncertainty which still hangs over them, to examine in great detail the minute histological changes associated with the diseases mentioned above. The most important facts known with regard to these changes have been already introduced into the chapter on the constituents of the central nervous system. It would be very interesting if we could prove in any single case whether the nervous elements of the spinal cord are the starting points of the disease (parenchymatous processes), or whether, on the other hand, the disease commences in the connective-tissue or blood-vessels (interstitial processes).

Suppose, for example, that we compare secondary degeneration of the lateral pyramidal tract with a similar lesion in disseminated sclerosis. The former degeneration is an injury to fibres; the sclerosed plaque, on the other hand, is formed by an overgrowth of connective-tissue, and only indirectly affects nerve-fibres; for a long time after the medullary sheath is destroyed, the axis-cylinder remains intact, and a certain amount of functional conductivity is retained; secondary degeneration proceeding from such a sclerosed spot is rare.

The occurrence in the spinal cord of fat granule-cells and amyloid bodies gives rise to appearances varying somewhat with the disease, but always characteristic. Chronic atrophic processes rarely occur without the appearance of numerous amyloid bodies; fat granule-cells are never wanting in dementia paralytica (*Westphal*); they occur also in other diseases. Fat granule-cells put in an appearance in the cord when other organs are diseased, but they are, in these cases, limited to the segment of the cord from which the nerve for the diseased organ comes off. Hardly ever are either fat granule-cells or amyloid bodies scattered equally throughout the whole cross-section. In secondary degenerations the former are almost always limited to the diseased columns of fibres.

Deficient development of the spinal cord is not rare, especially as regards the grey matter. Asymmetry of the two halves, prolongation of the anterior horns to the periphery, and other defects are found, but one must make quite sure that the distortion is not artificial. Especially remarkable are the cases in which one (*Bramwell*) or even both (*Fürstner* and *Zacher*) halves of the spinal cord appear to be double. Cases in which the two vesicular columns of Clarke are situate in the posterior commissure, nearly touching one another in the middle line, as was first described by *Pick*, have been repeatedly found. *Pick* also described a very rare heterotopia of grey gelatinous substance. In another case, *Musso* found in the posterior column a little heterotopic lesion, which did not quite correspond in structure with the column of Clarke, but was connected with this column of cells by a narrow grey neck.

SECTION V.—TOPOGRAPHICAL EXAMINATION OF THE BRAIN.

It is not, as a rule, necessary to obtain an unbroken succession when making a series of sections; at any rate when working with human brains or the brains of the larger mammals. Such an attempt would result in a great waste of apparatus, reagents, and time. It is sufficient to take somewhat thick pieces at different levels, from each of which the finest sections possible are again prepared. Care must be taken, however, not to discard material in such places as show important anatomical changes within a small area, as, for instance, in the region of the decussation of the trochlear nerves.

When, however, it is a question of establishing minute anatomical relations, as in tracing the course of fibres, we must be careful to prepare an unbroken series of sections. In such cases, as also for pathological work, methods like those recommended on p. 170 are often valuable.

We shall begin by examining a section obtained from the brain-stem in front of the anterior corpora quadrigemina. The drawings are all made from carmine-preparations. The sections are cut in a plane at right angles to the long axis of the medulla. By artificial stretching of a freshly removed brain-stem during the hardening process it is possible to bring the spinal cord and medulla (which naturally make a right angle with one another) into the same straight line. According to *Forel's* method the axis is called "Meynert's axis of section," and the perpendicular planes, to which our sections correspond, "Meynert's cross-planes." In lower animals this bending of the stem is less, and so the long axis of the spinal cord is more nearly a continuation of the brain-axis.

The comprehension of the complicated structure which the central nervous system presents above the medulla is facilitated by bearing in mind the arrangement, both as to form and course of fibres, which we have found in the spinal cord. Although there are many details in the arrangement of the fibres peculiar to each region, there are yet some general points of view of which mention may first be made:—

(1) The tracts of long fibres of the spinal cord can be followed for

210 PLANES IN WHICH SECTIONS ARE CUT.

a longer or shorter distance into the medulla, whence some go to the cerebellum, and others to the great brain; they are, however, to a greater or less extent, enveloped in other structures.

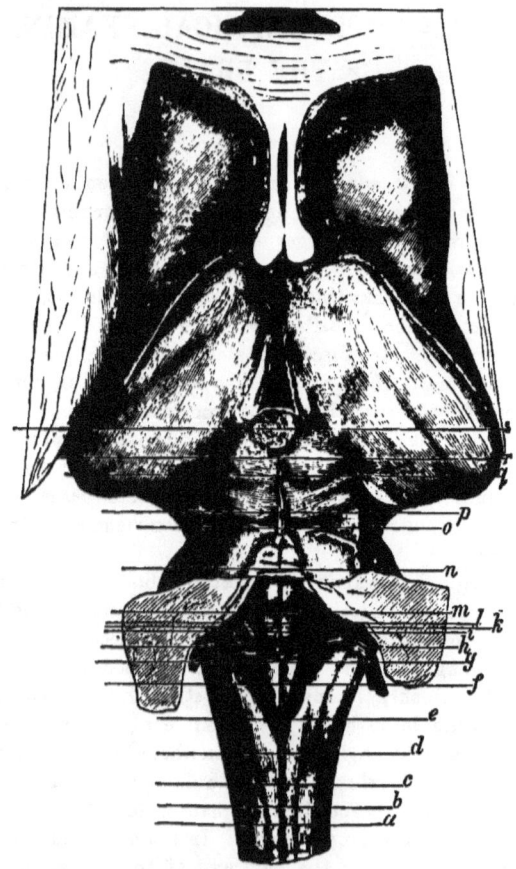

Fig. 117.—Serves to show the level at which the cross-sections, represented in figs. 118-136 (with the exception of 131) are made. It must be remarked that the lines drawn across this figure do not indicate the plane of the cross-sections to which they correspond, but only the situation of their most dorsal portions, cf. fig. 13.

(2) The same can be said for the grey matter of the cord, which, with many changes in form, but with no break in its continuity, takes part in the formation of the medulla oblongata.

(3) Several new grey masses appear with the fibres which belong to them, and introduce varying complications into the picture.

(4) A striking change is produced in the relation of the several constituents of the medulla by the opening out of the central canal into the fourth ventricle. Structures which were before dorsal to the canal, come to lie on each side of it.

We shall follow the structure of the medulla through a series of sections.

But in order not to lose sight of the level at which successive sections are made it is desirable first to give an account of its external form.

For this purpose it is well to have always at hand a well-hardened brain, and to follow in imagination the planes of the sections as we study them (as shown in fig. 117). It is also desirable to make at the same time cross-sections of a fresh brain, and to study in them the details which microscopic examinations of the prepared brain bring to light.

Changes in formation preparatory to its conversion into the medulla are already visible in the cervical cord at the level of the second nerves.

It is at about this level that we usually separate the cord in taking the brain out of the skull-case. The change in shape of the posterior horn consists in its assuming an almost cylindrical form, while the cervix cornu posterioris, *Ccp*, becomes thinner and the apex disappears. The caput with its substantia gelatinosa Rolandi, *Sgl*, is now separated from the surface by the longitudinally running fibres of the ascending root of the trigeminus, *Va;* it usually, however, makes a noticeable external prominence, the tuberculum Rolandi (not to be seen in fig. 118).

A strong development of the processus reticularis again makes its appearance in the lateral column. Individual bundles of fibres appear cut across at different angles. More and more obliquely-cut bundles are seen, especially traversing the central part of the anterior horn. Still farther brainwards one can see distinctly that large fasciculi from the lateral column pierce the anterior horn, cross the middle line, and add themselves to the anterior columns of the opposite side. These indicate the crossing of the pyramids (decussatio pyramidum), *DPy*. Farther up the quantity of fibres crossing from the lateral column of one side into the anterior column of the other becomes so immense that the tip of the anterior horn (Ca^1) is completely cut off from its central parts (Ca^2). At the same time the anterior longitudinal fissure (*fsla*) becomes much shallower. In some places it is almost completely filled up. Only that part of the lateral column which we have named the lateral pyramidal tract, shares in this crossing. The bundles which cross ascend obliquely upwards, forwards, and outwards. Thus it comes about that the anterior fissure is pushed sometimes to one side, sometimes to the other, or it may happen that the crossing of the pyramids is bounded below by a double sickle-shaped cleft, and presents a mammilliform process, processus mammillaris (as in fig. 120). The anterior commissure appears to be involved in the overwhelming mass of the crossing pyramids; as a matter of fact, however, it remains independent of the latter, and fibres homologous with it can be followed even as far as the mid-brain.

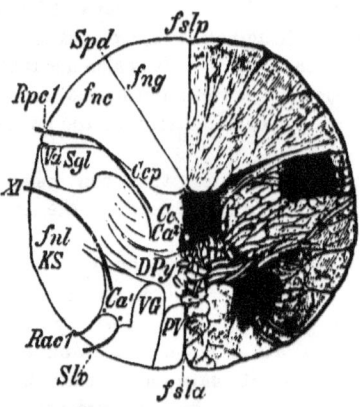

Fig. 118.—Section of the medulla oblongata represented as a in fig. 117. *Magn.* 4. —*fslp*, Fissura longitudinalis posterior; *Spd*, sulcus paramedianus posterior; *fng*, funiculus gracilis; *fnc*, funiculus cuneatus; *Rpc 1*, radix posterior cervicalis prima; *Va*, ascending root of trigeminus; *Sgl*, substantia gelatinosa; *Cep*, cervix cornu posterioris; *fnl*, funiculus lateralis; *KS*, lateral cerebellar tract; *Rac 1*, radix anterior cervicalis prima; *Slv*, sulcus lateralis ventralis; *fsla*, fissura longitudinalis anterior; *Cc*, canalis centralis; *DPy*, decussatio pyramidum; Ca^1 and Ca^2, peripheral and central parts of the cornu anterior; *VG*, anterior ground-bundle; *PV*, anterior pyramidal tract.

A transverse section through the region in which the greatest number of fibres cross shows the following changes. With the gradual increase in the cross-section, the central canal is displaced dorsally. The dorsal border of the central grey substance exhibits two swellings corresponding to the two portions into which the posterior column is divided. In the mesial of the two columns there appears an elongated club-shaped grey mass, the point of which rests against the mesial of the two swellings which we have mentioned; this is the nucleus funiculi gracilis, *Ng* (post pyramidal nucleus of *Clarke*, the postero-mesial accessory horn of *Reichert*). A little farther brainwards another lateral swelling makes its appearance as a broad elevation of the grey substance lying beneath the cuneate fasciculus (nucleus funiculi cuneati, restiform nucleus of *Clarke*, postero-lateral accessory horn of *Reichert*). Neither the nucleus gracilis nor the nucleus cuneatus forms a sharply-defined grey mass; both are made up of separate little groups of nerve-cells; an inconstant isolated group of cells, situate peripherally, is known as the outer nucleus of the cuneate fasciculus, *Nce* (fig. 121).

The lateral column becomes progressively smaller as its fibres cross the middle line to take up their position on the ventral side of the

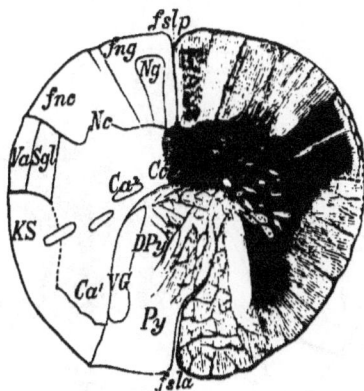

Fig. 119.—Section indicated in fig. 117 by the line *b*.—*Ng*, Nucleus funiculi gracilis; *Nc*, nucleus funiculi cuneati. Other lettering as in fig. 118.

medulla; the lateral cerebellar tracts are seen in figs. 118, 119, 120, lying almost unchanged in the lateral region, while the remainder of the lateral column is lost in a mass (staining light-red with carmine) which passes over into the portion of the anterior horn, which already, by the crossing of the pyramids, has been cut off and displaced laterally.

The farther we advance brainwards with our sections, the more indistinct becomes the lateral boundary of this portion of the anterior horn, until at last it loses itself in a mixed region (substantia or formatio reticularis grisea, *seu* lateralis) lying to the side of the ventral half of the medulla.

Mesially this region is bounded by a very distinct white bundle, the lowest root of the hypoglossal (fig. 120, *XII*), which starting in the neighbourhood of the central canal runs ventrally with an inclination sidewards towards the periphery. Lying on its inner side somewhere about the middle is to be seen a long, often interrupted, but very conspicuous, group of large nerve-cells, which may be called the nucleus of the ground-bundle of the anterior column, *Nfa* (nucleus funiculi anterioris). The ground-bundle of the anterior column retains its original position on the mesial side of the former anterior horn, appearing in cross-section as a fairly recognisable area rounded dorsally, but pointed on its ventral side (figs. 118, 119, 120).

As soon as the pyramids (*Py*) have taken up their position on the ventral side of the medulla, as great compact bundles, the crossing of the fillet, *DLm* (decussatio lemnisci, piniform decussation), makes its appearance in the middle line from the back of the pyramids to the central

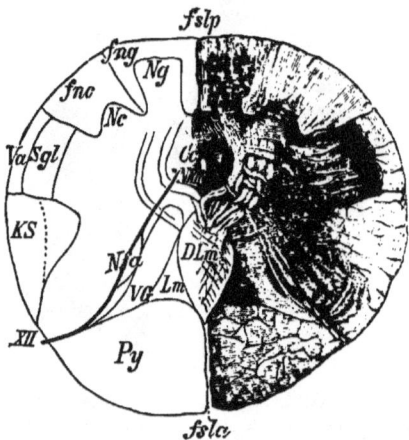

Fig. 120.—Section *c* in fig. 117.—*XII*, Nervus hypoglossus; *NXII*, nucleus of ditto; *DLm*, decussatio lemnisci; *Lm*, lemniscus; *Nfa*, nucleus of anterior column; *Py*, pyramids. For remaining lettering, see *supra*.

canal. Fairly thick white bundles extend in concentric curves out of the region of the posterior columns, where they are raised up by their two nuclei and surround the central canal. They cross on the ventral side of the canal at an acute angle, and take up their position on the dorsal side of the pyramids in the lemniscus layer (*Lm*).

The crossing of the fillet occurs immediately to the dorsal side of the crossing of the pyramids, so that in adult brains it is impossible to see a boundary between the two; whereas in the embryo the fillet-bundle is recognisable by its early myelination. The crossing of the fillet might be termed the sensory or upper pyramid-crossing.

The tract occupied by the crossing of the pyramids and fillet increases steadily in its dorso-ventral extension as the brain is approached, becoming at the same time narrower. For a long distance its median diameter is the greatest, and it hence appears as a fusiform area in section. All the way up to the third ventricle the median plane of the brain-stem is occupied by fibres which cross one another at an acute angle, but the area in which this decussation occurs is reduced to a vertical plate, termed the raphe, *Ra*, the dorsal edge of which abuts upon the central canal.

In the following sections the smaller groups of grey substance which constitute the nuclei funiculi gracilis et cuneati widen out, so that they make on the surface noticeable swellings; in the fasciculus gracilis the swelling is the cause of the formation of the clava; in

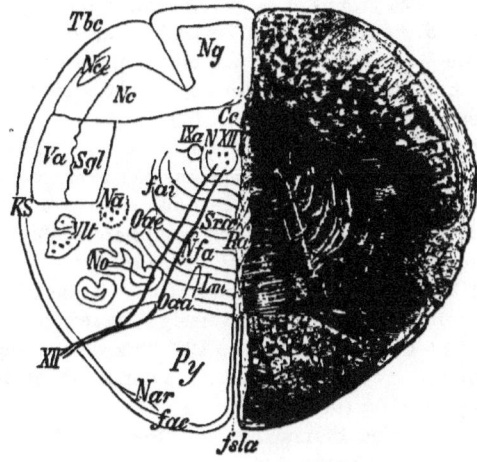

Fig. 121.—Section *d* in fig. 117.—*IXa*, Ascending root of glossopharyngeal; *Nce*, outer nucleus of cuneate fasciculus; *Nlt*, nucleus of lateral column; *Na*, nucleus ambiguus; *No*, olive; *fai*, fibræ arcuatæ internæ; *Sra*, substantia reticularis alba; *Ra*, raphe; *Oae*, external accessory olive; *Oaa*, anterior accessory olive; *fae*, fibræ arcuatæ externæ; *Nar*, nucleus arcuatus; *Tbc*, tuberculum cuneatum.

the fasciculus cuneatus, the tuberculum cuneatum is the external expression of the grey nucleus. The concentric fibres which before took part in the crossing of the fillets, now constitute thin bundles, all, or almost all, of which come from the posterior columns. Hence the radius of curvature of the outer arcuate fibres becomes constantly greater, and the portion of the medulla lying on the ventral side of the central canal is traversed by these bundles in a characteristic manner.

As these arching fibres no longer have the same meaning as the crossing of the fillets in a strictly literal sense, we simply call them fibræ arcuatæ internæ (*fai*). They traverse the substantia reticularis grisea, cross the roots of the hypoglossus, XII, and then becoming more distinct, divide this region into a number of little fields.

Only a few nerve-cells are scattered on the mesial side of the hypoglossal roots. This region, which extends dorsally to the level of the central canal, consists almost entirely of medullated fibres, and is termed the substantia or formatio reticularis alba (*Sra*).

Numerous large cells, analogues of the cells of the anterior horn, are scattered about the substantia reticularis grisea; a formation which we may look upon as derived, to a certain extent, from the anterior horn. In certain places the cells are united into compact clumps of grey substance. The groups of large cells which lie midway between the periphery and the central canal, before it opens out into the fourth ventricle in the lateral portion of the medulla, are described as constituting the nucleus ambiguus, *Na* (nucleus lateralis medius, motor vago-glossopharyngeal nucleus), figs. 121, 122, 123.

It is well to distinguish from this nucleus the numerous separated masses of small cells which lie nearer the periphery on the ventral side of the ascending root of the trigeminus, constituting the "nuclei of the lateral column;" often two such groups are visible, nuclei laterales anteriores et posteriores (figs. 121, 122, 123, *Nlt*).

The dorsal boundary of the pyramid is now formed, in its middle portion at any rate, by a long transversely disposed grey mass which is soon joined at an angle of 100° to 120° by a shorter sagittally disposed limb, *Oaa* (nucleus of the pyramid, anterior olive), figs. 121 and 122. The sagittal piece (antero-posterior with regard to the sections) has a greater extension brainwards than the horizontal limb (which lies transversely in the sections).

In a section intermediate between those represented in figs. 120 and 121, the nucleus of the pyramid and the nucleus of the lateral column lie very close together. A little farther forward a very characteristic formation of grey matter, the olive, *No* (nucleus olivaris), insinuates itself between them. The olive (figs. 121 to 125) appears in cross-section as a much folded dentate band curved upon itself with the convexity to the outer side, where it produces on the surface a well-marked swelling ordinarily known as the olive (or olivary body, also inferior olive), *Oi*, figs. 11, 122, 123, 124.

At the periphery of the sections various tracts of fibres are cut in the direction of their length; tracts, therefore, which have a more or less horizontal course. These are the arcuate fibres (fibræ arcuatæ, *seu* arciformes externæ), *fae*. They have various sources of origin. Many of them curl round the pyramid to join at the bottom of the anterior fissure with the raphe.

FIBRÆ ARCUATÆ. 217

On the ventral side of the pyramids, certain clumps of grey matter develop in the arciform fibres, the largest of which is in human brains, triangular and very strongly developed, the nucleus arcuatus triangularis, *nar* (anterior nucleus of the pyramid, one of the small pyramidal nuclei of *Stilling*), figs. 121, 122, 123. The number of these groups of cells, which we shall term nuclei arcuati, increases brainwards, and at last they go over into the nuclei of the raphe or into those great collections of grey matter which we shall come to know later on as the nuclei of the pons.

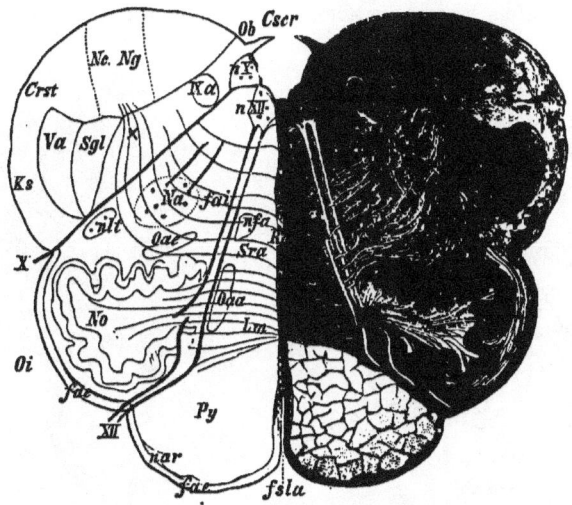

Fig. 122.—Cross-section, fig. 117, *e.*—*Cscr*, Calamus scriptorius; *Ob*, obex; *nX*, sensory nucleus of vagus; *X*, vagus; *Crst*, corpus restiforme; x, fibræ arcuatæ from the most external part of the nuclei of the posterior column; *Oi*, olivary eminence.

Superficially situated tracts of fibres are also found in figs. 121 and 122 in the dorsal portions of the sections. For the most part these belong to the direct (lateral) cerebellar tract which (passing the now quickly growing ascending root of the trigeminus, *Va*) comes into contact with the posterior column, and completely gives up its position in the lateral column.

The substantia gelatinosa Rolandi, *Sgl*, diminishes in amount as rapidly as the ascending root of the trigeminus grows; but it is to be recognised as the companion of the root of the trigeminus, on the concave mesial border of which it lies as far as the point of entrance of this nerve.

A small round bundle, lying on either side of the central canal in the sections through these planes, is still to be described. Farther forwards it becomes a conspicuous isolated tract, round in cross-section, the ascending root of the glossopharyngeal, *IXa* (figs. 121 to 124).

If the section falls not far above the level at which the central canal opens out into the fourth ventricle (fig. 122) at the calamus scriptorius (*Cscr*) the following points will be noticed:—The grey matter which lies dorsally to the central canal is pushed as the cleft opens to the outer side, being displaced farther and farther outwards as the floor of the fourth ventricle becomes flatter, while the grey matter of the anterior

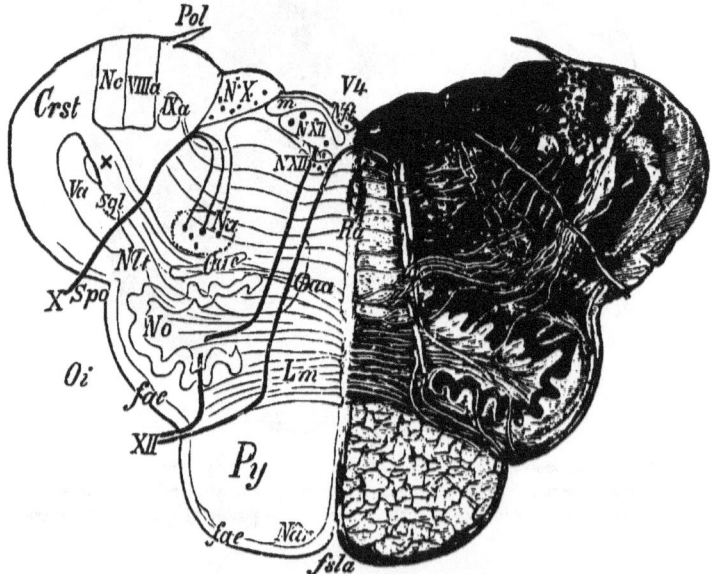

Fig. 123.—Cross-section, fig. 117, *f.*—*Pol*, Ponticulus ; *V4*, fourth ventricle ; *VIIIa*, ascending root of auditory nerve ; *m*, white matter covering the chief nucleus of the hypoglossal nerve (*NXII*); *NXII'*, small-celled nucleus of hypoglossal nerve; *Nft*, nucleus funiculi teretis; x, accession to the restiform body of fibres from the fibræ arcuatæ internæ (*Crst*); *Spo*, sulcus postolivaris.

horn, which lay on the ventral side of the canal, travels upwards and comes to occupy the mesial portion of the floor of the fourth ventricle.

The only remains to be found of the embryonal roof of the fourth ventricle are certain little plates of tissue (varying in form and development in different individuals), which are enclosed in pia mater and rest against the fasciculus gracilis with their free borders directed towards the inner side (*cf.* p. 59). An inconstant platelet [of white matter] which fills in the angle between the diverging fasciculi graciles is known as the obex, *Ob*, fig. 122. The symmetrical plates in front of this are the ponticuli, *pol* (alæ pontis), fig. 123.

The small collections of grey matter which we have learnt to know as the nuclei of the fasciculus gracilis and fasciculus cuneatus become constantly smaller. Their place is taken by a quickly growing field of fibres, the corpus restiforme, *Crst*, with which the lateral cerebellar tract blends. The restiform body slopes obliquely upwards and forwards around the outer side of the ascending root of the trigeminus. Only by embryological investigations can we make out the complicated elements out of which the restiform body is built up; attention must be called, however, to the considerable mass of fibres which passes on the mesial side of the substantia gelatinosa and ascending root of the trigeminus to join the restiform body, the fibræ arcuatæ internæ laterales, x (figs. 122 and 123).

By this time the olive has reached its greatest development, and is more conspicuous than anywhere else from the surface. On the dorsal side of the olive proper, an extended grey mass has made its appearance in the formatio reticularis, the upper or outer accessory-olive, *Oae* (nucleus olivaris accessorius externus *seu* superior), figs. 121 to 124.

The roots of the hypoglossal nerve, which spring from large nerve-cells, *NXII* (hypoglossal nucleus or chief nucleus), situate for the most part in the median grey matter in the floor of the fourth ventricle, are now at their greatest development (figs. 122, 123). They make a sharp boundary between the substantia reticularis alba and substantia reticularis grisea, and course for the most part between the sagittal limb of the pyramidal nucleus (*Oaa*) and the olivary nucleus (*No*). Often they seem to be connected with the latter, in reality they only cross through it or run within it, for a certain distance downwards towards the cord, and then bend horizontally and come out in the furrow between the olive and the anterior pyramid. The principal nucleus of the hypoglossal is still separated from the surface of the sinus rhomboidalis by a layer of fine medullated fibres, disposed for the most part in a longitudinal direction. On the mesial edge of the hypoglossal nucleus, and still more on its lateral edge, the cross-section of this white column assumes the shape of a club, *m* (fig. 123). These fibres give to the hypoglossal triangle on the floor of the fourth ventricle its striking white colour. Close beneath the ependyma and very near the raphe a small group of nerve-cells is cut through, known as the nucleus funiculi teretis, *Nft* (figs. 120 to 127).

Other tracts of fibres, less considerable than the roots of the hypoglossus, radiate outwards from groups of nerve-cells of medium size,

NX, lying in the grey substance of the floor of the ventricle. Not equally visible in all sections, they pass ventrally to the ascending root of the glossopharyngeal through the substantia reticularis grisea, and often pierce in a very striking manner the ascending root of the trigeminus, *Va* (figs. 123 and 124). These are the root fibres of the vagus and glossopharyngeal nerves; the grey mass from which they spring is, therefore, the vago-glossopharyngeal nucleus. A number of fibres originate in the nucleus ambiguus, *Na*, or group of large cells, which lies in the substantia reticularis grisea; first passing dorsally, many of them arch over and join the glossopharyngeal and pneumogastric roots. These groups of cells may be looked upon as the motor nuclei of *IX* and *X*. Other fibres from these nuclei incline medianwards towards the raphe.

Sections farther forward (figs. 124 and 125) differ in shape from

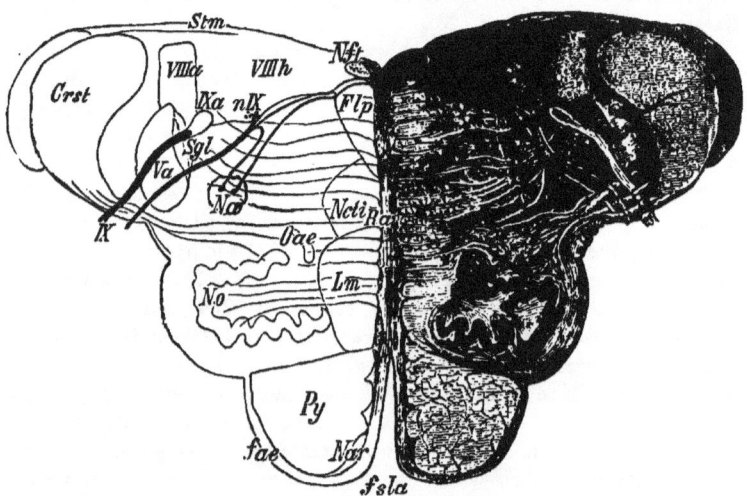

Fig. 124.—Transverse section, fig. 117, *g*.—*Stm*, Striæ medullares *seu* acusticæ; *VIIIh*, chief auditory nucleus; *IX*, nervus glossopharyngeus; *nIX*, glossopharyngeal nucleus; *Flp*, fasciculus longitudinalis posterior; *Ncti*, nucleus centralis inferior.

those already described owing partly to the flattening out of the broader floor of the ventricle and partly to the constant increase in size of the corpus restiforme which rises up above its dorso-lateral margin. The last traces of the posterior columns disappear.

At the level at which the root-fibres of the hypoglossal first disappear from the section, the ascending glossopharyngeal root, *IXa* (fig. 124), seems to bend horizontally outwards to make its exit parallel with the other root-fibres of the glossopharyngeal nerve. It is much the strongest bundle of root-fibres belonging to this nerve, and pierces the ascending root of the fifth, reaching the periphery on the ventral side of the corpus restiforme, *Crst*.

By this time the hypoglossal nucleus has disappeared from the floor of the fourth ventricle; only the last remnants of the nuclei of vagus and glossopharyngeal are left; but a great triangular grey field occupies the greater part of the floor of the fourth ventricle and reaches some distance down into the substance of the medulla oblongata, its apex directed towards the middle line; this is the "chief" nucleus of the auditory nerve, *VIIIh*. The commencement of this nucleus might have been looked for in fig. 123, in the region which extends from *NX* laterally as far as *VIIIa*. As it grows in size it presses the nuclei of *IX* and *X* downwards into the substance of the medulla, and at last when the chief nucleus of *XII* gives way to it, it extends itself inwards as far as the middle line. Between the auditory nucleus and the corpus restiforme is seen, in addition to the remains of the funiculus cuneatus (fig. 123), a nearly rectangular area of medullated fibres, transversely cut, embedded in a network of grey matter called the ascending root of the auditory nerve (*Roller*), figs. 123 to 127. Other fine bundles of fibres, which enter the fourth ventricle over the restiform body, also belong to the auditory nerve, the striæ medullares, *Stm* (*seu* acusticæ).

In the course of the striæ medullares are usually embedded larger or smaller masses of grey matter which sometimes make considerable eminences in the neighbourhood of the corpus restiforme (tæniola cinerea, tuberculum acusticum). When the striæ medullares are well developed it is easy to see that their fibres, just before reaching the middle line, bend ventrally, descending towards the pyramid in the outer margin of the raphe. In fig. 124 this is only shown to a small extent, at the spot where the letters *Nft* are placed.

No great change in the cross-section of the pyramids or of the substantia reticularis grisea and alba is exhibited by fig. 124. In the substantia reticularis alba, however, we begin to see a distinction between its most dorsally situate longitudinal fibres, *Flp*, which lie close to the floor of the fourth ventricle and the most ventral fibres, *Lm*. This distinction is due to the increasing rarity of the fibres in the intervening region and the accumulation in their place of grey substance, intercalated between the longitudinal and transverse fibres. This grey matter, nucleus centralis of *Roller*, *Ncti* (nucleus centralis inferior), is not sharply marked off from the substantia reticularis grisea.

The smaller division of longitudinal fibres, derived in part from the ground-bundle of the anterior column, retains its position close up against the raphe throughout the whole of the floor of the fourth

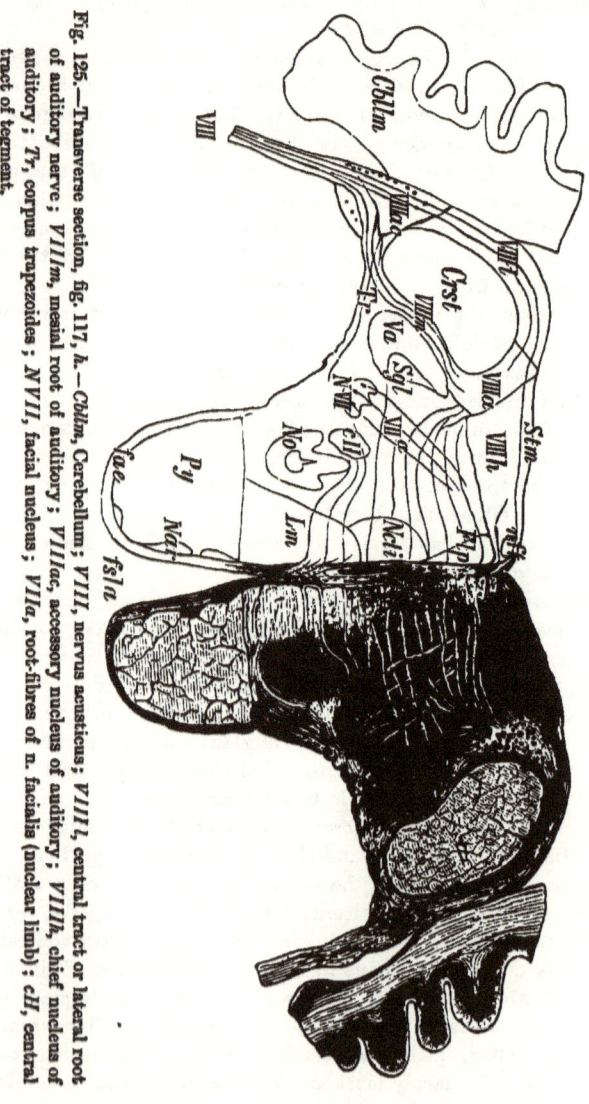

Fig. 125.—Transverse section, fig. 117, h.—*Cblm*, Cerebellum; *VIII*, nervus acusticus; *VIIIl*, central tract or lateral root of auditory nerve; *VIIIm*, mesial root of auditory; *VIIIac*, accessory nucleus of auditory; *VIIIh*, chief nucleus of auditory; *Tr*, corpus trapezoides; *NVII*, facial nucleus; *VIIa*, root-fibres of n. facialis (nuclear limb); *cH*, central tract of tegment.

ventricle and the aqueduct of Sylvius. It is known as the posterior longitudinal bundle, *Flp* (fasciculus longitudinalis posterior). The larger ventral collection of fibres—the continuation upwards of the inter-olivary layer—constantly changes its position in a manner to be presently described. It is known as the bundle of the lemniscus or fillet, *Lm* (mesial fillet).

In sections cut just below the pons, which may be regarded in the ascent of the system as the last sections of the after-brain, the upper convolutions only of the olive, *No*, are to be seen; the transverse diameter of the pyramids, *Py*, is now a little less, but their dorso-ventral diameter is proportionally greater. The chief nucleus of the auditory nerve, *VIIIh*, retains the same relation as before to the ascending roots of this nerve, *VIIIa*, and of the trigeminus, *Va*, as well as to the restiform body. The restiform body is surrounded more obviously than in fig. 124 by great bundles of fibres which, although they belong to the auditory nerve, are, nevertheless, not to be looked upon as root-fibres (the so-called lateral root of the auditory nerve, *VIIIl*), rather they constitute a connection with the great brain of the accessory auditory nucleus soon to be described. The mesial root courses down between the restiform body and the ascending trigeminal root. As well in the angle between the mesial and lateral roots as on the inner and outer sides of the conjoined roots appear collections of grey matter, the accessory nucleus of the auditory, *VIIIac;* the auditory roots are especially characterised by their richness in nerve-cells. Out of the accessory nucleus scattered bundles of fibres course transversely towards the median line; they belong to the corpus trapezoides, *Tr*, which only attains to its full development in sections higher up the axis.

The separation between the posterior longitudinal bundle and the fillet effected by the nucleus centralis inferior, *Ncti*, becomes increasingly distinct. After the nucleus of the lateral column has disappeared the groups of cells, which already in posterior sections formed the motor nuclei of the vagus and glossopharyngeal nerves, increase considerably in size, and as soon as the last fibres of these two nerves have been supplied with cells, other fine fibres belonging to another motor nerve, the facial, take their place, and are seen coursing dorsally and medianwards. This is the lower end of the facial nucleus, *NVII*, which is nothing more than the continuation of the nucleus ambiguus, and, therefore, indirectly of the cells of the anterior horn [or of the lateral horn].

If our sections are carried through the brain in more anterior planes, they take the form of rings, the ventral half of each of which is formed by the pons, the dorsal half by the cerebellum (*vide* p. 47). Through the ring thus formed, and in organic connection with its lower half, extend most of the structures hitherto described as taking part

224   ROOTS OF AUDITORY NERVE.

in the formation of the after-brain. It goes without saying, that the corpora restiformia, which are columns of fibres destined for the cerebellum, are excluded from this ring. The pyramids intertwining

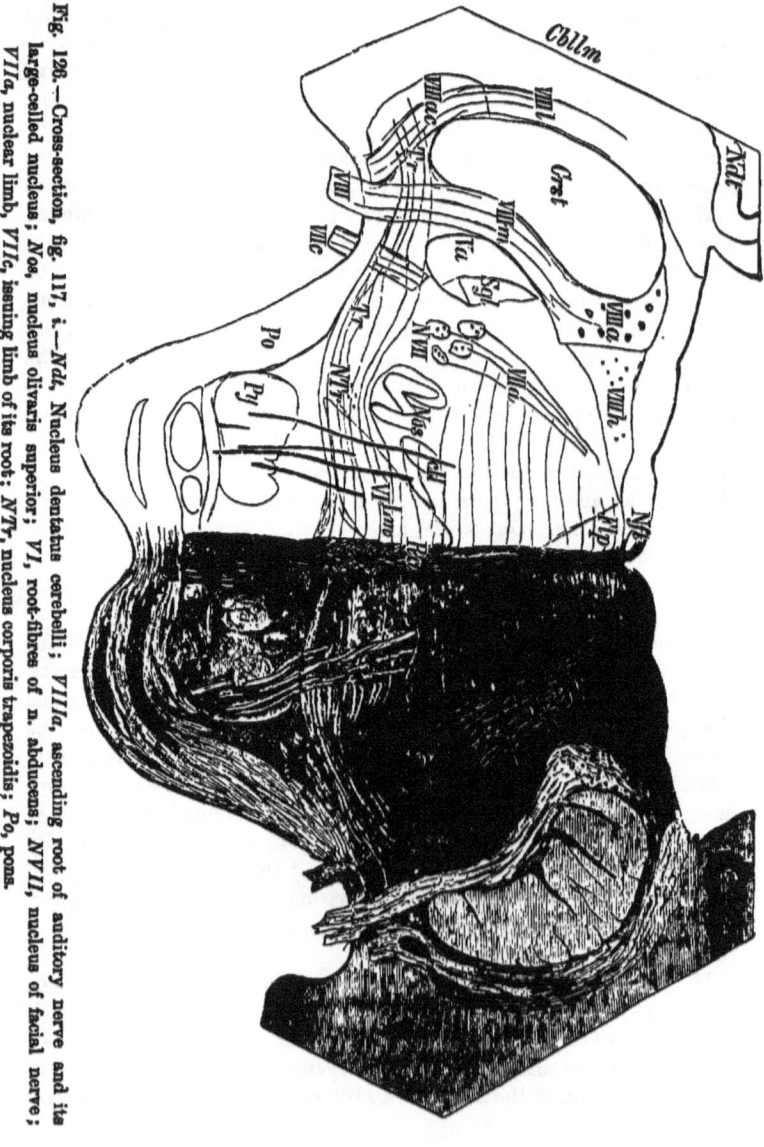

Fig. 126.—Cross-section, fig. 117, i.—*Ndt*, Nucleus dentatus cerebelli; *VIIIa*, ascending root of auditory nerve and its large-celled nucleus; *Nos*, nucleus olivaris superior; *VI*, root-fibres of n. abducens; *NVII*, nucleus of facial nerve; *VIIa*, nuclear limb, *VIIc*, issuing limb of its root; *NTr*, nucleus corporis trapezoidii; *Po*, pons.

with the fibres of the pons lose their distinctness, and take the opportunity, perhaps, of forming numerous connections with them. In examining the human brain it is better to cut off the cerebellum before hardening, leaving only the lingula in connection with the brachia pontis. In monkeys and small animals the cerebellum may be cut in the same sections as the pons (see fig. 17). Hence, we exclude the cerebellum at present from this description, and treat it subsequently by itself.

The most conspicuous difference between a section through this region (fig. 126) and sections through the after-brain is due to the appearance of the pons, *Po.*

The pons takes the form of great bundles of white fibres which start in the cerebellum, and running transversely across the middle line enclose amongst them irregular masses of grey substance, the nuclei pontis.

Every section through the pons is divided into two quite distinct portions—one ventral, the other dorsal. The latter contains the continuation upwards of the structures of the hind-brain with the exception of the pyramids; the ventral half contains, in addition to the proper formation of the pons, the continuation upwards of the pyramids, *Py.* The dorsal portion may well be called the tegmental field, since most of its longitudinal fibres appear later in the tegmentum of the crus cerebri.

In figs. 126, 127, and 129, an artificial boundary between the cerebellum and the pons is traced. The chief nucleus of the auditory nerve, *VIIIh*, already diminished in size, still lies beneath the floor of the ventricle; to its outer side the reticular formation of the ascending auditory root, *VIIIa*, has become thicker, and is distinguished, especially in many animals, by conspicuous large multipolar cells; this region is, therefore, known as the large-celled nucleus of the auditory nerve (Deiters' nucleus). The mesial auditory root, *VIIIm*, is seen to proceed from the region of the large-celled nucleus and the lateral and ventral angle of the chief nucleus, between the corpus restiforme, *Crst*, and the ascending root of the trigeminus, *Va*, taking its exit at the lateral part of the pons. The accessory nucleus, *VIIIac*, lies on the convexity of the corpus restiforme, and is traversed a little to the ventral side of this by the lateral root, *VIII l.* The transverse fibres which we have already described as proceeding for this group of cells form the largest part of the corpus trapezoides, *Tr.*

In the lateral part of the reticular substance the facial nucleus, *NVII*, becomes even more distinct. It takes the form of rounded groups of cells, from which separate bundles of fibres, *VIIa*, never

15

united into large tracts, wend their way obliquely in the direction of the dorsal surface and mid-line, towards the posterior longitudinal bundle as it seems. As they run at the same time somewhat forwards, it is only in later sections that we shall be sure that we have to do with the fibres of origin and the nucleus of the facial nerve. In the same section we see these same fibres traversing the pons obliquely near its margin, *VIIc*, close to the inner side of the ascending root of the fifth nerve, but this time in the form of a compact bundle. The two portions are united by a tract which undergoes various windings as subsequent sections (figs. 127 and 128) will reveal. The roots of the facial and trigeminal nerves towards their exit are distinguished by their course, the one on the mesial, the other on the lateral side of the ascending root of the fifth.

While in the most distal section through the region of the pons all the fibres of the pons surround the pyramids on their ventral side, we find in sections farther brainwards that scattered bundles of fibres and masses of grey substance, as well as the fillet, insinuate themselves between them. Farther forward still, scattered clumps of grey matter are found embedded in amongst the hitherto compact bundles of the pyramid; and, lastly, the farther forwards we make our sections, the more horizontal fibres do we find interlacing with the bundles of the pyramid, as well as lying to their dorsal side. The tracts which lie on the ventral side of the pyramid may be designated superficial bundles of the pons, those on its dorsal side, deep bundles, and those which traverse it, middle (or piercing) fibres.

In animals the pons is much less strongly developed than in Man, and consequently we find in the former, as a rule, that a considerable portion of the corpus trapezoides is left uncovered, and appears superficially on the ventral side of the medulla as a somewhat trapezoidal area, which occupies the whole space behind the pons and between the ventral margins of the cerebellum; [part of the corpus trapezoides runs over and part under the pyramids].

In the section shown in fig. 126 a number of fairly thick bundles of coarse fibres, *VI*, are to be remarked, which cross the tegment in a dorso-ventral direction, piercing also the fillet, the corpus trapezoides, and the pyramid. Neither their beginning nor their end is shown in this section. These are the fibres of the nervus abducens, the nucleus of origin of which, lying near the great brain, will be seen in fig. 127; while its exit from the medulla just behind the pons would be shown in a section taken between figs. 125 and 126 but not here delineated.

Between the facial nucleus and the roots of the sixth nerve is situate a somewhat ill-defined body of about the size of the facia nucleus, the superior olive, *Nos*. The superior olive descends almost into the corpus trapezoides, and presses its slender bundles close together. The appearance of this cup in which the superior olive lies

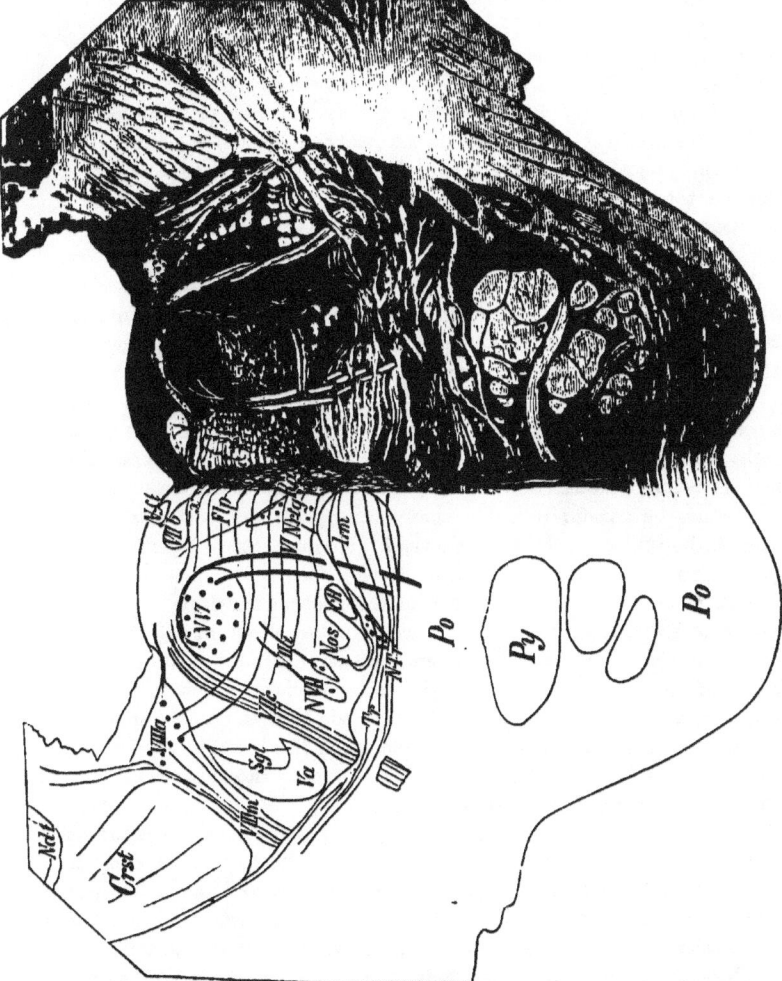

Fig. 127.—Transverse section, fig. 117, k.—*VIIb*, Ascending root of facial nerve; *x*, accession of crossed fibres to the nervus facialis; *NVI*, nucleus of nervus abducens; *Ntrg*, nucleus reticularis tegmenti.

helps us to recognise the body. The nerve-cells scattered about in the corpus trapezoides make up the nucleus corporis trapezoidis, *Ntr*. The fibres of the corpus trapezoides reach the raphe, in which, collected into slender bundles, they pierce the fillet.

Dorsal to the fibres of the pons lie a number of structures which we have already studied, but may with advantage recapitulate. In addition to the transverse trapezoidal fibres, we find in order from the middle line outwards :—(1) The raphe, (2) the fillet, (3) the roots of the nervus abducens, (4) the nucleus trapezoides, (5) the superior olive, (6) the nucleus nervi facialis, (7) the issuing fibres of the trigeminal nerve, (8) the ascending root of the trigeminal, (9) the mesial root of the auditory, (10) the restiform body, (11) and (12) the lateral root of the auditory nerve with the accessory nucleus of the same.

On the mesial border of the superior olive is found a small tract of fibres cut transversely, the central tegmental tract, cII (*Bechterew* and *Flechsig*). It is not, as a rule, sharply defined. Its fibres are supposed to take origin in the inferior olive.

In the next section (fig. 127) the fillet, which lies just dorsal to the fibres of the pons, is broader, its dorso-ventral diameter being diminished to a corresponding extent. It is traversed in the manner already described by the fine bundles of the corpus trapezoides.

In this section, as in all those behind it, fibræ arcuatæ are seen curving through all parts of the tegmental region, from the pons fibres right up to the floor of the fourth ventricle. They traverse the posterior longitudinal bundle, Flp, in their course towards the raphe. One must be careful to avoid confounding with the posterior longitudinal bundle a medullated nerve, $VIIb$, which for a time finds its place between it and the surface of the ventricle.

This nerve, the ascending limb of the root of the facial nerve, is easily distinguished from the posterior longitudinal bundle by the fact that it is not traversed by fibræ arcuatæ. It is better defined too. Most of the fibres arising from the facial nucleus, which is already much diminished in size, incline at first towards the raphe, applying themselves gradually to the nerve-root as it lies beneath the floor of the ventricle close to the middle line, while at the same time they assume a longitudinal direction.

In this section, too, we see for a greater distance the descending limb, $VIIc$, of the root of the facial nerve, laterally to its nucleus. So it comes about that the root of the facial nerve, $VIIa, b, c$, is three times met with on its course from its nucleus to the surface, without the connection between the three pieces being visible in any one section.

Close to the nucleus of the facial nerve in the bay formed by the fibres of the corpus trapezoides lies the superior olive, which takes the form here of a narrow riband more or less folded.

Near the olive the central tract of the tegmental region is usually but slightly marked, and next to it the bundles of the abducent nerve are conspicuous, as they form arches convex towards the raphe in their passage towards a grey mass, *NVI* (nucleus nervi abducentis), which lies near the middle line, not far below the floor of the ventricle. Owing to their oblique direction spinalwards, the fibres of the abducens are only seen in one portion of their course.

It should be mentioned that in this section the mesial auditory root, the fibres of which take origin in the large-celled nucleus, is still to be seen lying between the ascending root of the trigeminus and the corpus restiforme; the accessory nucleus has disappeared. The restiform body, as soon as it is set free from the bands of the so-called lateral root of the auditory nerve, begins to sweep off into the cerebellum. Scattered nerve-cells belonging to the nucleus reticularis tegmenti, *Nrtg*, are found far down in that portion of the substantia reticularis which lies near the raphe, between the fillet and the posterior longitudinal bundle.

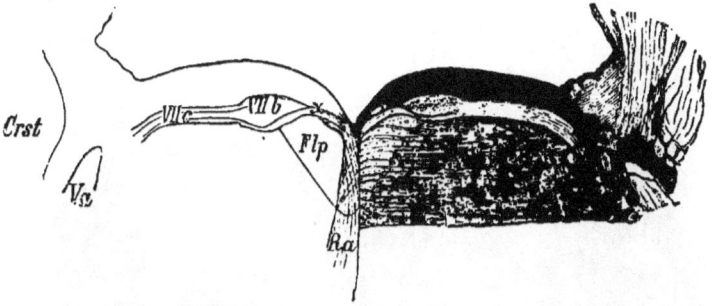

Fig. 128.—Transverse section corresponding to fig. 117, *l*; shows the bending over of the ascending limb of the facial nerve into its issuing limb.

In the next section (fig. 128) the nervus acusticus is almost wanting; numerous small masses of grey matter, composing the sensory nucleus of the trigeminal (which only reaches its full development in the next section) are seen enclosing the fibres of the transversely-divided nerve. In order that the relation to one another of the issuing and ascending limbs of the facial nerve may be well seen, only so much of the section as lies near the floor of the fourth ventricle is represented in the woodcut. The way in which the root changes its vertical for the horizontal direction is well seen, as is also the obvious accession at *x* of fibres from the opposite side.

Now we have reached the real region of origin of the trigeminus (fig. 129). The section shows us the posterior longitudinal bundle passing up to the situation which rightfully belongs to it, beneath the floor of the ventricle. The fillet spreads out farther sidewards until

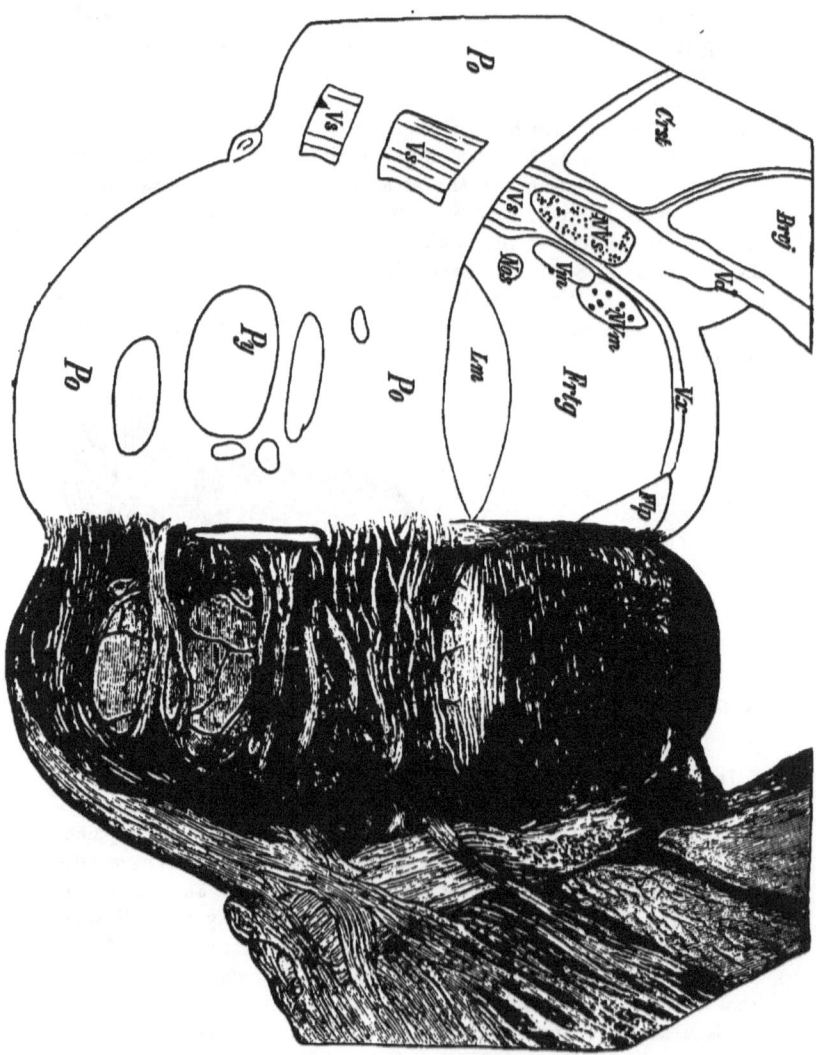

Fig. 129.—Transverse section, fig. 117, m.—*Brcj*, Brachium conjunctivum; *Vd*, descending root of trigeminus; *Vx*, crossed root of trigeminus; *NVs*, sensory trigeminal nucleus; *Vs*, sensory trigeminal root; *NVm*, motor trigeminal nucleus; *Vm*, motor trigeminal root; *Nos*, cerebral end of upper olive; *Frtg*, tegment.

it touches the cerebral end of the superior olive. Laterally to the olive lies the territory of the trigeminus. The clumps of grey substance already noticed farther spinewards in the substance of the ascending trigeminal root are obviously more numerous and larger. They make up the sensory nucleus of this nerve, NVs. Fibres from this spot are seen joining the ascending root; other fibre-bundles coming from the groups of cells also join the great sensory root which traverses the crus pontis (middle cerebellar peduncle) obliquely ventralwards and outwards, Vs. In this section and also in those farther forwards, the sensory root is cut obliquely.

On the mesial side of the sensory nucleus lies a compact roundish group of large nerve-cells, NVm, the motor nucleus of the fifth nerve. From both nuclei tracts of fibres can be followed which curve inwards towards the raphe, Vx; these belong to the crossed origin of the trigeminus. A number of coarse fibres, seen in the section as a conspicuously white tract, lie up against the ventral pole of the motor nucleus; this is the part of the motor root of the nerve which originates farther forwards.

The trigeminus receives a further accession of fibres which come from the neighbourhood of the lateral angle of the ventricle. They are only seen in sections farther forwards; the descending root of the trigeminus, Vd. At the lateral edge of the section the corpus restiforme is seen passing into the cerebellum; a great cross-cut field of fibres lies on its inner side, in the form of a curved club (the upper part of the club is cut away in fig. 129, but in fig. 130 the whole of it is seen). This is the brachium cerebelli ad cerebrum, $Brcj$ (brachium conjunctivum, superior cerebellar peduncle), which passes forwards from the cerebellum, sinking into the tegmental region as soon as the trigeminal nerve makes room for it.

As soon as the facial and abducent nerves have disappeared, the portion of the section which lies between the raphe and the trigeminus (formatio reticularis tegmenti, or tegmental region) begins to be traversed by arcuate fibres which are not well defined. In subsequent sections it rapidly diminishes in area.

Although in the following sections (figs. 130 to 132) the total cross-section of the pons is still of considerable size, owing to the direction in which it is cut, the entrance of the brachium pontis into the cerebellum is no longer seen, consequently no artificial section of the pons on either side appears in the drawings.

The mesial fillet, Lm, is now (fig. 130) displaced towards the margin of the section. The brachium conjunctivum, $Brcj$, the ventral point of which is distinctly curved, has descended somewhat ventrally. The fourth ventricle is fast narrowing into the aquæductus Sylvii, Aq

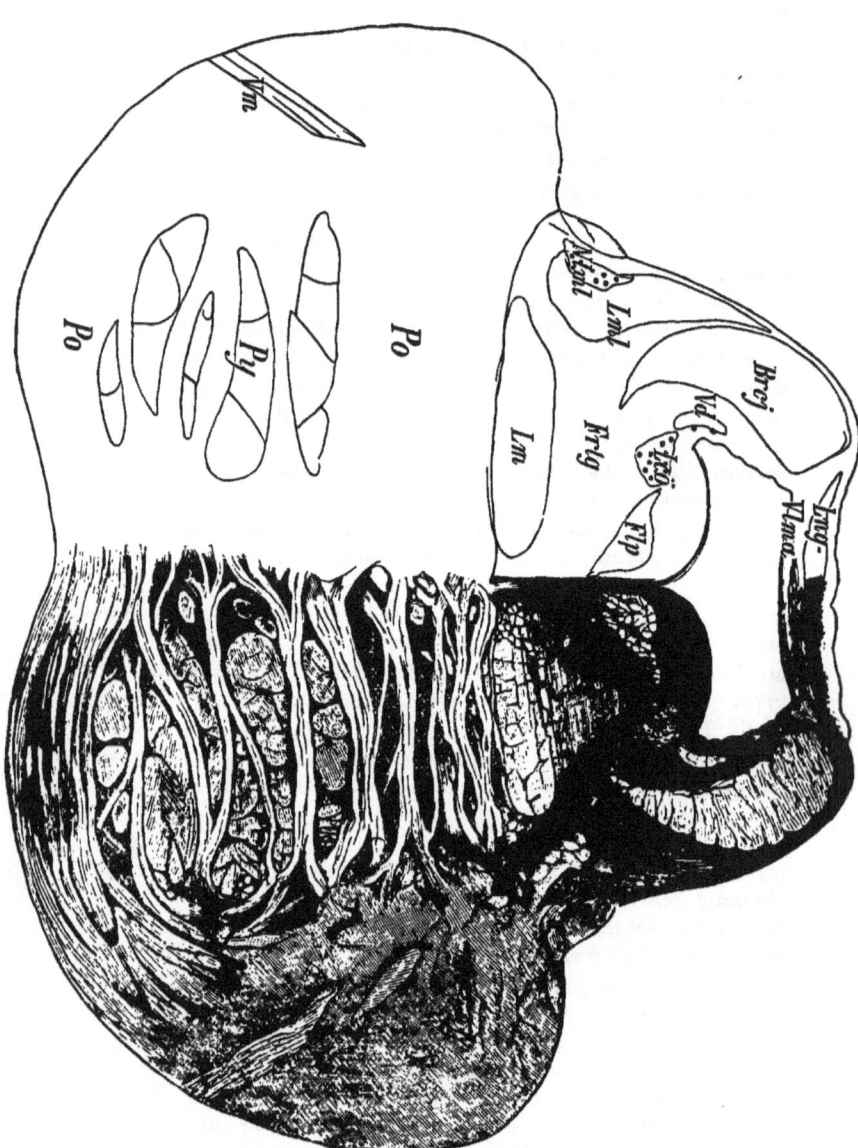

Fig. 130.—Transverse section, fig. 117, n.—*Vlma*, Velum medullare anterius; *Lng*, lingula; *Lcδ*, locus cæruleus [substantia ferruginea]; *Lml*, lateral fillet; *Nlm*, nucleus of lateral fillet; *Lm*, mesial fillet; *Vd*, descending root of trigeminus; *Vm*, motor root of trigeminus.

(figs. 131 to 134). For the first time the roof of the ventricle is represented by the velum medullare anterius, *Vlma*, carrying the lingula, *Lng*.

A somewhat triangular area is left between the pons and brachium conjunctivum. For the most part it is occupied with medullated fibres, *Lml*, which course obliquely dorsalwards, a small portion of them entering the velum medullare anterius, while the larger number can be followed forwards to the corpora quadrigemina. This is the tract of fibres which really constitutes the fillet, as seen from the outside, and it was to this that the name was originally applied. Although it may be looked upon as fillet *par excellence*, it is well to call it the lateral fillet, to distinguish it from the one which we have followed upwards from the spinal cord. To the latter we should now apply the name mesial fillet, *Lm*. Later on, it will be necessary to give some account of various confusing synonyms applied to its several parts.

Certain little groups of nerve-cells situate in this triangular cross-section of the fillet, and probably serving to give origin to some of its fibres (nuclei lemnisci lateralis, *Nlml*), deserve attention.

On the lateral side of the posterior longitudinal bundle lies a group of nerve-cells, which owing to its dark colour, dark enough to render it visible to the naked eye, is termed substantia ferruginea, *seu* locus cæruleus, *Lcö*. [Perhaps it is better to restrict the term locus cæruleus to the area in the floor of the ventricle, which receives a grey colour from the black substantia ferruginea which lies beneath it under a stratum of white fibres. The cells of this group are easily recognised as belonging to the visceral column or " vesicular column " of Clarke, owing to their well-filled outlines, round or oval form, and the small number of their processes, as well as by the closeness with which they are packed together. Wherever this intermittent column is present, whether in the sacral or dorsal regions of the cord, the nucleus of the vagus, or the substantia ferruginea, it is impossible to mistake it for groups of the motor cells of skeletal muscles.] Dorsally and laterally to the locus cæruleus, always occupying the vicinity of the angle of the ventricle, a narrow tract of fibres, somewhat prolonged in the dorso-ventral direction, is cut across, *Vd*, the descending root of the trigeminal nerve.

The fibræ arcuatæ which traverse the tegmental region become sparser, they no longer pierce the mesial fillet. Owing to the increasing depth of the median fissure in the floor of the fourth ventricle, these arched fibres are constantly being driven farther ventralwards. The fibres of the motor root of the fifth, *Vm*, are still seen just before their exit from the lateral border of the pons.

The next sections belong to the mid-brain, although a portion of the

CEREBELLUM OF MONKEY.

pons is still appended to them, for they exhibit nerves peculiar to this region.

We have not yet, however, described the **cerebellum**, which belongs to the hind-brain.

On account of the large size of this organ in Man it is convenient to make sections of the cerebellum of the monkey instead; if the human cerebellum is examined, it is well before preparing it for cutting frontally, to divide it by two sagittal sections, one corresponding to the lateral edge of the pons, and another 1 to $1\frac{1}{2}$ cm. away from it, on the other side. In this way the central nuclei are completely shown on the one side and a sufficient amount of the other side is left to exhibit its

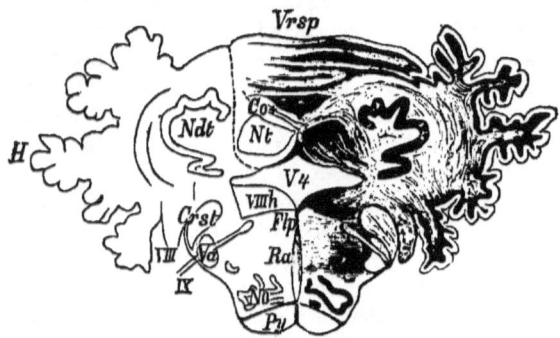

Fig. 131.—Frontal section through the cerebellum and medulla oblongata of a monkey. *Magn.* 2.—*H*, Cerebellar hemisphere; *Vrsp*, vermis superior; *Ndt*, nucleus dentatus; *Nt*, nucleus tecti; *Co+*, commissure; *V4*, fourth ventricle; *Crst*, corpus restiforme; *Py*, pyramid; *Flp*, fasciculus longitudinalis posterior; *Ra*, raphe; *No*, nucleus olivaris; *VIII*, nervus acusticus; *VIIIh*, chief auditory nucleus; *IX*, nervus glossopharyngeus; *Va*, ascending root of trigeminus.

relation to them. In this place we shall restrict ourselves to the consideration of a frontal section through about the centre of the cerebellum of the monkey, dividing it close behind the corpus trapezoides (fig. 131). The superior vermis, *Vrsp*, is seen in the middle line, a number of its convolutions being cut through one above the other. The inferior vermis does not reach so far forwards; the roof of the fourth ventricle, V_4, is no longer covered with cortex. On either side lie the hemispheres, *H*, divided into lobes, everywhere covered with cortex.

Amongst the central masses of grey matter are seen—(1) in the vermis the considerable, somewhat wedge-shaped "nucleus of the roof," *Nt*, its angle almost reaching the middle line; (2) in the hemispheres the corpora dentata cerebelli, *Ndt* (nuclei dentati), with their hila

ORIGIN OF TROCHLEARIS.

directed ventrally and mesially. Nucleus globosus and nucleus emboliformis of Man are not present.

Among the tracts of white fibres certain bundles which lie dorsally to the nuclei of the roof and cross one another in the middle line, taking part in the formation of the great commissure, $Co+$, are especially conspicuous. Certain of these fibres, however, dip down in

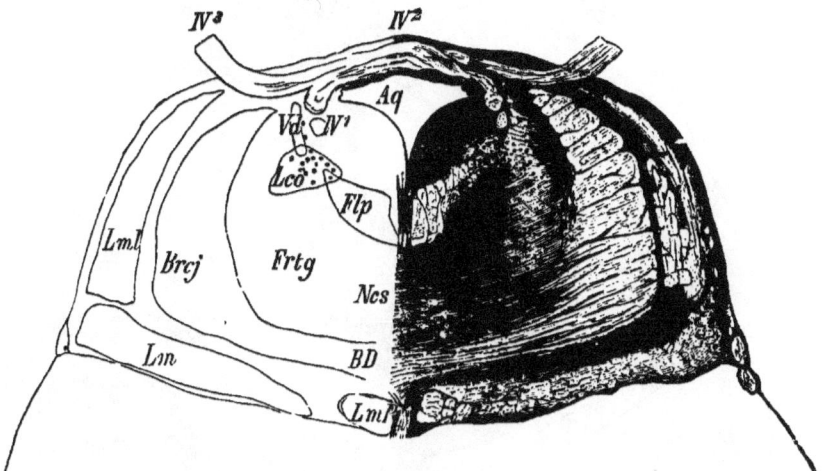

Fig. 132.—Transverse section, fig. 117, o.—IV^1, Descending root of the trochlear nerve; IV^2, trochlear decussation; IV^3, issuing root of n. trochlearis; Aq, aquæductus Sylvii; Ncs, nucleus centralis superior; LmP, bundle from fillet to crusta; BD, commencing decussation of brachia conjunctiva.

the middle line between the roof nuclei and form a kind of raphe. They probably run in a sagittal direction (brain- or spinewards) after crossing. Strongly marked concentric arches of medullated fibres are very distinctly seen on the outer sides of the corpora dentata.

If we commence our description of the mid-brain with fig. 132, it happens that the section shows the origin of the trochlear nerve, IV. Its fibres are seen distinctly crossing their fellows from the opposite side in the roof of the aquæductus Sylvii. Certain bundles of fibres lying on the inner side of the descending root of the fifth are cut transversely or obliquely, IV^1. These are the root-fibres of the trochlear nerve which have their nuclei of origin farther brainwards. On the dorsal side of the posterior longitudinal bundle, close to the raphe, lies a striking darkly coloured rounded group of the smallest cells, which also appears to give origin to fibres of the trochlear nerve.

NUCLEUS OF TROCHLEARIS.

It is the posterior nucleus of the trochlear nerve or Westphal's nucleus (not lettered in the figure).

The recognised (anterior) nucleus of the trochlear nerve lies immediately to the cerebral side of this group of cells. [If, as in all probability may safely be done, we accept pigmentation and

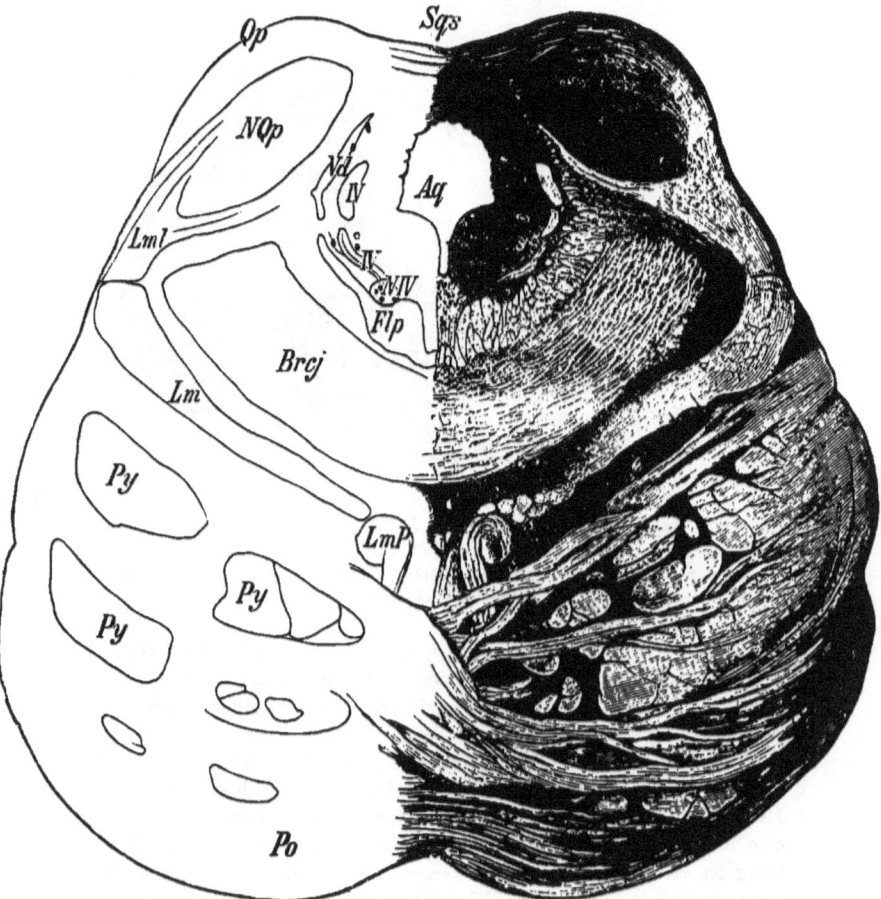

Fig. 133.—Transverse section, fig. 117, p.—*Qp*, Posterior corpora quadrigemina; *NQp*, nucleus of *ditto*; *NIV*, trochlear nucleus; *Sqs*, sulcus corporum quadrigeminorum longitudinalis.

shrinking in size as an evidence of atrophy, and look upon groups of strongly pigmented cells as the vestiges of nuclei no longer functional, this group may possibly represent the nucleus of an additional portion of the trochlear nerve, which there are good grounds for believing existed in early vertebrates. Both the third and fourth nerves take origin from spots in the mid-brain which contain deeply-pigmented cells, and both contain, mixed up with their nerve-fibres, masses of tissue which occur in no other nerves, but have all the characters of atrophied nerve-cells and cell-sheaths (*Thomson*), and may well be, as *Gaskell* believes, the vestiges of root-ganglia. Apart from these indications within the nerves and their nuclei, there are many reasons for thinking that the third and fourth nerves, which in all, or almost all, vertebrates are limited to fibres supplying certain muscles of the eye, had, before the consolidation of the vertebrate head, a wider range, and included sensory as well as motor elements.]

Laterally and ventrally to the aqueduct the descending root of the fifth, *Vd*, the locus cæruleus, *Lcö*, and the posterior longitudinal bundle, *Flp*, retain the same relative position as heretofore. The lateral fillet, *Lml*, lies on the outer side of the brachium conjunctivum, *Brcj*, the ventral limb of the cross-section of which joins the main body almost at a right angle. The ventral division of its fibres already reaches the middle line forming the commencement of the decussation of the brachia conjunctiva, *BD*. The tendency of the brachia towards the middle line is seen in the following sections, while the mesial fillet, *Lm*, inclines in the opposite direction, away from the raphe. Only the most mesial of its fibres remain behind in a rounded bundle lying dorsally to the pons, *LmP*. The nerve-cells which are seen near the raphe, between the posterior longitudinal bundles and the decussation of the brachia conjunctiva, belong to the nucleus centralis superior, *Ncs*. The two little bundles of fibres, cut across on the periphery between the fillet and the pons, constitute the ponticulus.

The fibres of the trochlear nerve take origin in large part, as the following sections show, from a rounded grey mass, the (anterior) nucleus of the trochlear nerve, *NIV*, which in part is embedded in a concavity on the dorsal side of the posterior longitudinal bundle. Now for the first time the corpora quadrigemina are seen in section, their united portions bridging over in the middle line, the aquæductus Sylvii, *Aq*, which is prolonged into a deep channel on the ventral side. The centre of each lobe of the corpora quadrigemina is occupied by an ill-defined grey mass, the nucleus of the posterior corpora quadrigemina, *NQp*. On the outer side of this is recognised the bundle of the lateral fillet, *Lml*, part being prolonged to the middle line and across it. A smaller part of the lateral fillet passes beneath the nucleus of the corpora quadrigemina, so that the grey matter is almost completely encapsuled in white. The mesial fillet continues its course laterally and dorsalwards. The brachia conjunctiva, *Brcj*, enter to a greater extent into their decussation, and moving downwards come to occupy the greater part of the formatio reticularis of the tegment. The pons fibres, *Po*, have split the pyramid, *Py*, into a great number of separate bundles; but, nevertheless, we find the pyramids in the sections just in front of the pons condensed into an immense close-set field of fibres, presenting a cross-section convex on its ventral side, *Pp* (pes pedunculi [crusta]).

Fig. 134 represents a section carried through the hinder part of the anterior corpora quadrigemina, *Qa*. A superficial indentation is seen in the middle of its dorso-lateral border, *Sqt*, this is the fissure (sulcus interbrachialis) which bounds the brachium corporis quadrigemini posterioris on its dorsal side, and it shows us, therefore, that we have passed the posterior and entered the region of the anterior tubercles of the corpora quadrigemina. The nucleus of the anterior tubercle is already visible although indistinct, *NQa*.

Between the pes pedunculi and the now no longer sharply-defined fillet, an ever enlarging grey mass insinuates itself, *SnS*; it is distinguished by possessing strongly pigmented cells, and assumes, therefore, to the naked eye a characteristic dark grey colour (substantia nigra Soemmeringi). Many bundles of fibres are seen to stream from the pes pedunculi into the substantia nigra. They cannot be followed farther.

On either side of the middle line, the brachia conjunctiva, beyond their main decussation, begin to form an oval field placed with its long axis vertically (the white nucleus of the tegment). These tracts of crossed fibres are reinforced by those still crossing.

The rounded bundles, *LmP*, which we had noticed (figs. 132, 133) as separating from the mesial fillet, place themselves, as soon as the fibres of the pons have disappeared, on the mesial side of the pes pedunculi, on the margin of which they soon spread out sideways. Hence they constitute the tracts from the fillet to the pes.

The space between the posterior longitudinal bundles and the aqueduct of Sylvius has considerably increased in its dorso-ventral diameter. It is occupied by a region rich in cells, the ventral part of which belongs, as later sections will show, to the nervus oculomotorius, *NIII*.

The large brown cells of the substantia ferruginea have completely disappeared, and the descending root of the fifth, *Vd*, is only picked out with difficulty under a low power; it can still be recognised,

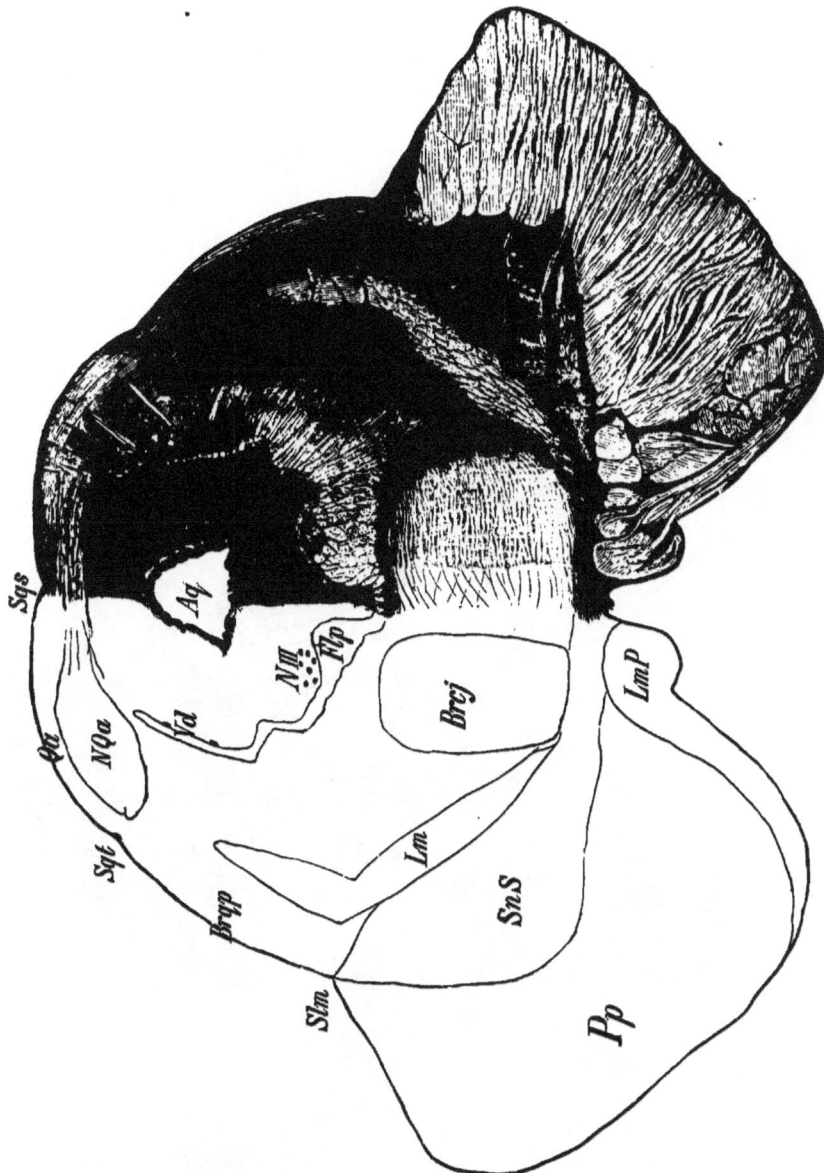

Fig. 134.—Transverse section, fig. 117, q.—*Qa*, Anterior corpora quadrigemina; *NQa*, nucleus of *ditto*; *Sqt*, sulcus corpor. quadrigem. transversus; *Brqp*, brachium corporis quadrigemina posterioris; *Slm*, sulcus longitudinalis mesencephali; *NIII*, nucleus of nervus oculomotorius; *SnS*, substantia nigra Soemmeringi; *Pp*, pes pedunculi.

240   FRONT OF MID-BRAIN.

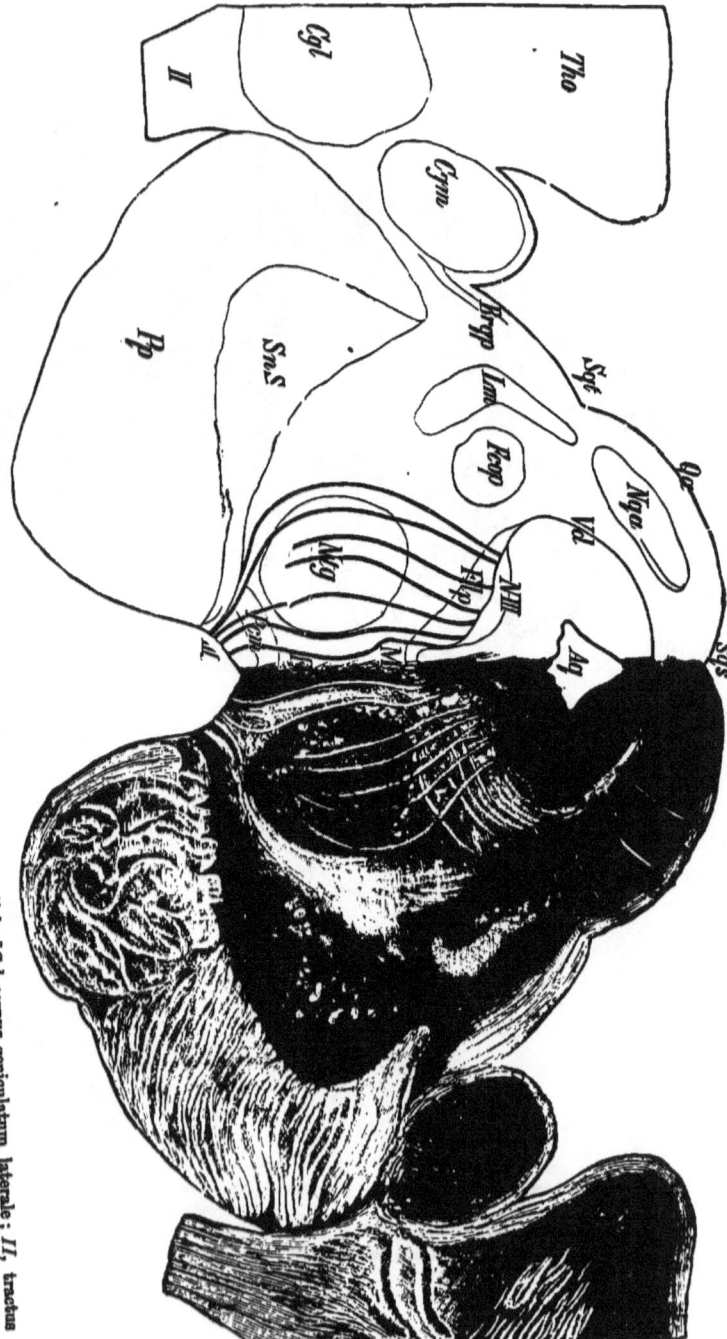

Fig. 135.—Transverse section, fig. 117, r.—*Tho*, Thalamus opticus; *Cgm*, corpus geniculatum mediale; *Cgl*, corpus geniculatum laterale; *II*, tractus opticus; *Pcp*, posterior commissure; *Ng*, nucleus ruber tegmenti; *III*, root fibres of n. oculomotorius; *Pcm*, pedunculus corporis mammillaris; *M*, Meynert's fontanal decussation of the tegment; *F*, Forel's ventral decussation of the tegment.

however, if one relies upon the scanty but very characteristic large nerve-cells which it contains.

The fibres which cross dorsally to the aqueduct are very conspicuous, they can be followed farther to each side in a well-formed arch, sweeping down towards the descending root of the fifth, as appears more clearly in the following sections.

A section carried through the summit of the anterior tubercles of the corpora quadrigemina shows certain well-marked changes (fig. 135). The sulcus corporum quadrigeminorum sagittalis, Sqs, is deep and well defined; while the sulcus interbrachialis which separates the anterior tubercle from the brachium of the posterior tubercle seems to have inclined farther ventralwards, Sqt. Many medullated fibres cross the middle line dorsally to the aqueduct of Sylvius; some come from the fillet, others belong to the central connections of the descending root of the trigeminus, Vd, others to the fibræ arcuatæ of the tegment which will soon be described.

The crossing of the brachia conjunctiva is complete, and instead of the brachia themselves a round area formed of reticular substance occupies their place on the dorsal side of the substantia nigra and not far from the middle line. This is the red nucleus, Ntg (superior olive of *Luys*, nucleus tegmenti).

On the dorsal side of the posterior longitudinal bundle, large nerve-cells are seen, $NIII$ (nucleus nervi oculomotorii). They give origin to bundles of fibres which sweep ventrally, first piercing the posterior longitudinal bundle; and pass some on the inner side, some on the outer side, and others through the red nucleus to reach the surface at the furrow between the two crura cerebri, III (root fibres of the oculomotor nerve). Ventrally to the red nuclei the fibres of the oculomotor nerve pass through a region, Pcm, occupied by fibres running from the corpus mammillare to the tegment (pedunculus corporis mammillaris). Numerous small nerve-cells lie in the space between the nucleus of the oculomotor nerve already described and the aqueduct of Sylvius; probably they, too, are in relation with this nerve. On each side these cells can be distinctly classified in two groups, separated by fibres. The mesial group lies dorso-ventrally, the lateral group extends transversely (*Westphal's* mesial and lateral oculomotor groups).

The distinctly diminished mesial fillet, Lm, appears as an inconspicuous, semilunar area, stretching towards the corpora quadrigemina. It takes part in the crossing in the roof of the aqueduct. A hardly recognisable clear patch, $Fcop$, on the mesial side of the fillet, contains fibres streaming from the posterior commissure into the tegment (*Wernicke*). Fine fibres are seen crossing in the raphe from the level of the posterior longitudinal bundles to the basis cerebri. The dorsal portion of these crossing fibres must be distinguished from the ventral (*Forel*). The fibres which cross in the dorsal segment of the raphe come from the roof of the aqueduct; sweeping in fine curves around the outer side of the descending root of the fifth, they curl in beneath the posterior longitudinal bundle, and so traverse the tegment towards the middle line. *Meynert*, thinking that these fibres took origin in the cells of the descending trigeminal root, called them the tracts of the fifth (Quintusstränge). *Forel* substituted the name "fountain-like or Meynert's decussation," *M*. The fibres which cross in the ventral portion of the raphe form *Forel's* ventral tegmental decussation, *F*.

The most remarkable feature of this section is the fact that a large number of new structures, which are connected with the optic nerve, have joined it on its lateral borders. A great white column cut obliquely lies up against the crusta, the tractus opticus, *II*. Dorsally it passes into a peculiar mass of alternating grey and white matter, the ganglion or corpus geniculatum laterale, *Cgl*. A smaller part of the optic fibres can be followed on the surface of the crusta farther dorsally to another grey body of oval form and almost the same size as the nucleus ruber tegmenti, the ganglion, *seu* corpus geniculatum medialè, *Cgm*. The lateral geniculate body is situate in the sulcus lateralis mesencephali; it is enveloped all over in bundles of fibres, and gives some bundles to the corpus quadrigeminum posterius. Lastly, the section has already cut into the thalamus opticus, *Tho*, which appears as a great grey mass lying on the dorsal and lateral sides of the structures described above.

Still another section must be made through the anterior border of the corpus quadrigeminum anterius, so that it cuts the posterior commissure, *Cop* (fig. 136). By this time the optic thalamus occupies an extended area. Each of the anterior corpora quadrigemina is united with the thalamus by a conspicuous tract of white fibres, the brachium corporis quadrigemini anterioris, *Brqa*, which lies in the furrow between them. The considerable tracts of the posterior commissure cross above the aqueduct (already opening out into the third ventricle); their most ventral fibres extend downwards on either side the aqueduct in the direction of the posterior longitudinal bundles, *Flp*, which are already indistinct. The dorsal fibres of the commissure separated from the ventral fibres by the recessus subpinealis, *Rsp*, can be followed farther sidewards into the thalamus. The most anterior portion of the nucleus oculo-motorius, *NIII*, is still to be seen near the raphe.

Fibres stream outwards from the lateral edges of the red nuclei, *Ntg*. Many bundles of fibres belonging to the thalamus take a similar course

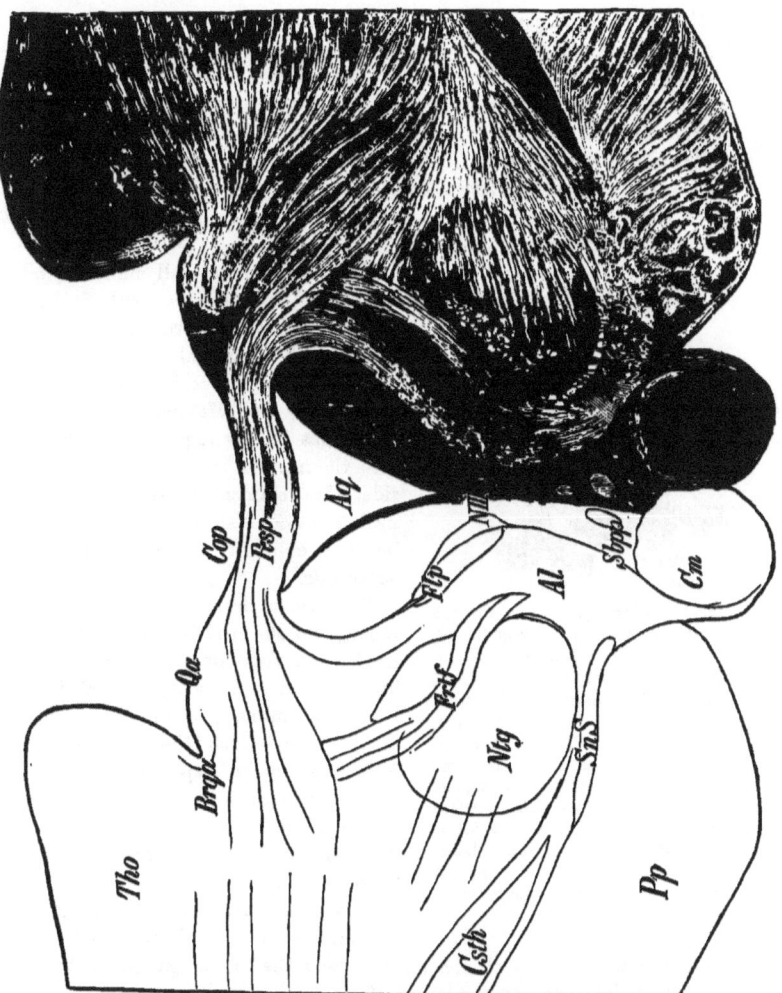

Fig. 136.—Transverse section, fig. 117, s.—*Cop*, Commissura posterior; *Rsp*, recessus subpinealis; *Brqa*, brachium corporis quadrigemini anterioris; *Qa*, front of anterior corpus quadrigeminum; *Aq*, aquæductus Sylvii where it opens into third ventricle; *Frtf*, fasciculus retroflexus; *Csth*, corpus subthalamicum; *Flp*, posterior longitudinal bundle; *Al*, ansa lenticularis; *Sbpp*, substantia perforata posterior; *Cm*, corpus mammillare; *SnS*, anterior end of substantia nigra Soemmeringi; *Ntg*, red nucleus and fibres streaming from it; *Pp*, pes pedunculi cerebri.

in the most lateral part of the section. The substantia nigra Soemmeringi, *SnS*, has disappeared with the exception of a little mesial portion. Its place is taken by a lenticular body, the corpus subthalamicum, *Csth*, which, as we shall see later on, ought to be apportioned to the 'tweenbrain. It is surrounded by a white capsule. The two corpora mammillaria, *Cm*, are squeezed in between the crura cerebri beneath the substantia perforata posterior, *Sbpp*.

On the mesial side of the red nucleus lies a region, *Al*, rich in fibres belonging to the ansa lenticularis.

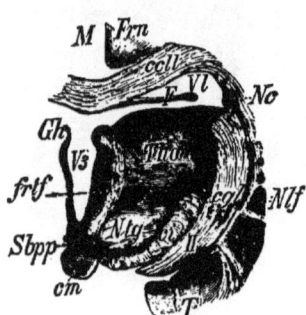

Fig. 137.—Frontal section through brain of monkey. *Magn* 2.—*M*, Incisura pallii; *Frn*, gyrus fornicatus; *ccll*, corpus callosum; *F*, fornix; *Vl*, ventriculus lateralis; *Nc*, nucleus caudatus; *Gh*, ganglion habenulæ; *V3*, ventriculus tertius; *Tho*, thalamus opticus; *Sbpp*, substantia perforata posterior; *cm*, corpus mammillare; *Ntg*, nucleus ruber tegmenti; *ci*, capsula interna; *cgl*, corpus geniculatum laterale; *Nlf*, nucleus lenticularis; *T*, temporal lobe; *frtf*, fasciculus retroflexus.

Dorsally it is not well marked off from the field of cross-cut fibres which lies on the ventral side of the posterior longitudinal bundle. A tract of coarse fibres, the fasciculus retroflexus, *frtf* (Meynert's bundle), discharges in this region. It enters into the nucleus ruber on its inner side. Its commencement and termination are not visible in this section, but it is evidently coming from the outer side.

The human brain is inconveniently large for the study of the **'tween-brain and fore-brain.** They are best studied in cross-sections of the brains of animals, especially small monkeys, in which the departures from the human type are not of great importance.

The following account is derived from the study of the cerebrum of Cercopithecus.

If sections are to be made of the human brain it is well before hardening to cut away all the parts beyond the "great ganglia at the base" and the cortex of the island of Reil, otherwise the sections are unmanageable. It is desirable, too, to change a little the plane of the sections, making them truly frontal—*i.e.*, perpendicular to the longitudinal axis of the great brain (*cf.* p. 209).

As will be shown directly, the anatomy of certain parts of the 'tweenbrain presents peculiar difficulties, which are increased by the fact that to certain tracts of fibres no distinct physiological purpose can yet be assigned, and so we are obliged to be contented with dry and often doubtful anatomical data. The very existence of the tracts in question is sometimes rather taken for granted than demonstrated. In fact, the gaps in our information so often felt in the study of many parts of the brain are experienced acutely in the case of the 'tween-brain.

A section carried in front of the posterior commissure (fig. 137) shows us structures with most of which we are already acquainted.

The aqueduct of Sylvius is fully enlarged into the third ventricle, $V3$. The optic thalamus shows us its two free surfaces; the mesial looking into the third ventricle, the upper one belonging to both the third and lateral ventricles. The edge between the two surfaces is marked by a little swelling, the ganglion habenulæ, Gh, from which the fasciculus retroflexus, $frtf$, extends downwards to the mesial side of the red nucleus, Ntg, at the basis cerebri. In Man the fasciculus retroflexus causes an indenting of the red nucleus. The two round bodies, the corpora mammillaria (*seu* albicantia), Cm, which in most animals are fused into a single rounded body [the corpus mammillare], lie beneath the substantia perforata posterior, $Sbpp$.

The external boundary of the thalamus is partly formed by the crus cerebri, which on entering the brain-substance is transformed into the "internal capsule," ci. The optic tract, II, on its road to the lateral corpus geniculatum, cgl, embraces the crus. The posterior end of the lenticular nucleus, Nlf, is placed almost directly upon the tractus opticus and the corpus geniculatum laterale. In the upper part of the preparation we must mention the corpus callosum, $ccll$, the band-like fornix, F, as well as the tail of the nucleus caudatus, Nc.

Fig. 138 represents a section carried through the principal mass of the optic thalamus and the optic chiasm, Ch.

The thalamus, again, exhibits its two surfaces, but the edge between them is no longer formed by the ganglion habenulæ, but only by the insignificant tænia ventriculi tertii, $Tv3$.

The two thalami have fused together over a great part of their mesial surfaces. This united portion corresponds to the grey commissure, Cm, of Man. The thalamus is divided by the lamina medullaris medialis, Lmm, into a smaller medial, Nm, and a larger lateral nucleus, Nl. Numerous white fibres enter the lateral nucleus on its outer side, giving to it for a certain thickness a peculiar reticulate appearance; hence it is known as the stratum reticulatum, str. On the lateral border of the thalamus these fibres are collected into a thin boundary layer, the lamina medullaris lateralis, Lml. On the outer side of this comes the internal capsule, ci.

Not everything, however, which lies between the internal capsule and the third ventricle belongs to the optic thalamus; the basal portion of the region, which cannot, it is true, be sharply defined from the thalamus, is known as the regio-subthalamica (stratum intermedium of *Wernicke*). It must be remarked that only in this section, on account of the different plane in which the brain is cut, does this region, which has already (fig. 136) been examined in its posterior part, appear in its full development.

Our attention is arrested by a somewhat lenticular body which lies above the internal capsule, the corpus subthalamicum, *Csth* (nucleus

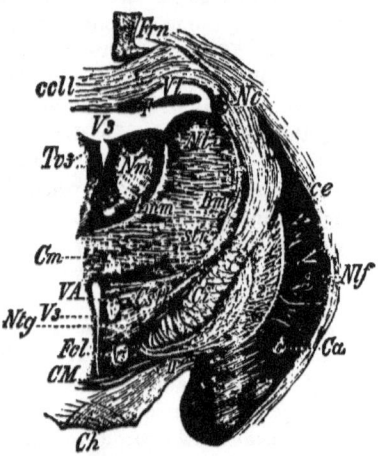

Fig. 138.—Frontal section through brain of monkey, passing through the middle of the optic thalamus. *Magn.* 2.—*Tv3*, Tænia ventriculi tertii; *Cm*, commissura mollis; *Nm*, nucleus medialis thalami optici; *Nl*, nucleus lateralis thalami optici; *Lmm*, lamina medullaris medialis; *Lml*, lamina medullaris lateralis; *Csth*, corpus subthalamicum; *str*, stratum reticulatum thalami optici; *Ci*, capsula interna; *ce*, capsula externa; *Nlf*, nucleus lenticularis, *1, 2, 3*, its three segments; *Ca*, anterior commissure; *VA*, Vicq d'Azyr's bundle; *Fcl*, anterior crus of fornix; *Ntg*, anterior end of nucleus ruber tegmenti; *Ch*, chiasma nervorum opticorum; *II*, tractus opticus; *CM*, Meynert's commissure. Other lettering as in fig. 137.

amygdaliformis, nucleus of Luys, nucleus of Forel, bandelette accessoire de l'olive supérieure). In Man it is much more sharply defined than in many animals. The corpus subthalamicum is, with the single exception of its mesial angle, shut in by a thin but distinct capsule of medullated fibres, capsula corporis subthalamici. The ventral lamella of its capsule separates it from the crus, or internal capsule, as the case may be; the dorsal lamella separates it from the region which is in connection dorso-laterally with the fibres of the lateral nucleus of the thalamus, and gains on the ventral and mesial side a greater importance by blending with the region, *Ntg*, in which we must look for the fibres

ANSA PEDUNCULARIS.

which stream out of the dorsal portion of the red nucleus. This area is continued almost to the wall of the third ventricle, but it must be pointed out that our views with regard to its definition are by no means concise. *Forel* distinguishes the ventral portion of this area which lies next to the corpus subthalamicum (zona incerta) from the upper more abundantly medullated one (*Forel's* field H).

Further, fibres are seen streaming from the lateral and ventral part of this region beneath the inner segment of the lenticular nucleus (for this nucleus is already divided into its three segments, *Nlf*, 1, 2, 3), and so curving round into the internal capsule, the ansa peduncularis.

It consists of a variety of fibre-tracts, but in this place it is composed of bundles to which the name ansa lentiformis is especially applied.

Near to the wall of the ventricle are seen two bundles cut across; one, the anterior limb of the fornix, *Fcl*, coursing backwards towards the corpus mammillare; the other, which lies farther dorsalwards, is the bundle of Vicq d'Azyr, *VA*, on its road from the corpus mammillare to the anterior nucleus of the thalamus. Beneath the lateral segment of the nucleus lenticularis lies another bundle of transversely cut white fibres; this is the anterior commissure, *Ca*, directed backwards in this part of its course.

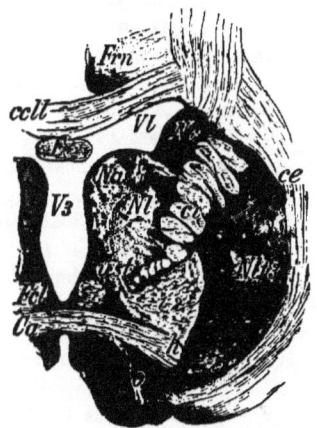

Fig. 139.—Frontal section through monkey's brain at level of front of thalamus. *Magn.* 2.—*Na*, Anterior nucleus of thalamus; *ust*, inferior peduncle of thalamus; *h*, hemispheral portion of anterior commissure; *o*, olfactory portion of *ditto*. Other lettering as in figs. 137 and 138.

On the base of the brain lies the optic chiasm, above which, in the narrow interval between it and the third ventricle, certain thick fibres cross the middle line, Meynert's commissure, *CM*.

The next drawing (fig. 139) represents a section which passes through the anterior commissure, *Ca*. The commissure is seen dividing into two parts, the principal one, derived from the hemisphere, passes laterally beneath the globus pallidus, and then, when it reaches the putamen (fig. 138), inclines backwards, and so passes to the occipital lobe, and perhaps to the temporal lobe also. The smaller olfactory portion, *o*, of the anterior commissure turns basally and slightly forwards, to end in the tractus olfactorius and the neighbouring portions of the cortex. [It is not improbable that the two parts of the anterior commissure here described belong to the same system, the descending anterior portion *coming from* the olfactory tract and crossing over into the larger posterior portion of the opposite side.] The thalamus opticus now decreases as fast as the nucleus caudatus grows in cross-section. We still recognise in the optic thalamus the lateral nucleus with its stratum reticulatum. The nucleus anterior, Na, comes to an end at the upper mesial angle, and fibres from the base of the brain stream into the no longer well-defined nucleus medialis. These fibres constitute the inferior peduncle of the thalamus, *ust* (inferior and inner peduncle of *Meynert* and *Wernicke*). This peduncle is also a constituent of the ansa peduncularis, and like the ansa lentiformis sweeps downwards and outwards beneath the anterior end of the inner capsule. Possibly its fibres take origin in the two inner segments of the nucleus lenticularis. The other fibres derived from these two segments of the nucleus lenticularis, always more numerous than those just described, may have quite a different destination.

The bridges of grey matter connecting the outer segment of the nucleus lenticularis with the nucleus caudatus must be especially pointed out. The anterior crus of the fornix, *Fcl*, lies just above the anterior commissure, almost exposed on its ventricular side, although still covered with a thin layer of grey matter.

In more anterior segments the head of the nucleus caudatus will be found to have completely displaced the optic thalamus. The third segment of the nucleus lenticularis alone keeps it company, the two being connected by numerous wide bridges. From the lower surface of the corpus callosum the septum pellucidum extends downwards on either side the middle line.

If sections are made still nearer to the frontal pole, first the nucleus lenticularis and then the nucleus caudatus disappears. The bending over of the corpus callosum (its genu) is next encountered; and still farther forwards nothing is left but the sections of the two frontal lobes, now completely separated from one another.

SECTION VI.—COURSE OF FIBRES.

A. TRACTS IN THE SPINAL CORD.

Hitherto our attention has been directed to the larger topographical changes in constitution which a continuous series of cross-sections of the central nervous system shows. Such a series of sections from the filum terminale to the front of the great brain affords material for the study of the course of fibres on the one hand and minute structural relations on the other.

We will attempt, in the first instance, to distinguish the separate tracts of fibres in the cord and then to follow them as far brainwards as possible, bearing in mind what has already been said (p. 160, *et seq.*) about fibre-routes and roads. It must, however, be remarked that it is not our intention to quote in this place all the fibre-paths which have been described, especially when their course is of but little importance.

1. Pyramidal Tracts (fig. 140).—We have learnt to recognise the lateral, PyS, and anterior pyramidal tracts, PyV, in the spinal cord.

The **lateral pyramidal tract** increases in cross-section almost constantly, from the caudal end of the cord upwards. We are bound to suppose that some of the fibres which are seen issuing from the lateral border of the anterior horn enter the lateral pyramidal tract and add to its size. Since we may take for granted that these fibres have their origin in the large cells of the anterior horn [or are connected with these cells indirectly through the medium of the plexus in the grey matter; for while it may almost be said to be proved that a motor-root fibre is connected with each anterior horn cell, no second axis-cylinder process has been demonstrated as arising from it]; it follows that in the lateral pyramidal tract fibres course brainwards which are the indirect (since they are interrupted in the cells of the anterior horn) continuations upwards of the anterior roots, fig. 140, p^1, p^3. It may also be accepted that the anterior roots of the opposite side are represented, though to a much smaller extent, in the lateral pyramidal tract. At any rate we have seen that certain fibres pass through the anterior white commissure to end in the mesial group of cells of the anterior horn. Each lateral tract must, therefore, consist of a large number of fibres destined for the more caudally situate muscles of its own side, and a smaller number for the muscles of the opposite side; the two kinds of fibres are, however, mixed together.

Concerning the **anterior pyramidal tracts**, it has already been

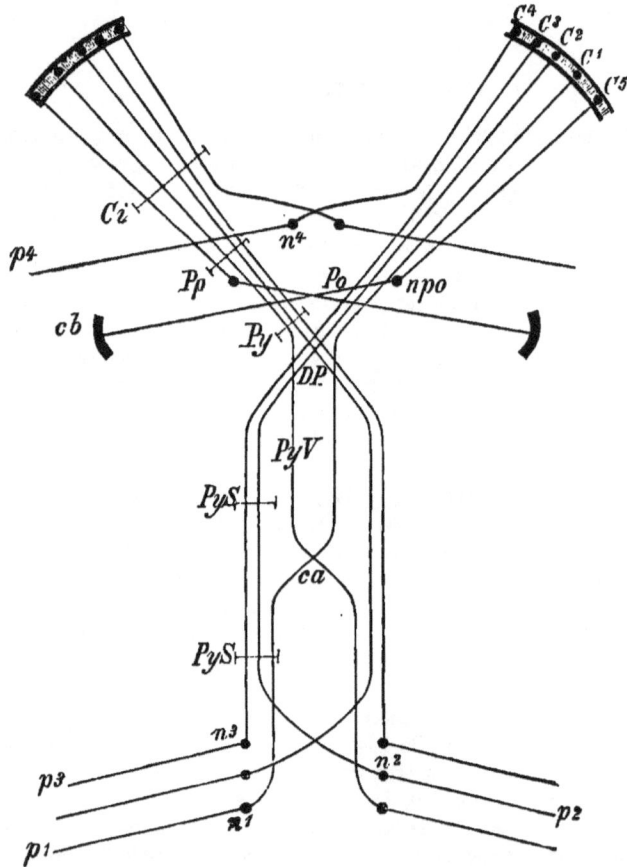

Fig. 140.—Scheme of pyramidal tracts.—p^1, p^2, p^3, Periphery of body; n^1, n^2, n^3, spinal nuclei of origin; PyS, lateral pyramidal tract; PyV, anterior pyramidal tract; ca, anterior commissure of spinal cord; DP, decussation of pyramids; Py, pyramids; Pp, pes pedunculi cerebri; Ci, internal capsule; Po, pons; npo, nuclei pontis; cb, cerebellum; p^4, periphery supplied by cranial nerves; n^4, nucleus of origin of a cranial nerve; C^1 to C^5, cortex cerebri.

stated that they consist, for the most part, of fibres, p^1, derived from the lateral tract of the opposite side, which have crossed the middle line in the anterior commissure, *ca;* the decussation of the pyramids may thus be prepared for throughout a considerable part of the spinal cord. The possibility of a further direct accession by the anterior pyramidal tract of fibres from the cells of the anterior horn of the same side must be taken into account.

The crossing of the lateral tracts begins at the level of the second cervical nerves, **decussatio pyramidum,** *DP.* Histologically, it is characterised by the fact that the fibres which, ascending forwards and medianwards, cross one another in this decussation, do not cross as isolated fibres but in bundles; this gives rise to a peculiar appearance in cross-section (figs. 118, 119).

Owing to the crossing of the lateral tracts they and the anterior tracts are henceforth mixed together in the **pyramids,** *Py.* There are many reasons for thinking, however, that quite a small portion of each lateral tract does not cross, but runs directly into the pyramids of its own side.

The behaviour of the pyramidal tracts within the spinal cord and at the decussation is subject to numerous individual differences. *Flechsig* has made it the subject of detailed communications. Both anterior and lateral tracts are found in the majority of spinal cords (75 per cent.); the lateral is so much the larger, however, that beneath the decussation it usually gets 91 to 97 per cent. of all the pyramidal fibres, while the anterior tract only gets 3 to 9 per cent. Nevertheless, this relation is exceedingly variable; it may happen that all the pyramidal fibres cross (total decussation in 11 per cent. of all spinal cords), in which case no anterior tract comes into existence; on the other hand, this total crossing may only affect the pyramidal fibres of one side. Further, it may happen that nine-tenths of the fibres of the pyramid remain on the same side in the anterior tract, and only one-tenth passes across the middle line to the opposite lateral tract. In the latter case, the opposite lateral pyramidal tract appears abnormally small, while the anterior tract of the same side is conspicuous for its great size. A symmetrical disposition of the two tracts on each side occurs in only 60 per cent. of all cases, in the remaining 40 per cent. the one pyramid is not split into anterior (direct) and lateral (crossed) tracts in the same proportion as the other.

The pyramids extend brainwards along the ventral side of the medulla as compact columns, as far as the pons, where fibres begin to cover them, and later on split them into numerous separate tracts. The huge bundle of fibres which appears as the continuation of the pyramids on the proximal side of the pons, the **crusta,** *Pp* (pes pedunculi cerebri), so greatly exceeds the pyramids in size that we are bound to conclude that the pyramidal tract has received a great accession of fibres while within the pons. A direct accession can be proved in the case of (1) the bundle added to the crus from the fillet (fig. 141). This bundle curls around the outer side of the crus as far as its lateral border (faisceau en écharpe, *Féré*); it usually remains intact when other parts of the crus degenerate downwards, and can then be distinguished from the grey degenerated columns upon which it lies as a conspicuous white band. In many animals this bundle attains to a very striking development relatively to the slender crus, and it can be seen that it only turns cerebralwards after reaching its lateral border, *LmP* (figs. 132–134).

(2) An increase in the number of fibres may be expected to occur in connection with the motor nerves (hypoglossus, vagus, glossopharyngeus, facialis, abducens, trigeminus), which take origin in or near the region of the pons, for there must exist a connection of these nerves with the continuation of the pyramidal tracts similar to that of all spinal nerves. Once the decussation of the pyramids is over, however, some other crossing over the middle line must be looked for in the case of the greater number of the nerves just mentioned. This is provided for by the raphe. The fibres coming from the motor nuclei extend in the raphe ventralwards, cross one another at an acute angle, form the most internal of the tracts of longitudinal fibres in the region of the pons, and join the crura on their mesial borders. The bundle which they now constitute has been termed the faisceau géniculé—a name which it deserves on account of the position in which we shall find it in its further course in the knee of the internal capsule (fig 142, *2*).

(3) The chief portion of the most internal (mesial) fibres of the crus have an unknown course spinewards; we shall be able to follow them, however, towards the front of the great brain as the frontal pontine tract (faisceau corticobulbaire, anterior cerebro-pontine tract). After disease of the frontal lobe or of the anterior part of the internal capsule, this tract degenerates as far as the pons, but not farther. Generally, however, a thin tract of fibres on the inner border of the crus is exempt from degeneration, so that we must allow that it has a different course unknown as yet.

(4) The lateral part of the crus is usually considered to contain sensory tracts, but their course spinewards through the pons is also unknown. They take their rise in posterior portions of the hemisphere, the parietal, occipital, and temporal lobes. Usually they, too, are spared in descending degeneration; in exceptional cases, however, they are drawn into an extensive degeneration (*Rossolymo*), a cir-

cumstance which may be regarded as opposed to their sensory interpretation.

The denotation of these tracts as Türck's bundle, sometimes adopted, should be avoided, since the name is usually applied to the anterior pyramidal tract.

(5) The layer limiting the crusta on its dorsal side towards the substantia nigra consists of thin fibres supposed by *Meynert* to take origin in the grey mass just named, and therefore termed by him pedunculus substantiæ nigræ. They extend downwards towards the pons, in the tegmental region of which they are lost.

(6) The portion of the crusta which remains as the proper continuation upwards of the lateral and anterior pyramidal tracts occupies its middle third (according to *Charcot* its two middle fourths).

Owing to the far from parallel course of the fibres in the crusta— they tend to diverge outwards in their course brainwards—it may easily happen that many degenerated bundles remain hidden in its depth; further, it often happens that in descending degeneration of the pyramidal tracts only a triangular grey-coloured degenerated area shows from the surface; the apex of the triangle pointing towards the pons, its base covered by the optic tract.

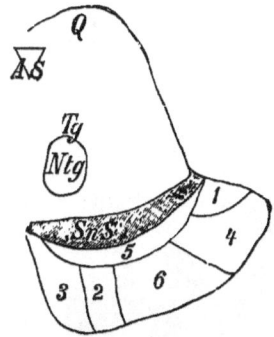

Fig. 141.—Diagram showing the constitution of the crus cerebri. —*AS*, Aquæductus Sylvii; *Q*, corpus quadrigeminum; *Tg*, tegmentum; *Ntg*, red nucleus of the tegment; *SnS*, substantia nigra Soemmeringi; *1-6*, pes peduncili; *1*, fasciculus from the fillet to the crusta; *2*, central tract of the motor cranial nerves which have their origin farther spinewards; *3*, frontal pontine tract; *4*, sensory portion of the pes pedunculi; *5*, dorsal boundary layer of the pes; *6*, pyramidal tract

Lastly, we must mention a seventh constituent of the crusta which perhaps more than any other contributes to its cross-section, namely, the fibres derived from the pons.

If a cross-section through the pons be examined, numerous clumps of grey substance very rich in medium-sized nerve-cells are seen lying amongst the cross-cut bundles which are ascending from the medulla to the pons, as well as amongst the fibres from the cerebellum to the pons, which are here cut lengthways. It is now fairly ascertained that many of the fibres imported from the cerebellum by the pons, cross the middle line in the pons to join with its nerve-cells, *npo* (fig. 140), from which ascending fibres are continued brainwards by some, as yet unknown, route. Thus is provided a crossed connection between the cerebellum and the cerebrum.

After all this account can apply to part of the pons fibres only, for it must be remembered that the cross-section of the crus pontis [middle cerebellar peduncle] is rather greater than that of the crus cerebri. We are still rather in the dark as to the fate of the other fibres of the pons. *Bechterew* has proved from embryonic brains that the fibres of the pons are not all of equal value, since they obtain their myelin-sheaths at very different times, and as a matter of fact not all the fibres of the pons cross the middle line. Some of the fibres coming from the cerebellum to the pons turn dorsalwards, and so are prolonged through the raphe pontis into the raphe tegmenti, and are supposed to find their destination for the present in the groups of nerve-cells which lie on either side the raphe (nuclei reticularis tegmenti pontis), *Nrtg*—fig. 127. We must now follow the pyramidal tracts into the great brain, and at the same time we will take into consideration the further course of the other constituents of the crus.

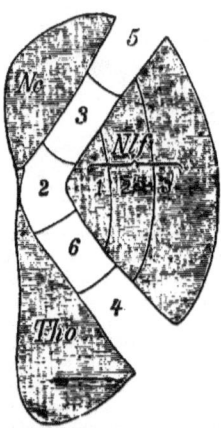

Fig. 142.—Horizontal section through the internal capsule. — *Nc*, Nucleus caudatus; *Nlf*,*1'*,*2'*,*3'*, the three segments of the nucleus lenticularis; *Tho*, thalamus opticus; *2*, tract of motor cranial nerves; *3*, frontal pontine tract; *4*, sensory tract; *5*, anterior peduncle of the thalamus; *6*, pyramidal tract.

It has been shown that the crus cerebri as it passes between the grey masses of the 'tween-brain and fore-brain becomes the internal capsule, *Ci* (fig. 140). No proper investment of the fibres occurs thereby. We can picture to ourselves the displacement which the fibres undergo by imagining that the whole crus is slightly twisted, its inner fibres appearing in horizontal section the most anterior, its outer fibres occupying the posterior part of the internal capsule (*cf.* figs. 141 and 142). Only the back part of the anterior segment of the internal capsule consists, however, of crural fibres, its whole anterior half being occupied by a tract of fibres connected with the optic thalamus, the anterior peduncle of the thalamus, *s*. Behind this comes, *3*, the frontal pontine tract; next in the neighbourhood of the knee of the capsule, the cerebral connection of the motor nerves of the brain, *2* (fig. 140, p^4, n^4, C^4); then the continuation of the pyramidal tracts in the strict sense of the word, 6; and, lastly, the posterior third of the hinder segment of the internal capsule is devoted to the conduction of sensory impressions, 4. In the last-mentioned region of the internal capsule, which corresponds to the outer border of the crus, are, perhaps, to be found fibres belonging to the optic and olfactory nerves, which two nerves are represented in the crus, if they are represented there at all, in a different way to other sensory nerves, arising lower down the system. Since various sensory tracts meet together in this region of the inner capsule, it has been termed the "carrefour sensitif."

Other tracts which are also present in the internal capsule will be mentioned later on. *Flechsig* calls attention to the important fact that the individual fibre-tracts traversing the internal capsule are inconstant in their relation to its "knee;" field 2, for example, does not always correspond to the knee itself, as represented in fig. 142.

As soon as the parts of the internal capsule, which are squeezed together so long as they occupy the narrow defile between the central grey masses of the brain, come out into the open field of the centrum semiovale Vieussenii they stream away from one another on all sides. They never again assume a distinctly stratified arrangement, but are, on their way to the several parts of the cortex, scattered about as constituents of the corona radiata Reilii.

It is impossible at present to say how the bundle coming from the fillet (fig. 141, 1) is disposed; perhaps it occupies the back of the internal capsule. Just as little is known with regard to the seventh element of the crusta, namely, the fibres from the pons.

The fibres of the frontal pontine tracts pass forwards to the frontal lobe and nucleus caudatus; the pyramidal tracts end in the central convolutions, the lobulus paracentralis, and the anterior part of the parietal lobes [collectively the Rolandic area]; while the hindmost fibres of the capsule turn backwards to the occipital lobe (optic radiations of *Gratiolet*, sagittal fibres of the occipital lobe), and also ventrally to the temporal lobes.

Thus it is seen that the pyramidal tract is a long unbroken fibre-route between the cortex of the great brain (and especially that part of it to which we attribute motor functions) and the cells of origin of the motor nerves. For the larger part this connection is a crossed one, some fibres, however, run without crossing. [According to *Sherrington* a considerable number of the fibres, especially at the level of the cervical and lumbar enlargements, cross back again to the side from which they started, "recrossed fibres."] The cortico-muscular connection consists, therefore, of two segments or divisions—(1) the pyramidal tract, $C-n$; (2) the peripheral motor-nerves, $n-p$; between the two divisions of every fibre is inserted at least one anterior-horn cell, n (or its homologue, a motor-cell of the medulla oblongata). It is possible that the connecting link may be more complicated than this; may consist of more than one nerve-cell, or possibly of a nerve-plexus. All the nerve-cells of the anterior horn are also, by means of their numerous processes, connected with one another (by a fine network only, however), and also with those nerve-routes which bring them into relation with the cerebellum, the grey central ganglia of the great brain, and with sensory regions.

Meynert has called attention to the fact that in Man, as in all other mammals, the cross-section of the crusta greatly exceeds in extent that of the tegment, a fact of great importance which must be mentioned in this place, since the pyramidal tracts constitute a considerable portion of the crusta. *Spitzka* has found that not only has the dolphin, which is destitute of hind limbs, rudimentary pyramids, but the same condition obtains also in the elephant and armadillo.

Again, attention must be called to the fact that the pyramidal tracts first become medullated in the centrum semiovale, and that the myelination proceeds from above downwards, taking several weeks to reach the lumbar cord.

2. The Posterior Columns and the Tracts derived from them.

—A great part of the fibres of the posterior columns stand in direct relation with the posterior roots; numerous fibres run into Burdach's column, both in a curved direction, as seen in transverse section of the cord, and also with an ascent brainwards. The presence of long tracts in the posterior columns is denied by some anatomists, but the fact that Goll's columns always degenerate upwards as far as their nuclei in the medulla oblongata certainly tells in favour of their existence.

The crossing, or at any rate partial crossing, as indicated by physiological experiments, of the fibres of the posterior roots before they come into relation with the posterior columns can hardly be accounted for except by supposing that it occurs in the posterior grey commissure, and probably also by fibres traversing the septum posterius.

In the medulla oblongata, as we know, the posterior columns swell out, owing to the deposition in them of certain grey masses (nuclei funiculi gracilis et cuneati), *Nc*, *Ng* (figs 119, 120, 121), which nuclei, considered together, may be termed shortly the nuclei of the posterior columns. They must, according to the data already given, be looked upon as sensory nuclei for the muscle-sense of the extremities. Burdach's nucleus is supposed to be in relation with the upper limb; Goll's nucleus with the lower limb.

The fibres coming out of these nuclei, the indirect connections of the posterior columns—that is to say, go partly to the corpora quadrigemina and the great brain *viâ* the fillet; partly to the cerebellum *viâ* its inferior peduncle, the corpus restiforme. We must, therefore, leaving out of account some less well-known connections, consider these two separately.

(a.) **The fillet.**—The term fillet (lemniscus, laqueus, ruban de Reil) was originally applied to the triangular area on the surface of the crus, which extends downwards and backwards from the posterior tubercle of the corpora quadrigemina. Latterly the term fillet has been made to include also other allied fibre-tracts. This composite system has been divided in several ways without anything like uniformity in nomenclature. The difficulties which beset the subject are due as much to a confusion of names as to a complexity of structure. We are still, however, far from understanding the origin and destination of all its fibres. The part of the fillet best established is that which is in connection with the posterior columns of the spinal cord, this is our reason for considering it here.

We have seen that arcuate fibres extend from the nuclei of the posterior columns ventrally towards the middle line. The greater number of these fibres are to be found at the spinal end of the medulla. They commence in the funiculus gracilis, and arching round the central canal, take up their position on the dorsal side of the pyramids in the opposite interolivary region or "fillet layer," where they are joined by fibres from the anterior column (*Homen, Spitzka*). The fibræ arcuatæ, which originate farther cerebralwards in the nuclei of the posterior columns, sweep round in finer bundles and wider curves. They may, according to *Darkschewitsch* and *Freud*, be divided into two groups; the dorsal group is collected partly out of the proper interolivary layer into the mesial region of the medulla oblongata (middle part of the substantia reticularis alba), in which it extends brainwards; the other group of fibres keeps the horizontal direction for a greater distance, and, as we shall presently see, joins the corpus restiforme of the opposite side.

The cross-section of the mesial fillet lying in the ventral part of the tegment, where it is pierced by fibres of the corpus trapezoides, *Tr*, can be followed into the mid-brain. One can recognise a steady increase in the size of this field, which enlargement must be attributed to the accession of new fibres of doubtful origin. The lateral fillet joins the mesial fillet farther brainwards. According to *Roller's* observations, most sensory nuclei have connections with the fillet, and this, even apart from its origin in the posterior columns, would mark it as a sensory tract. Near the middle of the cross-section of the fillet little clumps of nerve-cells occur (called by Roller the "fillet-flock" nuclei lemnisci mediales), which may be regarded as the centres of origin of fillet-fibres. *Bechterew* finds a double accession of fibres from his nucleus reticularis tegmenti pontis, *Nrtg* (fig. 127); one set of bundles joins the lateral fillet, and the other, distinguished by its smaller fibres, is supposed to swell the mesial fillet.

Many other fibres belonging to the mesial fillet have been described. We have seen (fig. 133) that near the corpora quadrigemina the fillet is disposed in three divisions :—(1) The most mesial bundle joins the crusta, *LmP* ; (2) the mesial fillet, *Lm* ; (3) the lateral fillet, *Lml*, which, as it covers the brachium is the part of the fillet visible on the exterior, extends to the posterior corpora quadrigemina, and crosses in part above the aqueduct ; it is also called the inferior fillet, while the mesial fillet, which can invariably be followed into the anterior corpora quadrigemina and thalamus, is also known as the superior fillet. The nucleus of the lateral fillet (nucleus lemnisci lateralis, fig. 130, *Nlml*, and fig. 143, *Nll*), yields numerous fibres to the lateral fillet, as does also the upper olive, *Os* (fig. 157). It receives, too, a considerable addition from the corpus trapezoides. The lateral-fillet-nucleus corresponds in position to the upper olive, the cerebral end of which it almost reaches. The fibres from the nucleus reticularis, as already mentioned, join the lateral fillet, and as this nucleus is connected with fibres from the lateral column, we may consider that a connection between the lateral column and posterior corpus quadrigeminum is thus established.

The principal part of the upper or mesial fillet turns dorsally under the anterior corpus quadrigeminum to form its white matter just in the same way as the lateral fillet has been seen to do with regard to the posterior tubercle. Probably a part of the fibres lying above the aqueduct are continued across the middle line to the tubercles of the opposite side, but whether or not they extend into the brachia corporum quadrigeminorum is uncertain. A small remnant of the fillet is to be traced still farther brainwards, on the outer side and a little dorsally to the red nucleus, as a feebly marked half-moon-shaped bundle, mixed with the fibres which stream out from the nuclei of the regio subthalamica. It must be accepted that many of these fibres end in the thalamus (fig. 143), *Th*, and, perhaps, in the inner segments of the nucleus lenticularis. Some of the fibres of the fillet are supposed to reach the parietal part of the cortex by turning outwards in the subthalamic region in the ansa lenticularis, and traversing the two inner segments of the nucleus lenticularis, the cortex-fillet, *C*. *Edinger* describes bundles which, coming out of the fillet, are to be met with on the upper and outer side of the nucleus ruber on their way to the cortex of the upper part of the parietal lobe *viâ* the internal capsule

DEGENERATION OF FILLET. 259

—avoiding thus the nucleus lenticularis. They form a part of his so-called tegmental system, which will be described in detail later on.

Secondary degeneration of the fillet has been repeatedly observed; usually descending. Ascending degeneration (*P. Meyer*), and even

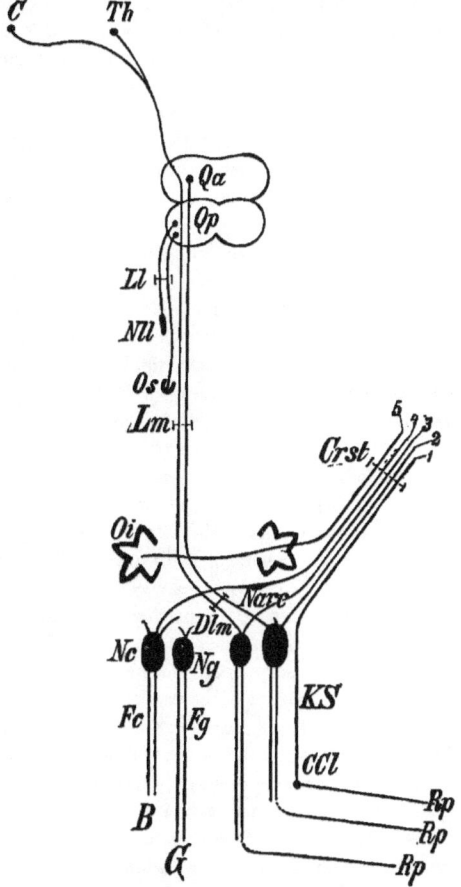

Fig. 143.—Plan of the central connections of the posterior columns.—*Rp*, Radix posterior; *B*, Burdach's column; *Fc*, funiculus cuneatus; *G*, Goll's column; *Fg*, funiculus gracilis; *Nc*, nucleus of f. cuneatus; *Ng*, nucleus of f. gracilis; *Dlm*, decussatio lemnisci; *Narc*, nucleus arcuatus; *CCl*, Clarke's vesicular column; *KS*, lateral cerebellar tract; *Oi*, inferior olive; *Crst*, corpus restiforme; *Lm*, mesial fillet; *Ll*, lateral fillet; *Os*, superior olive; *Nll*, nucleus of lateral fillet; *Qa*, *Qp*, anterior and posterior tubercles of the corpora quadrigemina; *Th*, thalamus opticus; *C*, cortex cerebri.

degeneration in both directions (*P. Meyer, Spitzka*), has also been observed. From the latter circumstance the conclusion may be drawn that both sensory and motor fibres (*Mendel*), or at any rate different kinds of fibres, traverse the fillet.

Although in many cases degeneration of the fillet has been found to be associated with atrophy of the inferior olive, this may be attributed to the fact that other bundles in the tegmental region are also involved, as the relations of the fillet to the lower olive sometimes referred to (*Roller*) are certainly only of a subordinate character.

(*b.*) **The Inferior Peduncle of the Cerebellum.**—The connection of the posterior column with the cerebellum is effected by the inferior peduncle of the latter (corpus restiforme). The passage over from the posterior column to the peduncle of the cerebellum is not, however, so simple as a superficial examination of the medulla would lead one to suppose.

Into the constitution of the cerebellar peduncle enter—(1) fibres from the spinal cord, notably from its lateral column, although those coming from the posterior column must not be overlooked; (2) fibres from the inferior olivary nucleus (olivary cerebellar tract).

(1) The fibres derived from the lateral column constitute the lateral cerebellar tract, KS (fig. 143), which we will again mention later on. It is supposed that the restiform body receives a further accession of fibres from the nucleus of the lateral column which lies fairly near to the lateral cerebellar tract (figs. 121-123), Nlt.

(2) The constituent of the restiform body derived from the posterior column is a very considerable one; partly direct, partly crossed. *Darkschewitsch* and *Freud* have again pointed out the importance of the uncrossed connection, which had for a long time been overlooked. They have shown that the corpus restiforme receives a much more considerable accession of fibres from the nuclei of the posterior column of the same side, especially from Burdach's nucleus, than it does from the arcuate fibres; the latter extend a short distance on the posterior aspect of the periphery of the medulla, and connect the restiform body with Goll's nucleus (fibræ arcuatæ externæ posteriores of *Edinger*). At the higher levels of the nuclei of the posterior columns, it can be shown that they decrease as fast as the corpus restiforme grows; the latter occupies the place of the successively disappearing clumps of grey matter (*cf.* figs. 122, 123, 124), and so the fibres of the posterior columns, after interruption in the cells of the nuclei, continue their course in the restiform body with little change in their direction.

The restiform body receives a further reinforcement of fibres from the posterior column by a round-about way through the fibræ arcuatæ internæ (figs. 121, 122, Fai) which constitute, as has already been briefly explained (p. 257), the proximal continuation of the crossed fillet. They do not remain in the interolivary layer but turn ventralwards in the raphe, crossing one another at an acute angle. The periphery reached, they sweep onwards over the surface of the pyramid of the opposite side and over the olive to the corpus restiforme. These are the fibræ arcuatæ externæ anteriores (fig. 143, *Crst 4*). Thus they connect the posterior column with the corpus restiforme of the opposite side. Certain small collections of grey matter, as well as the larger nucleus pyramidalis anterior (nucleus arciformis), are imbedded in the course of these fibres (fig. 121 and the following figs., *Narc;* fig. 143, *Nar*). It may also be pointed out that great numbers of these fibres, while they are still traversing the medulla as fibræ arcuatæ internæ, enter the olives (fig. 123). *Edinger* has shown that they pass through the olives without forming connections with them.

(3) The olivary portion of the corpus restiforme is also constituted in a complicated way.

The **olivary nucleus** (inferior olive, figs. 121-125) appears on crosssection as a double sinuous plicated band, of which the two segments are united laterally, but are open towards the middle line. As a whole, the inferior olive may most readily be compared to a bag, the mouth of which is only partly drawn together, the hole into it (hilum) being directed medianwards. The thickness of the band is almost uniform throughout, between 0·3 and 0·4 mm. With slight magnification one can see that numerous nerve-bundles, especially the considerable hypoglossal root, traverse the substance of the olive. The nerve-cells of the olive are round or somewhat fusiform, slightly pigmented, and almost all of the same size (12 to 20 μ in diameter). They are fairly evenly distributed throughout the grey band, although occasionally some cells lie outside the grey substance. Besides the bundles of fibres which pierce the grey band in a horizontal direction, or otherwise, bundles of nerve-fibres, running longitudinally, as well as a rich network of medullated nerves, can be shown to exist within the grey substance of the olive. The two accessory olives exhibit a similar structure.

Numerous groups of fibres come out of the hilum (they constitute the peduncle of the olive); others envelop its outer side, the individual fibres having a horizontal direction (stratum zonale). Lastly, a considerable number of fibres extend from the stratum zonale to the restiform body, passing by the outer border of the ascending root of the trigeminus (figs. 122, 123).

It is not possible by anatomical methods to distinguish between the several sets of fibres described above. We must be influenced in this matter by pathological observations, the most notable being that when one side of the cerebellum atrophies the opposite olive atrophies also.

It would seem that the course of the fibres connecting the olive with the corpus restiforme is as follows:—The fibres which come out of the hilum cross those of the opposite side in the middle line, thence they extend to the opposite restiform body, piercing the olive for the most part, and also taking part in the formation of its stratum zonale, *Crst 5*, (fig. 143).

Bechterew and *Flechsig* have described a connection between the inferior olive and the nucleus lenticularis *viâ* the central tegmental tract. This bundle is gradually formed on the lateral and dorsal surface of the inferior olive, *cH* (fig. 125), passes between the mesial fillet and the superior olive (figs. 126 and 127), continues its course towards the cerebrum on the outer side of the posterior longitudinal bundle, and, lastly, enters the ansa lenticularis. The central tegmental tract is rarely distinctly marked in adult brains. Various other connections of the olive with the brain and cord must surely exist, but at present they are unknown.

The inferior peduncle, formed by the union of all the tracts of fibres above described, soon enters the substance of the cerebellum, in which its further course can only be followed by embryological methods.

According to *Edinger*, the spinal portion of the restiform body is destined for the vermis, whilst the olivary elements form the bundles of fibres which surround the corpus dentatum as its "stratum zonale." Further details will be mentioned in connection with the cerebellum. The tracts of fibres which pass from the fifth and eighth nerves into the cerebellum are also frequently reckoned to the corpus restiforme.

3. **The Lateral Cerebellar Tract.**—The facts known with regard to this may be recapitulated here in a few words (fig. 143). The lateral cerebellar tract receives its fibres from the direction of Clarke's vesicular column, and thus, in all probability, from the posterior roots, *Rp*. When it reaches the medulla the lateral cerebellar tract inclines obliquely across the ascending root of the fifth towards the dorsal surface (figs. 122, 123); gradually the other constituents of the inferior cerebellar peduncle apply themselves to it; and its conspicuously-large fibres finally end, after a fairly-simple course, in the vermis, *Crst 1* (fig. 143). The lateral cerebellar tract is, therefore, an uncrossed path between the posterior roots and the cerebellum. The fact that it degenerates upwards indicates that we are to look upon it as a centripetal conducting system. A portion of the lateral cerebellar tract is supposed not to enter the corpus restiforme, but to go on brainwards almost as far as the corpora quadrigemina, and then, near the fillet, to turn backwards, and so,

lying on the surface of the brachium conjunctivum, to stream into the cerebellum (*Löwenthal*).

4. Gowers' Tract.—This bundle is regarded as formed of fibres derived from the posterior roots, which having crossed in the posterior commissure and been interrupted in nerve-cells, collect together in the lateral column for their cerebral course (fig. 108, *25*). One portion of the fibres is stated to disappear in the upper cervical cord, another portion ends in the nucleus lateralis of the medulla oblongata (*Bechterew*). These tracts of fibres are not indicated in the scheme (fig. 108) owing to the uncertainty which still invests them.

5. The Rest of the Anterior and Lateral Columns.—In this paragraph we shall include all those tracts which have not found a place hitherto. As far as a division into short and long tracts is allowable, we may say that we have here to deal with short tracts, fibres which come out of the grey matter to enter it again after a short longitudinal course, forming in this way connections between segments of the spinal cord at various heights.

All the several constituents of the cord which are here described can be followed at any rate as far as the proximal end of the midbrain within the substantia reticularis of the tegment; not that we wish to imply by this statement that each individual fibre has anything like such an extensive course as this; rather are the several tracts made up of fibres frequently disappearing to be replaced by new ones, so that no essential alteration in the cross-section of the tracts need necessarily occur at any particular height.

The ground-bundle of the anterior column is the one most easily followed brainwards. We have already seen that this bundle, VG, is a little displaced by the crossing of the pyramids (fig. 118, *et seq.*). Farther forwards the interolivary layer made by the crossing of the fillets presses the anterior ground-bundle together with a portion of the lateral column dorsally, the three together forming the substantia reticularis alba (formatio reticularis medialis). The most ventral portion of the substantia reticularis alba (the interolivary layer) has already been traced upwards in the fillet. The middle portion corresponds to the part of the lateral column just mentioned, to it certain bundles of fibres originating in the nuclei of the posterior columns join themselves (p. 257), while the most dorsal section of the substantia reticularis alba, which is sharply marked off from the grey matter on the floor of the fourth ventricle, is developed from the anterior ground-bundle. It may be mentioned here before hand that the middle portion which is formed out of the remains of the lateral column seems to end, above the origin of the hypoglossal nerve, in those grey masses (nuclei centrales inferiores of *Roller*, figs. 124, 125,

Nct), which lie up against the raphe on either side the middle line, and separate the fillet from the continuation of the anterior ground-bundle, *VG*, which henceforth receives the name "posterior longitudinal bundle."

The posterior longitudinal bundle, *Flp* (fig. 124, *et seq.*), can be followed as far as the anterior corpus quadrigeminum. It forms a bundle, very distinct in cross-section, which lies beneath the grey matter of the fourth ventricle and the aqueduct of Sylvius. Its ventral edge is never sharply defined, for it cannot be separated from the other longitudinal bundles of the tegment with which it mingles. It is very difficult to trace the posterior longitudinal bundle beyond the oculomotor nucleus; not improbably it ends at this level (*Flechsig, Edinger*).

We do not believe that the posterior longitudinal bundle arises in the nucleus lenticularis or its surroundings, or in the cortex as has been repeatedly stated. In disproof of the latter origin, *Spitzka* advances the very telling fact that the posterior longitudinal bundle is particularly strong in amphibia and reptiles in which the fore-brain is only feebly developed, except in the members of these classes in which the eye is atrophied. On this he bases the theory that the posterior longitudinal bundles connect the anterior tubercles of the corpora quadrigemina, greatly developed in these animals as lobi optici, with the nuclei for the eye-muscle nerves, and beyond them with the nuclei of the nerves that innervate the muscles by which the head is moved. The posterior longitudinal bundles are exceedingly small in the mole (*Forel*).

As already mentioned, we may conclude that the posterior longitudinal bundles consist for the most part of short fibres connecting together the motor nuclei which follow one another from the spinal cord up to the brain.

It is not impossible that the root-fibres of peripheral nerves run for a certain distance in these bundles before crossing the middle line— *e.g.*, fibres from the oculomotor nerve may reach in this way the nucleus of the abducens. The fact that the larger part of the posterior longitudinal bundle myelinates very early, simultaneously with the peripheral nerves, accords with this view.

Less is known as yet about the continuation upwards of the rest of the lateral column. Part of the lateral column (after accounting for the lateral cerebellar tract, the lateral pyramidal tract, and Gowers' bundle) forms, as we have already learnt, the middle portion of the substantia reticularis alba, and seems to end somewhere about the level of the most anterior roots of the hypoglossus, in the nucleus centralis inferior. All the remaining bundles attain the substantia reticularis grisea, and, consequently, take part in the formation of the tegment. Here are found numerous scattered nerve-cells, which may be looked upon as the preliminary terminations of the fibres ascending from the spinal cord. *Bechterew* claims for this purpose the upper olive especially, as well as the nucleus reticularis and its prolongation forwards, the nucleus centralis superior. He looks upon the nucleus reticularis as one of the most important nodal points in the central nervous system; we have already called attention to its connection with the pons, as well as its manifold relations to the fillet. *Monakow* designates a tract, already described by *Meynert* and others, as the "aberrant bundle" of the lateral column; it originates in the peripheral portion of the lateral column, lies up against the corpus trapezoides between the facial nucleus and ascending root of the fifth nerve, and finally passes over into the fillet.

In the region of the corpora quadrigemina where the brachia conjunctiva force themselves into the tegment, taking up a great part of its area in Man, only a very small number of longitudinal fibres, as a matter of fact, remain over from the formatio reticularis, apart from the posterior longitudinal bundle and the fillet. An ill-defined small bundle of medullated nerves may be pointed out on the lateral side of the posterior longitudinal bundle, *Fcop* (fig. 135). According to *Wernicke's* researches this bundle bends towards the middle line in front of the corpora quadrigemina, crossing over in the roof of the most anterior portion of the aqueduct of Sylvius. After helping to form the posterior commissure, it reaches the optic thalamus of the opposite side, in which it ends.

Throughout the whole extent of the cross-section of the tegment, the longitudinal fibres, which are early myelinated, run in separate small bundles only. Numbers of these fibres cross the middle line in the vicinity of the anterior corpora quadrigemina, some near the basis cerebri on the ventral side of the red nucleus (*Forel's* ventral tegmental decussation), others more dorsally beneath the posterior longitudinal bundle (*Meynert's* fontanal tegmental decussation), F and M (fig. 135).

B. THE CEREBRAL NERVES.

1. Nervus Olfactorius.—The central apparatus of the sense of smell may be regarded in Man, not only as a relatively feeble organ, the development of which has been arrested, but as an organ affected in the adult by a distinct retrogressive atrophic process in addition to its genetic inferiority. In its want of development it resembles the corpus callosum of lower mammals which may be almost or completely absent.

[The corpus callosum or great commissure of the secondary forebrain is present in all cephalota (*Osborn*). Since it is, however, essentially the commissure of the mantle (cortex) its development varies with this part of the brain. An indubitable brain-cortex is first met with in reptiles (although some parts of the mantle in amphibia and dipnoans contain cortex-formations), and here it is that the corpus callosum first acquires considerable proportions.]

Not only is the olfactory bulb in Man undeveloped, but its atrophy is evidenced by the numerous amyloid bodies to be found in the course of the cerebral connections of the olfactory nerves.

In studying the central organs of the sense of smell, it is well to employ not only the human brain, but also the brains of animals in which the sense of smell is well developed; *e.g.*, carnivora and rodents.

The olfactory organs are very ill-developed in apes as well as in the aquatic predatory mammalia; in many cetacea, the dolphin, for example, it is absolutely wanting.

[The nature of the sense served by the olfactory membrane in the several classes of vertebrates offers much room for speculation. In fishes the membrane and its central connections are well developed. In lacertilia and ophidia not only is the olfactory membrane highly organised, but it also presents a further specialised portion, the organ of Jacobson, of much greater sensitiveness than the rest (*Beard*). The relative development of the membrane is fairly constant throughout the four lower classes of vertebrates, whether aquatic or terrestrial; as soon, however, as a mammal takes to the water its olfactory organs dwindle. In the otter they are very ill-developed; in sirenia still more rudimentary; in cetacea they are practically absent in the adult. A consideration of the alteration in character which the sense of smell must undergo to adapt it from a power of appreciating the quality of substances in solution in water, to a power of recognising substances suspended in air, raises a doubt as to whether the sense is fundamentally the same in the two cases. As is well known the olfactory membrane of the mammal is quite insensible to the action of the most strongly-odorous bodies when presented to it in solution in water. On the other hand, an air-breathing animal, when under water, would be incapable of using its olfactory organ without such an adaptation of the apparatus as would allow of the renewal of the water in contact with the olfactory membrane without its passing into the lungs. Since this arrangement has not come into existence the olfactory membrane is useless.]

Broca divides the mammalia into osmatic and anosmatic animals, according as the sense of smell is well developed, or ill developed or absent.

The peripheral nerves of smell originate in the pigmented regio olfactoria of the Schneiderian membrane; they are non-medullated, and pass through the perforations of the cribriform plate into the interior of the skull-case, where they attach themselves to a greyish-yellow, rounded body of small size in Man, the **bulbus olfactorius**, *Bol* (caruncula mammillaris, lobe olfactif), fig. 144.

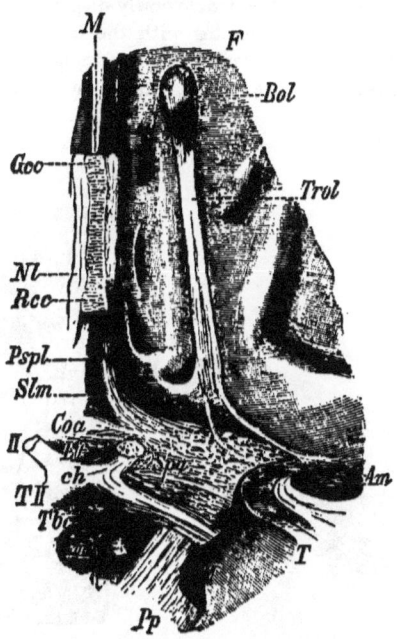

Fig. 144.—A portion of the base of the left hemisphere in front of the optic chiasm. The apex of the temporal lobe is cut away.—*Pp*, Pes pedunculi; *Cm*, corpus mammillare; *Tbc*, tuber cinereum; *T II*, tractus opticus; *ch*, chiasma; *II*, nervus opticus; *T*, temporal lobe; *U*, uncus; *Am*, nucleus amygdaleus; *Spa*, substantia perforata anterior; *Lt*, lamina terminalis; *Coa*, bulging forward of the grey commissure of the floor produced by the anterior commissure; *Pspl*, pedunculus septi pellucidi; *Slm*, sulcus medius subst. perf. ant.; *Rcc*, rostrum corporis callosi; *Gcc*, genu corp. callosi; *Nl*, nervus lancisii; *M*, incisura pallii; *F*, frontal lobe; *Bol*, bulbus olfactorius; *Trol*, tractus olfactorius.

The olfactory bulb lies on the orbital surface of the frontal lobe, at the front of the sulcus olfactorius. It is free on all sides, with the exception of its attachments to the olfactory nerves, and a strong stalk or peduncle which runs backwards to join with the rest of the brain, the tractus olfactorius, *Trol.*

The fine anatomy of the olfactory bulb is best studied in sagittal sections through this structure in the dog (figs. 145 and 146). On slight magnification we see, if the section runs through the middle of the bulb, *b*, and the tract, *t*, that a fine canal, *V*, traverses the tract almost as far as the front of the bulb. In frontal section this canal proves to have the form of a transversely-disposed slit (ventriculus bulbi olfactorii). It communicates with the lateral ventricle of the brain [by a narrow slit-like opening on the inner side of the head of the nucleus caudatus]. The bulbus olfactorius covers the tract as with a hood.

The bulbus olfactorius exhibits a complicated stratification, the

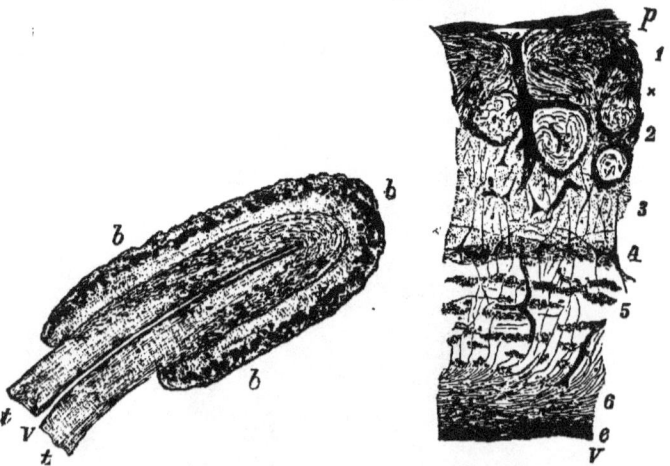

Fig. 145.—Sagittal section of the bulbus olfactorius of the dog. *Magn.* 4.—*b*, Bulbus olfactorius; *t*, tractus olfactorius; *V*, ventriculus olfactorius.

Fig. 146.—Portion of a sagittal section of the olfactory bulb of the dog.— *P*, Pia mater ; *1*, layer of peripheral nerve-fibres ; *2*, stratum glomerulosum, at × fibres are seen streaming out of the first layer into a glomerulus ; *3*, stratum moleculare ; *4*, nerve-cell layer ; *5*, stratum granulosum ; *6*, medullary substance ; *e*, ependyma ; *V*, ventricle.

meaning of which is only revealed by stronger magnification (fig. 146). First comes the enveloping pia mater, *p*, which does not, however, appear as a continuous layer, as shown in the picture, but is rather torn into many pieces by the numerous olfactory fibres which enter the bulb. Large vessels from the pia mater sink into the bulb. The first nervous layer is made up of the very fine bundles of the olfactory nerve (1), which after passing the pia mater run, as a rule, some distance in a sagittal direction, so that they are cut across in transverse section.

The second layer (stratum glomerulosum) is already very conspicuous with a low power. It is formed of peculiar globular masses, 0·05 to 0·30 mm. in diameter, fairly closely packed together. They stain but little in carmine. It is very difficult to make out the finer constitution of these glomeruli. Not rarely bundles of fibres from the first layer are seen to enter the glomerulus (as at x), but here they lose themselves in the finely-granular mass which constitutes the glomerulus. Scattered connective-tissue nuclei only are apparent in the irresolvable substance of the glomerulus.

In Man it is especially easy to gain the impression that the olfactory fibres are in a certain sense coiled up within the glomerulus. The finely-granular mass is better developed in animals, and covers up the nerve-fibres. This mass can hardly be comprehended in the term connective-tissue; nor is it like the neuroglia found in other parts of the nervous system. It is distinguished amongst other things by its different behaviour towards staining reagents.

The large vessels which enter from the pia tend to apply themselves closely to the glomerulus to which they give off fine branches. The glomeruli are separated from surrounding structures by a more or less broad layer of nuclei of the sort met with in many other places (in the cerebellar cortex, for instance); [not, however, arranged in a continuous stratum, but in overlapping plates]. The third layer or stratum moleculare (stratum gelatinosum), *3*, is about 0·3 mm. thick. It consists of a finely-granular ground-substance in which are scattered stellate connective-tissue cells, free nuclei, and a fairly-close meshwork made up partly of medullated, but principally of non-medullated nervefibres; the medullated fibres run, as a rule, perpendicularly to the surface of the bulb. It is easy to see in preparations stained according to Weigert's hæmatoxylin-method that these fibres come without exception out of the inner layer of the bulb (the medullary layer), and lose the medullary sheaths at a greater or less distance from the glomeruli, with which they join company as non-medullated fibres. Finally, this layer contains scattered large nerve-cells, usually triangular in shape.

As fourth layer, *4* (nerve-cell layer), follows a narrow strip, not more than 0·04 mm. in thickness, which appears in carmine preparations, when only slightly magnified, as a dark line. This layer consists of thickly-packed granules, amongst which lie large triangular nerve-cells, usually arranged in a single row. These cells have a diameter of 30 to 50 μ, and give off a process towards the periphery, and also another process directed obliquely inwards towards the deeper layers.

The next layer, stratum granulosum, *5*, which is not marked off sharply from the sixth, is broadest at the apex of the bulb (1 to 1·5 mm. in diameter); towards the hinder end it disappears altogether. It is especially characterised by its closely-packed granules [nuclei] arranged in several rows parallel to the surface, between which bundles of nerve-fibres course in the same direction. This layer is in addition pierced by a number of radiating medullated fibres, which coming out of the medullary layer of the bulb lose their myelin, some in this layer, some (as already mentioned) in the third layer.

The innermost, or sixth layer (*6*), the medullary centre of the bulbus olfactorius, consists of nerve-fibres which run parallel to one another with a somewhat undulating course. This layer gives off at right angles fibres to the superficial layer, and so diminishes towards the apex of the bulb in the same proportion that the fifth layer grows in thickness. It is limited towards the ventricle by an ordinary ependyma, *e*, with ciliated epithelial cells.

It appears, therefore, from the above account that cells of decidedly nervous character occur in the bulbus olfactorius only as scattered cells, or collected into a sheet in the fourth layer.

In the human olfactory bulb nerve-fibres and glomeruli are present, and, as stated above, the fibres in the glomeruli are more easily recognisable than they are in animals. The third and fourth layers are not sharply defined; genuine nerve-cells occur but very sparsely. The granular and medullary layers are distinctly recognised. The ventricle is wanting, but its situation is indicated by gelatinous substance in the centre of the bulb. The layers above mentioned are found on the ventral side of this gelatinous substance only, usually the dorsal portion consists of nothing but medullary substance. The numerous amyloid bodies which occur in Man have been already pointed out.

The olfactory nerves find their first interruption in the olfactory bulb; the bulb is comparable to the nuclei of origin of most other nerves, or in some sense to the ganglion-cell layer of the retina perhaps also to the spinal ganglia; but in no sense to the cortex (fig. 148, *p.* and *Bo*).

[The similarity between certain of the elements contained both in the retina and olfactory bulb has attracted the attention of several anatomists; but it appears to the *translator* that the homology is much deeper than has been hitherto supposed, and that both similarities and differences throw a great deal of light upon the morphology of the central nervous system. If sections of the olfactory bulb and retina of any given animal (there is none more suitable for observation than the rat) be compared with one another, after similar staining in osmic acid, it will be found that the resemblance between the "granules" of the bulb and the "nuclei" of the inner layer of the retina, and between the "gelatinous" substance of the bulb and the inner "molecular" substance of the retina, respectively, is so great that it is almost impossible to distinguish between them. It is difficult at the first moment to recognise the homology of the "nuclei" of the retina, the "granules" of the bulb, the "cells" of the ganglion spirale of the ear, and the "cells" of the ganglia on the posterior roots of the spinal nerves; but it might on *a priori* grounds be supposed that the same plan would be adopted in the connection with the central nervous system of all sensory nerves; and certain considerations with regard to the development of the nervous system give the key to this plan.

The nervous system is first formed in the animal kingdom as a connection between certain cells on the surface, well situate for the purpose of acquiring information with regard to the environment, and contractile, muscular cells, or cell-processes, the action of which adapts the animal to its environment. The specialisation of spots on the surface into sense-organs is due to the favourable position of these spots and to the sensitive character of the cells, whether pigmented or containing crystals of carbonate of lime, or otherwise adapted to receive impressions. When these cells are collected into sense-organs they need long filaments to connect them with the contractile elements. Further than this, as soon as the cells of the sense-organs are able to distinguish between impulses of different strength and kind, a mechanism is needed for the purpose of distributing the impulses they receive. These distributive cells or nerve-cells are derived, as R. and O. Hertwig have shown, from sense-cells, which, having lost their receptive properties, have sunk down from the sense-organ into the subjacent mesoblast, and serve henceforth for the distribution to appropriate muscles of the impulses received from their more favoured sisters. The central nervous system consists, in the first instance, of clumps of deposed epithelial cells and the plexus formed by their processes; the clumps lie, therefore, immediately beneath sense-organs.

The next step in the evolution of the nervous system consists in its withdrawal in part to a more central sheltered situation. Its local origin is always marked, however, by the presence of nerve-cells in the vicinity of the sense-organs (the nose, the eye, the ear), or the situation formerly occupied by sense-organs (neighbourhood of the spinal ganglia).

The central connections of the nose and the eye differ from those of the more posterior sense-organs. Less of the nervous labyrinth has been withdrawn from its original local situation in the neighbourhood of the sense-organ, to take part in constituting a cerebro-spinal axis, than in the case of the ear and segmental organs corresponding to the spinal ganglia. In the olfactory bulb and retina are found bipolar cells (granules or nuclei), plexus (gelatinous or molecular substance), and associating nerve-cells. In the ganglion spirale and spinal ganglia, bipolar cells alone are found, the plexus (substantia Rolandi) and associating cells (? cells at the base of the posterior horn) are withdrawn into the axis (see fig. 57).

This is not merely a morphological speculation. It is hopeless to attempt to trace the central connections of the olfactory and optic tracts until the relation of the nervous elements in the bulb and the retina to the rest of the cerbero-spinal axis has been determined.

Looking at the nervous system from the *translator's* point of view, it is seen to be composed of sense-organs and grey matter, united by nerve-fibres. Processes of the cells of the grey matter stretch out to and innervate muscle-fibres. The grey matter is essentially a plexus providing alternative routes for impulses originating in the sense-organs. In it several different kinds of elements are found. The basal processes of sense-cells never terminate directly in the plexus, but first pass through bipolar cells. On leaving the bipolar cells they break up into a plexus, the processes of which are associated into nerve-fibres by nerve-cells. The bipolar cells are the granules or nuclei, and the cells of the spinal ganglia. The plexus is the gelatinous or molecular substance.]

In Man the pedunculus bulbi or **tractus olfactorius**, *Trol* (called formerly, by mistake, the olfactory nerve), which runs backwards towards the substantia perforata anterior, *Spa*, is essentially triangular in form.

The region immediately in front of the anterior perforated space is also termed the tuber olfactorium. The free superficial layer of the tract consists of white matter; its upper angular portion embedded in the sulcus olfactorius rises abruptly behind, and blends with the mesial wall of this sulcus. Another convolution passes from the tractus obliquely backwards and outwards, closing in this sulcus posteriorly. At the hinder end the superficial visible white fibres of the olfactory tract also separate into several bundles, all of which course outwards and backwards, the outer or **lateral** olfactory **root.** One of these bundles, the most lateral [stria externa], is always distinctly visible; it disappears in the gyrus uncinatus, near the nucleus amygdaleus, *Am.* Of the other bundles one or more, not always distinct, pass outwards and backwards, skirting close by the large holes of the substantia perforata; they cannot (with the naked eye) be followed into the temporal lobe.

No white **mesial root** [stria interna], as commonly described, is to be seen; nor does a middle grey root, in the common meaning of the term, exist.

The cross-section of the tract in Man is, as a rule, triangular with rounded corners and slight concave sides (fig. 147). A layer of fine medullated nerves, about 0·3 mm. thick, occupies the basal surface, and extends round the lateral angles. Next to this follows a layer, 0·1 to 0·3 mm. thick, which consists for the most part of connective-tissue, and corresponds to the obliterated ventricle, while all the rest of the tract is derived from modified cortex. On its dorsal surface it is covered by a distinct stratum of medullated nerves, and it contains small, irregularly-disposed nerve-cells which are more numerous and more definitely pyramidal towards the hinder end of the tract. In almost all adults, and especially in old people, the basal nerve-layer contains numerous amyloid bodies. The middle layer, which corresponds to the ventricle, may be completely filled with these bodies, while, at the same time, the cortical layer also shows them in small numbers in its white stratum. Their presence enables one to trace the olfactory tract in its further course, especially after staining with hæmatoxylin or after rapid dehydration (*Tuczek*).

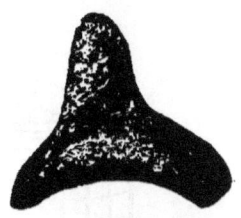

Fig. 147.—Transverse section of the human olfactory tract. *Glycerin preparation. Magn.* 15. The bundles of nerve-fibres appear dark.

In mammals endowed with a good sense of smell the tractus olfactorius is large enough to justify our designating it a distinct lobe of the brain (lobus olfactorius). In sagittal sections, stained with gold or with Weigert's hæmatoxylin-method, it is seen that a not inconsiderable number of fibres, in their course backwards, enter the grey layer of the tract, *cto* (fig. 148). This also is, therefore, a cortical centre for olfactory fibres, *1* and *2*. Such preparations show, too, fibres which, coming out of the cortex of the tract (*5* and *6*), turn backwards towards the brain and so represent the fibres which, having originated in the bulb, made their way into its cortex.

Returning to the human brain, we are able with the help of the amyloid bodies to follow the course of the olfactory tract farther back on the free surface of the substantia perforata anterior. Amyloid bodies affect especially the lateral white root. It is possible to follow this root some distance beyond the substantia perforata into the brain-substance on both sides of the corpus striatum. On the lateral surface of this nucleus we meet with a quantity of large, round, or spindle-shaped cells almost completely filled with light yellow pigment, 30 to 60 μ in diameter. They, too, are probably to be reckoned to the central apparatus of olfaction.

A strong bundle (*3, 6*), easily seen from the surface, goes from the tractus olfactorius into the temporal lobe—to the nucleus amygdaleus and the cornu Ammonis. In animals with a well-developed sense of smell a considerable tract (*5*) extends towards the anterior commissure. It is feebly developed in Man and apes.

The **anterior commissure** (*5, 7*) may be looked upon as supplimentary to the corpus callosum. Its function is to unite together identical points on the two hemispheres. It provides for those portions

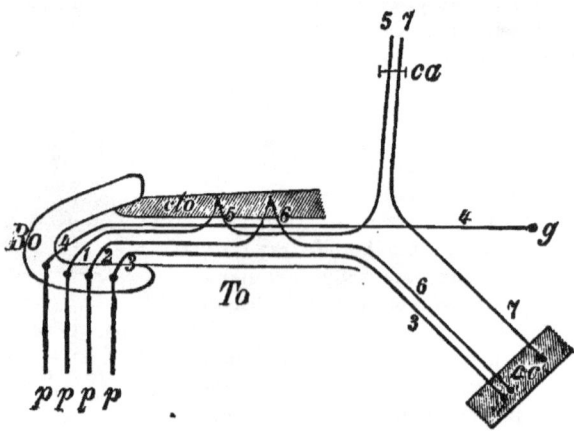

Fig. 148.—Scheme of the central apparatus of smell.—*Bo*, Bulbus olfactorius; *To*, tractus olfactorius; *p*, Schneiderian membrane; *cto*, cortex of the olfactory tract; *cc*, cortex cerebri; *g*, central brain-ganglia; *ca*, commissura anterior; *5*, olfactory portion of the anterior commissure; *7*, hemispheral portion of *ditto*.

of the cortex which are not supplied by the corpus callosum—part of the temporal lobes and perhaps also of the occipital lobes, as well as the cortex of the lobus (tractus) olfactorius. In Man the anterior commissure is seen on the under side of the nucleus lenticularis after it has forced itself into the substance of the hemisphere, having crossed in front of the ascending pillars of the fornix. Beneath the nucleus lenticularis it turns backwards and downwards, and so extends into the temporal lobe. This is its hemispheral portion (pars temporalis of *Ganser*). The olfactory portion of the anterior commissure is unimportant in the human brain in correspondence with the low development of the sense of smell. In all animals with a well-developed olfactory organ the olfactory portion of the anterior commissure is correspondingly considerable. It is relatively small in the monkey (fig. 139, *o*). There is present in Man a slender tract of fibres which, separating from the bundle which passes towards the anterior commissure, streams into the under border of the internal capsule and so reaches the front of the optic thalamus. It has been determined by *Ganser* that the anterior commissure contains commissural fibres alone and no decussating fibres [of the olfactory tracts]. Four kinds of medullated fibres may, therefore, be distinguished in the olfactory tract—

(1) Those from the bulb to the cortex of the tract (fig. 148, *1*, *2*).

(2) Those from the bulb which run in the tract, without coming into connection with its cortex, backwards towards other portions of the cortex (*3*), or to non-cortical ganglia (*4*, *9*).

(3) Fibres arising from the cortex of the tract, and extending *viâ* the anterior commissure to the cortex of the opposite side (*5*).

(4) Fibres from the cortex of the tract which run to other parts of the cortex or elsewhere in the brain (*6*).

It cannot be stated whether the very strong root of the olfactorius which passes to the nucleus amygdaleus and cornu Ammonis, consists of fibres of class 2 or class 3.

Besides the anatomical connections above named, there exist others which are either less easily determined in Man, or perhaps restricted entirely to the brains of certain animals. *Broca* describes a tract which courses backwards to the crus cerebri, and another, or upper root, which bends directly upwards to the frontal lobe.

A tract of fibres which extends from the temporal lobe obliquely forwards and inwards across the substantia perforata anterior towards the lower end of the gyrus fornicatus, is to be reckoned in with the central olfactory apparatus. This tract was first described by *Broca* as "la bandelette diagonale de l'espace quadrilatéral." It is only exceptionally seen in Man, in atrophied brains, as, for example, in old persons and in dementia paralytica.

If it is asked—To what portions of the cortex of the great brain do the olfactory nerves stand in direct relation?—The cortex of the olfactory tract itself should be mentioned first. To this can be added, in all probability, the nucleus amygdaleus, the anterior end of the cortex of the gyrus hippocampi, as well as, perhaps, the frontal end of the gyrus cinguli.

Gudden's extirpation-experiments have shown us that when the olfactory bulb is removed the gyrus uncinatus of the same side atrophies; so that there can be hardly any doubt about its relation to the olfactory centre. By comparing together the brains of different animals, *Broca* and *Zuckerkandl* have shown that it is very probable that the portions of the gyrus hippocampi attached to the uncus, as well as the anterior part of the gyrus cinguli, belong to the cortical centres of the olfactory nerve.

In those animals in which the olfactory organs are well developed, the gyrus hippocampi swells into a very large pear-shaped lobe on the base of the brain, and is elevated under these circumstances to the rank of a proper lobe of the brain, the lobus pyriformis.

The lobus pyriformis, which is smooth in most animals, but slightly fissured in the horse, tapir, and rhinoceros (*Zuckerkandl*), [even in Man the gyrus uncinatus is strikingly flatter than the convolutions to its outer side, clearly exhibiting its homology with the lobus pyriformis], is separated from the rest of the hemisphere by the scissura limbica. At least a trace of this fissure is to be recognised in most human brains (86 per cent., *Zuckerkandl*); it starts on the side of the island of Reil and runs in between the temporal pole and the uncus. The rudiment of this scissura limbica is shown [as a notch on the upper border of the temporal lobe] in figs. 24 and 33, *cf.* also 34.

The extent to which the cornu Ammonis is in intimate physiological connection with olfactory centres, is shown by the fact that it is quite rudimentary in the dolphin (*Zuckerkandl*) [and all other cetacea], and small in Man; whereas in animals with well-developed organs of smell it is large, and runs in company with the fornix far beneath the corpus callosum.

When we consider that sensations of smell, of taste, and of touch, as conveyed by the trigeminal nerve, are almost capable of being fused into a single perception in a way which is not possible with the other senses, such as sight and taste, we are prepared to believe (although direct anatomical proof is not yet obtainable) that the cortical terminations of the olfactory trigeminal and glossopharyngeal nerves either lie in the same neighbourhood or are, at any rate, very intimately connected by associating fibres.

[A consideration of the almost exclusive part played by sensations of smell in the daily life of most carnivorous animals would prepare us to expect that a very large portion of the cortex of their brains must be devoted to their reception. If the brains of a large number of different carnivores, such as is exhibited in the fine collection in the Hunterian museum, are compared together, it is obvious without need for measurement that the temporal lobe is very much larger in the dog, wolf, jackal, and other animals which track their prey with the nose, than in other mammals. Felines detect the proximity of their victims in forests and jungles rather by listening for broken twigs and crackling leaves and sticks than by sniffing along their trails; in carnivores of this habit, which might be distinguished as "springing" animals, the temporal lobe projects forwards to a less extent than it does in the "running" hunters. A comparison, however, of the aquatic otter with its terrestrial congeners is most instructive. The otter trusts to its sensitive whiskers for guidance amongst the snags and stones in the pools of brown water which the salmon frequent. Its sense of smell is extremely deficient, and, corresponding with this, its temporal lobe is reduced to very small proportions.

Herbivorous animals rely upon their sense of sight for safety. As far as possible they feed in open ground, keeping watchful guard. Doubtless they quickly discover any taint in the air when their enemies depart so far from their usual practice as to hunt from windward, but the use which they make of the nose to escape the enemy is not comparable in intensity or specialisation to the following of a trail which crosses and recrosses countless other lines of scent. On the other hand, they make selection of favourite herbs, and avoid poisonous ones with the aid of smell; so that the difference between the two classes of vegetable and animal feeders is one of degree rather than of kind, whilst carnivores are "osmatic" *par excellence*, herbivores cannot be justifiably termed "anosmatic." Indeed these terms are more likely to lead to confusion than to introduce order. Man, some quadrumana, and all marine mammalia are very deficient in olfactory apparatus. The sense of smell varies greatly amongst remaining mammals. But while it is impossible to speak in antithetical terms of the two divisions into which, from this point of view, they fall—the predatory and preyed-upon—as, the one, osmatic, and, the other, anosmatic, the relative preponderance of smell-perceptions as a substratum of mental processes must be very different in the two groups. Nor is it difficult to recognise the brain-characters upon which this difference depends. In carnivora the fissure of Sylvius is very oblique, and its margins are pressed close together; the temporal lobe projects a long way forwards beneath the frontal. In herbivora the fissure of Sylvius is more nearly vertical; its margins fall quickly apart, sweeping away from one another in easy curves; the temporal lobe does not project forwards. It is curious to notice the intermediate position taken up by the brain of the omnivorous root-hunting pig.

A study of the comparative anatomy of the brain throws much light upon questions of cortical localisation, and will probably be the ultimate tribunal to which all experimental evidence will be submitted. The outlines of the three brains given below (figs. 149, 150, 151) are tracings taken from the drawings in Gratiolet's atlas (published before localisation of function in the cortex was thought of), reduced by the pantograph to about the same size. They show the relative development

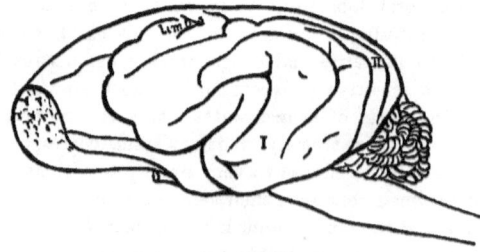

Fig. 149.—Brain of dog.

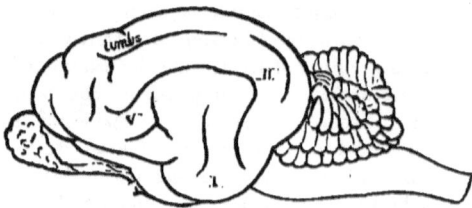

Fig. 150.—Brain of cat.

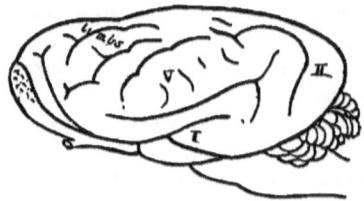

Fig. 151.—Brain of otter.

Tracings of the pictures in Leuret and Gratiolet's Atlas. The Roman numbers indicate approximately the cortical areas, the development of which in different animals varies as the cross-section of the several sensory nerves, to which they correspond.

of the temporal lobe in animals with an acute, moderate, and feeble sense of smell respectively.

Broca included the gyrus fornicatus in the cortical field of smell. That this is an error is shown by the fact that the gyrus fornicatus is well developed in the marine mammalia.]

Absence of the olfactory tract on one or both sides has been repeatedly observed in otherwise normal brains.

Kundrat associates together all forms of defect of the olfactory nerves under the name of arrhinencephalia, allowing, however, that other extensive defects in the structure of the brain are associated therewith.

2. Optic Nerve.

—Only the exceedingly short fibres which lead from the rods and cones to the nerve-cells of the retina can be termed peripheral nerves of sight in the proper sense of the word. The retina, as well as the fibres which originate in it, must be regarded as parts of the central nervous system. As is well known, both retina and optic nerves originate from a vesicular outgrowth of the fore-brain, which appears very early in development (primary optic vesicle). The column of fibres which constitutes the optic nerve differs from a peripheral nerve in that, if it is cut, its two ends will never grow together. This appears to be a differential character of all central tracts of fibres.

We will not in this place treat of the minute structure of the retina (fig. 152, R); but will confine our attention, in the first instance, to the optic nerve which, composed of thin medullated fibres, leaves the orbit as a round column, flattening out a little after it has entered the cavum cranii. It runs towards the basis cerebri, and forms, in front of the tuber cinereum, the optic chiasm with the nerve of the opposite side, *Ch* (fig. 152). From the optic chiasm the "optic tracts," *Tro*, extend backwards and outwards. According to *Salzer's* measurements, the optic nerve of Man has an average cross-section of about 9 sq. mm., reduced to 8 sq. mm. by deducting the space occupied by connective-tissue septa. The number of nerve-fibres averages about 438,000, a number which can only be understood if their great tenuity is borne in mind.

The optic nerve-fibres are collected into irregular bundles, round or polyhedral in form, which are separated from one another by thicker or thinner septa derived from the sheath of pia mater which surrounds the nerve. Secondary septa, rich in nuclei, enter the substance of the separate bundles.

The peripheral bundles which lie nearest to the pia, as well as the central bundles which border the arteria centralis, invariably atrophy to such an extent that the fibres are found (except in new-born children) to have completely disappeared; only the empty supporting connective-tissue remaining behind (*E. Fuchs*).

It is beyond doubt that there are three kinds of fibres in the optic chiasm :—

(1) Fibres from the lateral halves of the retinæ, which occupy the

OPTIC CHIASM.

lateral borders of the chiasm, and go to the optic tract of the same side.

(2) Fibres from the mesial halves of the retinæ, which cross in the chiasm to the tract of the opposite side.

(3) Fibres which occupy the back of the commissure and extend

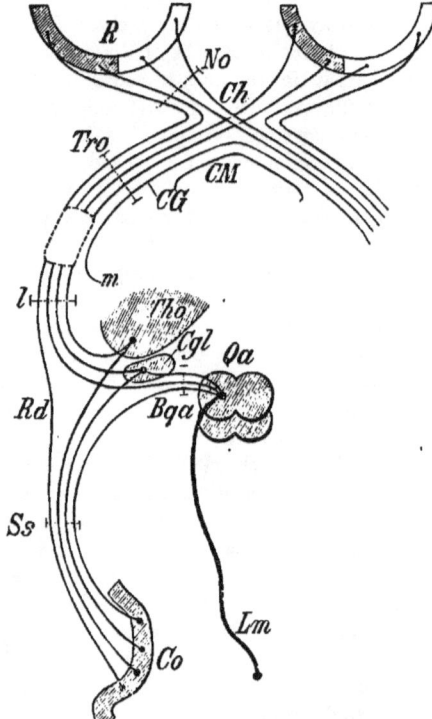

Fig. 152.—Scheme of the central apparatus of vision.—*R*, Retina, dark on the side connected with the left, light on the side connected with the right hemisphere; *No*, nervus opticus; *Ch*, chiasm; *Tro*, tractus opticus; *CM*, Meynert's commissure; *CG*, Gudden's commissure; *l*, lateral division of the tract; *m*, mesial division of the tract; *Tho*, thalamus opticus, *Cgl*, corpus geniculatum laterale; *Qa*, anterior tubercle of the corpora quadrigemina; *Bqa*, brachium of *ditto*; *Rd*, direct cortical root; *Ss*, sagittal fibres of the corona radiata to the occipital cortex [optic radiations]; *Co*, cortex cerebri (of the cuneus); *Lm*, mesial fillet.

from one tract to the other. They are distinguished by their fineness, *CG*, Gudden's commissure (commissura inferior, commissura arcuata posterior).

Even if other kinds of fibres are present in it, the three kinds just mentioned constitute the bulk of the chiasm.

Many anatomists regard it as certain that there exists an anterior commissure lying in the anterior angle of the chiasm and connecting the two retinæ. The relation to one another of the crossed and uncrossed portions of the tract varies exceedingly in different animals. It appears that in Man the uncrossed fibres preponderate, whilst in the lower mammals more crossed fibres are present. In many fishes uncrossed fibres are totally wanting. In the mole the optic nerves are rudimentary, and contain only a few poorly medullated fibres; the white inferior commissure is consequently very obvious.

In Man and other mammals the interweaving of the fibres in the chiasm is so intimate that very little light is thrown upon their relations by making sections; the degeneration-method first helped to clear the matter up. *Ganser* delineates a human brain in which the uncrossed bundle on the right side runs from optic tract to optic nerve as a distinctly isolated column.

In lower animals, and especially in fishes, the nerves cross in coarse bundles. In many fishes the optic nerves simply lie across one another without entering into a chiasm at all.

The **optic tract,** *Tro*, starts from the chiasm. At first it lies close up against the grey basal substance of the brain, but subsequently it rests upon the crus cerebri, around the most anterior free portion of which it winds.

It is easy to convince oneself that the optic tract splits in the human brain into two roots. The lateral (anterior) root runs towards the lateral geniculate body. The mesial (hinder) root runs towards the mesial (internal) geniculate body (figs. 11 and 12).

(1) Part of the **lateral root** (fig. 152, *l*) enters the lateral geniculate body, *Cgl*.

In Man, and still more in apes, the **external geniculate body** is heart-shape in horizontal section, the apex being directed forwards. So deeply is it split that in certain frontal sections it often happens that we see two separate pieces, while only the sections farther forwards show the segments united into one.

The structure of the corpus geniculatum laterale, *Cgl*, is so characteristic that it is always easy to recognise. It consists of layers of grey and white matter irregularly rolled in one another (fig. 135). The white strata are formed for the most part of fibres of the optic tract. The grey layers are of two kinds, some of them consisting of large round nerve-cells, others of closely-agglomerated little cells.

A considerable portion of the external root, however, does not enter the lateral geniculate body, but passes on to the optic thalamus, *Tho*, or to the anterior corpus quadrigeminum, *Qa*. Many bundles of fibres slip under the external geniculate body, so as to reach the back part of the thalamus, the pulvinar, producing its radial striation. Other fibres extend farther forwards on the surface of the geniculate body, and take part in the formation of the white layer which covers the thalamus (stratum zonale thalami). Very little is known about the ending of these fibres, especially those last mentioned.

Lastly, fibres sweep over the geniculate body into the brachium anterius, Bqa, and so to the anterior quadrigeminal body of the same side, Qa. Thus it comes about that the external root of the tract is in connection with the optic thalamus, the external geniculate body and the anterior corpus quadrigeminum. These three grey masses have this in common, viz., they all give fibres to the corona radiata. The fibres join the sagittal medullary tract of the occipital lobe (*Wernicke*), Ss (fig. 21), which comes from the posterior third of the posterior limb of the internal capsule skirts the outer side of the posterior horn of the lateral ventricle and runs to the cortex of the hinder portion of the great brain.

From the corpus quadrigeminum anterius the fibres of the corona are carried to the sagittal medullary tract by the brachium anterius. The cortical ending of the optic tracts will be treated of later on.

(2) The **mesial root** of the optic tract is easily followed to the **mesial geniculate body** in which some of its fibres end. This is a grey oval body united below the surface with the thalamus. Nerve-cells of medium size are scattered through it, fairly uniformly; they are somewhat more closely packed in its ventral portion than elsewhere. The fibres which enter the corpus geniculatum mediale are continued by the brachium posterius, Bqp, to the posterior quadrigeminal body. Another portion of the fibres of the internal root go over the mesial geniculate body to the anterior tubercle of the corpora quadrigemina; while still another set of fibres go directly into the posterior tubercle, perhaps without interruption in the mesial geniculate body. In the posterior brachium fibres extend towards the great brain attaining to its cortex, so far as it is possible to judge. The bundle of fibres which passes from the mesial geniculate body to the hemisphere seems to attain to the temporal lobe, for *Monakow* found atrophy of the mesial geniculate body after extirpation of this portion of the cortex.

J. Stilling describes as a third or middle superficial root those fibres which run between the two geniculate bodies to the anterior tubercles of the corpora quadrigemina.

The fibres which branch off to the crus cerebri in front of the geniculate bodies, might be designated a deep root. They remain for a short distance in the outermost part of the crus (*Wernicke*), and then join themselves to the sagittal medullary tract of the occipital lobe as a direct cortical root of the optic tract, Rd (direct hemispheral bundle of *Gudden*). Probably the direct cortical root contains fibres from both optic nerves. *J. Stilling* asserts that he has followed a portion of these fibres spinalwards in the crus as far as the crossing of the pyramids (radix descendens). *Darkschewitsch* says that this descending bundle receives its myelin sheaths considerably earlier than the proper optic fibres, and is, therefore, to be distinguished from them.

In the tuber cinereum, and in the portion of the anterior perforated substance over which the optic tract extends, lie large yellow pigmented nerve-cells, first described by *J. Wagner*. Nerve-fibres characterised by their considerable calibre take origin in these cells, and course backwards in the optic tract, *CM* (Meynert's commissure). In Man they are separated from the optic tract by a thin layer of grey matter.

These fibres soon leave the vicinity of the tract, traverse in curves the pes pedunculi, and seem to end in the corpus subthalamicum.

A root of the optic nerve, not to be overlooked, passes from the chiasm directly into the central grey matter of the third ventricle.

When both optic nerves degenerate, a large part of both optic tracts also comes to grief, as well as the lateral geniculate body, the anterior tubercle of the corpora quadrigemina, and the back of the thalamus (the pulvinar). Portions of the tract, however—namely, Meynert's and Gudden's commissures—remain intact; since they have, as these experimenters proved, nothing to do with the optic nerve itself, for they play no direct part in the act of seeing. Since the mesial geniculate and posterior quadrigeminal bodies do not suffer, we must accept it for a fact that the fibres of the commissura inferior (Gudden's commissure) run in the inner root of the optic tract, although they are not the only fibres which it contains. Very little is definitely known with regard to these remaining elements of the inner root.

Gudden's tractus peduncularis transversus, *Tpt* (fig. 12) is a portion of the brain which atrophies after degeneration of the optic nerve. It begins in the anterior quadrigeminal body, passes obliquely across the crus cerebri, and finally streams into the same; very little is as yet known concerning its anatomical connections.

Finally, *Darkschewitsch* finds that after extirpation of one eyeball, a bundle atrophies which leaves the tract on the side opposite to that on which the operation has been performed in the neighbourhood of the lateral geniculate body, extends through the thalamus and the pedunculus conarii to the pineal body, and so having crossed over to the side of the operation again is supposed to reach the oculomotor nucleus through the ventral portion of the posterior commissure. It possibly takes part in reflexes of the pupil.

After what has been said it will be understood that the lateral geniculate and anterior quadrigeminal bodies, as well as the thalamus (and, indeed, the ganglion-cells of the retina ought also to be included), constitute the primary centres of the optic nerves. These grey masses mediate between the optic nerves and the cerebral cortex through the sagittal medullary layer of the occipital lobes; they also serve to bring together other parts of the brain—*e.g.*, the corpora quadrigemina connect the optic nerves with the nuclei of the eye-muscle nerves. A direct connection between optic nerves and cerebral cortex is also found in the direct cortical root of the tract.

The parts of the cortex which are to be regarded as the terminals of the optic fibres, the **cortical visual centres,** *Co*, are already fairly well known. *Ferrier* and *Yeo* localise the visual centre in the occipital lobe and the angular gyrus as the result of their experimental investigations. *Séguin* feels justified in asserting that the optic radiations terminate chiefly in the cuneus. *Exner* comes to the conclusion that the most concentrated or active portion of the cortical field of vision is to be looked for at the upper end of the gyrus occipitalis primus.

Despite the diversity of opinion, one will probably not go far wrong in placing the cortical visual area in the occipital lobes, and of these lobes the most probable seat of vision is the cuneus. At the same time the fact must never be lost sight of that each visual centre is connected in a partially crossed manner with both eyes.

It is impossible to explain the physiological relation between the corpus quadrigeminum posterius, the mesial geniculate body, and the inferior commissure of one side, and the proper central apparatus of vision on the opposite side. Just as little is known regarding the function of Meynert's commissure.

A few words concerning the minute structure of the corpus quadrigeminum remain to be written.

In the **anterior tubercle** a number of layers are to be distinguished, but not clearly, in carmine-preparations at any rate.

. We have already called attention to the very distinct arch of medullated fibres which is seen in transverse sections sweeping through the anterior quadrigeminal bodies over the aqueduct (*cf.* figs. 134, 135). The central grey matter which surrounds the aqueduct is cut off fairly sharply from the region which lies on its dorsal and lateral sides, reaching as far as the brachium posterius, *Brqp*, and belonging to the anterior corpus quadrigeminum.

Proceeding from without inwards (fig. 135) we meet with :—

(1) A thin peripheral layer of white fibres which probably originates directly in the optic nerve (stratum zonale or superficial medullary layer). In most mammals this layer is so thin that the corpora

quadrigemina appear not white, as in Man, but grey, owing to the underlying grey matter showing through.

(2) A not very thick layer of grey substance, the nerve-cells contained in which are small and few (peripheral grey layer, cappa cinerea, stratum cinereum).

(3) Grey substance with small nerve-cells and numerous sagittally running fine nerve-fibres which originate in the brachium anterius (strato bianco-cinereo superficiale, *Tartuferi*). *Ganser* divides this layer into three, the outer and inner containing more fibres, the middle more grey matter. This region corresponds to the proper nucleus of the anterior tubercle, Nqa, but it is difficult to delimit it from the second layer.

(4) The fourth layer, which is sharply marked off from the central grey substance around the aqueduct (strato bianco-cinereo profondo, deep medullary layer, layer of the fillet), consists of grey substance with cells like those of the preceding layer, and nerve-fibres which become closer and closer the greater the depth from the surface, and arch around the grey matter of the aqueduct. Probably a large number of them arise in the fillet (see p. 257). The innermost fibres of this layer have no further relation to the corpora quadrigemina, but belong to the descending root of the fifth, and are recognised by the occurrence amongst them of occasional large vesicular cells which cannot be confounded with the other cells of this region. Besides the fillet and the fibres of the fifth nerve a bundle of tegmental fibres is also found in this place, which courses towards the middle line and enters the fountain-like tegmental decussation.

The crossing above the aqueduct in the anterior quadrigeminal area is formed of fibres of the fifth nerve, the fillet, and, perhaps, also the tegmental fibres of the fontanal decussation. Other elements, such as inter-quadrigeminal commissural fibres, are, in all probability, also present.

In immediate proximity to this decussation sagittal sections carried farther forward exhibit the **posterior commissure.** We have already found in this commissure (fig. 136) a tegmental tract which passes to the thalamus of the opposite side, as well as a tract which passes through the thalamus and the pineal body to the oculomotor nucleus of the opposite side. The remaining more considerable mass of the posterior commissure is not properly understood. It seems as if fibres of the fillet, perhaps also fibres from the posterior longitudinal bundle and the brachium anterius, enter into this commissure. It is always necessary to distinguish, as *Darkschewitsch* does, a dorsal and a ventral part of the posterior commissure. In the former, fibres from the deep medullary layer of the corpus quadrigeminum are supposed to extend to the cortex of the opposite side.

A slight radial striation is sometimes recognisable in the anterior quadrigeminal body on slight magnification. This is due to the entering vessels which take this direction, and also to numerous fibres which extend from the fillet radially outwards towards the superficial layer. Other radial nerve-fibres are described by *Meynert* and *Tartuferi* as passing from the corpora quadrigemina into the central grey substance around the aqueduct, and so making a connection with the nuclei of the eye-muscle nerves which lie there.

The great number of "spider" cells, each very rich in processes, which lie in this region of the anterior corpora quadrigemina, and are supposed to give to it its relative firmness and hardness, must also be pointed out.

The anterior quadrigeminal body is certainly in connection with the following parts of the brain:—

(1) Directly with the optic tract through the anterior brachium.

(2) With the lateral geniculate body, and so indirectly with the optic tract.

(3) With the cortex of the occipital lobe through the anterior brachium and the sagittal medullary layer.

(4) With the spinal cord (posterior column) through the mesial fillet.

(5) With the nuclei of the eye-muscle nerves.

In the rabbit, according to *Darkschewitsch*, the fibres for the optic tract come from the anterior two-thirds of the anterior corpus quadrigeminum of the same side, and principally from the outer part of its surface, while its mesial side gives origin to the fibres for the cortex cerebri. When the occipital lobe is destroyed in a new-born animal atrophy affects the same ganglionic masses as when the optic nerves are destroyed—*viz.*, the corpus geniculatum laterale, corpus quadrigeminum anterius, and part of the thalamus opticus on the same side (*Gudden, Monakow, Ganser*); but, in addition, the tractus opticus and tractus peduncularis transversus of the same side also atrophy.

After destruction of the optic nerves, it is the third layer of the corpora quadrigemina which comes to grief. In the mole and bat it is badly developed. We may, therefore, conclude that this layer stands in direct connection with the optic tract, while the inner, more deeply-lying, medullary part of this layer is connected with the occipital cortex by way of the internal capsule; these latter fibres also atrophy after destruction of the just-mentioned portion of the cortex (*Ganser*).

In the **posterior tubercle** of the corpora quadrigemina, as in the anterior, a stratum zonale is recognised, under which lies (in Man) a biconvex grey body, the ganglion of the corpus quadrigeminum posterius. For a considerable distance the ganglia of the two sides are continuous with one another in the middle line above the aqueduct. They contain but few large nerve-cells. Ventrally they reach almost as far as the descending root of the trigeminus. Fibres extend into the posterior brachium from the anterior and lateral portions of these grey masses, and thence, probably, reach the great brain; while other fibres, which come out in the posterior brachium, presumably form the principal connection of the inner root of the optic tract. Fibres from the lateral fillet are seen to enter the ventral and lateral portions of the ganglion. Above the aqueduct in this region also there is a decussation into which part of the fillet enters.

The connections of the posterior tubercle are far less clear than those of the front one :—

(1) An indirect connection with the inner root of the tract through the mesial geniculate body. Possibly there is a direct connection also.

(2) With the cortex cerebri. This, as well as (1), is effected by the brachium posterius.

(3) With parts of the system lying spinewards (the auditory centres especially) *viâ* the lateral fillet.

3. Oculomotor Nerves (Third Pair).

—The root-bundles of the nerve for the eye-muscles in general, originate in several groups of nerve-cells, which lie in the mid-brain below the anterior tubercles of the corpora quadrigemina (and, perhaps, a little farther forward than these tubercles in the floor of the third ventricle). The nuclei lie to the dorsal side of the posterior longitudinal bundles, $NIII$ (figs. 134, 135, 136). The whole oculomotor nucleus extends about 5 mm. in a sagittal direction.

That portion of the oculomotor nucleus which lies near the middle line, immediately dorsal to the posterior longitudinal bundle, is called also the chief nucleus. It is distinguished by its large cells.

Quantities of smaller cells lie between this nucleus and the aquæductus Sylvii, especially in the anterior half of the oculomotor region. These cells are united into groups, of which one, lying close to the middle line and occupying a considerable space in a dorso-ventral direction, is constant—the mesial oculomotor cell-group (*Edinger*, *Westphal*). It is well defined on its lateral side by a descending bundle of medullated fibres (fig. 135). The cells which lie laterally to this are associated into a less distinct group disposed transversely (*Westphal's* lateral group).

The upper oculomotor nucleus of *Darkschewitsch* is probably identical with these groups of small cells.

The way in which the fibres of the oculomotor roots curve through the tegment towards their point of exit has been already described.

In the most distal (posterior) sections in which root-bundles of the oculomotor nerve are still to be seen, they are usually found far to the side, leaving a large interval between themselves and the raphe.

The point of exit of most of the oculomotor fibres is to be looked for, as we know, in the trigonum interpedunculare, and especially in the sulcus oculomotorius. Not rarely single bundles traverse the crusta. This always happen in the case of the bundle which is sometimes present as an abnormal lateral root (p. 63).

Gudden has proved that in the rabbit the origin of the oculomotor nerves is a half-crossed origin. In this animal the nucleus on each side is divided into two clumps. The ventral group of cells belongs to the nerve of the same side, the dorsal group gives origin to fibres which cross to the opposite side. The ventral group lies a little farther forwards than the dorsal one, and is itself divided into two groups lying one behind the other.

All motor nerves have as already remarked a double origin; part crossed, part uncrossed. The crossed origin has not yet been proved for the oculomotor in Man, although the eye-muscles exhibit remarkable bilateral symmetry in action. The trochlear nerve comes almost wholly from the opposite side of the body, but it does not simplify matters to look upon the two nerves as one, since they supply different muscles, whereas, we must suppose that in the typical arrangement, each muscle is governed from both sides of the brain. We are, therefore, driven to believe that the crossed origin of the oculomotor described by *Gudden* in the rabbit, holds good for Man also. This we may do with the greater confidence if we remember that there are many fibres crossing the middle line in the oculomotor region.

Numbers of fibres commissural between the two oculomotor nuclei are seen crossing the middle line in the brains of kittens. These are most numerous in the posterior part of the region. They myelinate early (*Nussbaum*). Perhaps some of these fibres are decussational not commissural.

Duval and *Laborde* have shown that the oculomotor nerve of one side is connected, by means of the posterior longitudinal bundle, with the nucleus of the abducens nerve on the other side. It appears that the fibres take origin from the anterior pole of the abducens nucleus, sink somewhat ventrally in their course through the tegment, and not far behind the oculomotor nucleus, go across to the opposite side in the dorsal tegmental decussation (*Nussbaum*). Here they meet with root-fibres of the oculomotor nerve, with which they join company, on the mesial side. This is an anatomical datum for explaining the harmonious working of the external rectus muscle of one side and the internal rectus of the other.

Hensen and *Völker's* experiments on the dog show that the individual terminal branches of the oculomotor nerve originate in different portions of the nucleus, which are arranged one behind the other in physiological order, although, anatomically, they are separated imperfectly. Farthest brainwards lies, in the dog, the nucleus of origin for the nerves of accommodation, behind this the centres for the sphincter iridis, for the rectus internus, rectus superior, levator palpebræ, rectus inferior, and, last of all, for the obliquus inferior.

[It is easy to trace the harmony between the arrangement of these nuclei in anatomical sequence from behind forwards, and the several stages in the act of searching for an object and concentrating the gaze upon it. The head being first turned in the required direction, with adaptive movements of the oblique and other muscles of the eye, the object is searched for on the ground near the feet, whence the eyes are directed outwards over the plain (superior recti), the lids being lifted from before the pupils. As soon as the object is found the eyeballs are converged upon it (by the internal recti), the size of the pupil is regulated to the amount of light (by the sphincter iridis), and, lastly, the lens is focussed for the distance. Although these several actions occur simultaneously as far as we can tell, it is clear that the position of the several nerve-centres coincides with the order in which the movements have been evolved.]

It appears that the fibres for the uppermost portion of the facial nerve, especially for the part supplying the orbicularis palpebrarum, originate in the most posterior part of the oculomotor nucleus (*Mendel*). The observations of *Kahler* and *Pick* that the pupillar fibres of the oculomotor nerve run in its anterior bundles, agrees well with the above experimental results.

The posterior bundles of fibres are regarded as destined for the outer eye-muscles; they are divided into a lateral group (for the levator palpebræ, rectus superior, and obliquus inferior, which have a close functional connection), and a mesial group (for the rectus internus and rectus inferior).

It must be allowed that the oculomotor nucleus is closely connected on the one side with the central mechanism of sight, and on the other side with motor regions of the cortex; our knowledge on these points is still, however, very imperfect. As coming within the former category must be mentioned the radial fibres which stream from the anterior quadrigeminal body into the central grey matter of the ventricle, in which the oculomotor nucleus is embedded (p. 284). *Darkschewitsch* finds that fibres extend from his "upper oculomotor nucleus" to the ventral portion of the posterior commissure, and farther on through the pineal gland and its peduncle, reach the region of the lateral geniculate body and the optic tract (p. 281). According to *Bechterew's* view, which is not as yet supported by sufficient anatomical data, the fibres of the oculomotor nucleus which subserve pupillar movements are supposed not to extend backwards in the optic tract, but to leave at the chiasm for the brain-substance, entering the central grey matter of the ventricle and extending to the oculomotor nucleus of the same side.

The connection between the oculomotor nucleus and the cortex cerebri may be looked for in all probability in the fibres which pass from the nucleus to the raphe, cross one another at an acute angle, and extend ventrally into the pes pedunculi, on the mesial side of which they place themselves. It has not yet been settled to what part of the cortex cerebri these fibres pass in the corona radiata. The same has to be said with regard to the relation of other eye-muscle-nerves to the cortex. In some cases of cortical disease (especially of syphilitic origin) ptosis is the only symptom present as far as the eye-muscle-nerves are concerned, so that it appears that the cerebral path of the fibres for the levator palpebræ, in its course towards the cortex, separates from the other eye-muscle-tracts. The cortical centre for the levator palpebræ has been looked for in the gyrus angularis, since circumscribed disease of this part of the cortex is sometimes associated with paralysis of the opposite eyelid.

4. Trochlear Nerve (nervus patheticus, fourth pair).—

The origin of the nerve which supplies the superior oblique (anterior trochlear nucleus) is to be looked upon as the distal continuation of the nucleus of the oculomotor nerve, from which it is not, as a rule, sharply defined. It also lies on the dorsal side of the posterior longitudinal bundle and partly also in a groove in the same, NIV (fig. 133). Since the nucleus of the trochlear nerve lies in the plane of the front of the posterior quadrigeminal body, while it takes its exit from the brain much farther back, at the front of the velum medullare anterius, it follows that its intracerebral course must be of considerable length. Its course spinewards within the brain is somewhat complicated. The root-fibres which originate from the lateral part of the nucleus slope outwards across the dorsal surface of the posterior longitudinal bundle (crus of origin, nuclear crus); they are then collected on the mesial side of the descending root of the trigeminal nerve into two or three round bundles, which next bend spinewards and somewhat dorsally (the middle piece or descending crus), IV^1 (fig. 132). When they reach the front of the velum medullare they lie close up against the dorsal edge of the descending trigeminal root; here they turn abruptly over towards the opposite side in the roof of the aqueduct of Sylvius, IV^2, and take their exit by the side of the brachium conjunctivum (crus of exit), IV^3. As they curve round the proximal angle of the fourth ventricle they are almost the only things to be seen in the section. Nothing in brain-anatomy is more certain than the crossing of the trochlear nerve in the velum medullare anterius; nevertheless, it is by no means impossible that a certain number of the fibres of the trochlear nerve take their exit with the issuing root of the same side. If this be the case the trochlear nerve is no exception to the rule with regard to the partial crossing of motor nerve-roots, although it would still be exceptional in the preponderance of its crossed root-bundles. *J. Stilling* describes a fine root which comes out of the cerebellum, runs brainwards through the lingula and joins itself, perhaps without crossing, to the trochlearis. A rounded group of the most minute nerve-cells, which lies immediately to the spinal side of the proper (anterior) trochlear nucleus (*cf.* fig. 132, in which it is not lettered) is considered by *Westphal* to belong to the trochlear nerve (Westphal's trochlear nucleus or posterior trochlear nucleus).

We may suppose that the connections of the trochlear nerve with the great brain (*viâ* the raphe) with the corpora quadrigemina anteriora and with the posterior longitudinal bundle, is the same as that already described for the oculomotor nerve.

When the intracerebral course of the trochlearis is studied in animals certain important differences, especially as regards its relations to the descending root of the fifth nerve, attract one's attention. In the monkey, in which the nerve is comparatively well developed, we find the same conditions as in Man. In the cat and the dog its descending portion lies to the outer side of the descending root of the trigeminus. In the horse it lies so close to the outer side of the root of the trigeminus that its bundles, as seen in transverse section, are disposed, not in a straight line, but in an arch convex mesially; the curve is due to its course medianwards towards the velum medullare. Its fasciculi pierce the trigeminal root and interlace with its fibres to such an extent that some of the easily-recognised cells of the trigeminal root get displaced into the bundles of the trochlearis. In lower animals (rodents) the interlacing of the trigeminal and trochlear roots is usually still more intimate. In all mammals, and also in birds, the decussation of the trochlearis can be easily demonstrated.

It is not impossible that the posterior longitudinal bundle may enter into connection with the nucleus or with the root-fibres of the trochlear nerve. *Nussbaum's* observations on kittens' brains afford no support to the idea that there is a crossed relation between the root-fibres of the trochlearis and the abducens nucleus (as taught by *Duval* and *Laborde*).

5. Nervus Abducens (nervus oculomotorius externus, sixth pair).—The nerve to the external rectus muscle of the eye is only considered in this place on account of its relation to the other eye-muscle-nerves.

We met with the nucleus of origin of the nervus abducens, NVI (facialis abducens nucleus, upper facial nucleus, fig. 127), in the tegmental region in the vicinity of the knee of the facial nerve. It appeared as a fairly well-defined almost globular nucleus, its vertical axis only being somewhat extended, made up of large stellate cells. The separate bundles of the abducens nerve traced dorsalwards through the tegment in gently curving arches, apply themselves to the mesial side of the nucleus, curl round its dorsal side, and extend in some cases as far as its lateral side, sinking successively into its substance, to unite with the axis-cylinder processes of its cells. A very small and easily overlooked portion of the abducens turns medianwards beneath the nucleus, extends to the raphe, traverses it, as it appears, as far as its dorsal edge, and then, passing beneath the ascending crus of the facial nerve, enters the abducens nerve of the opposite side.

Since the facial root, during a great part of its course, lies on the abducens nucleus, and fibres as a matter of fact seem to come out of the nucleus and apply themselves to the facial nerve, it is easy to make pictures which give an illusory appearance of a partial origin of the facial nerve in this group of cells. *Gudden* and *Gowers* have, however, proved that the facial nerve has no connection with the grey mass in question, and that the fibres of the facial nerve which seem to come out of it really only cross it.

The abducens nucleus may be connected with the great brain by fibræ arcuatæ, which cross in the raphe and continue their course ventralwards to the pyramidal tract. We have already pointed out the connection between the abducens nucleus and the posterior longitudinal bundle, and the part which it, by this means, takes in the formation of the opposite oculomotor nerve.

It is supposed that a direct obvious bundle of fibres goes from the abducens nucleus to the upper olive, *Ost* (stalk of the upper olive, fig. 155); since the latter stands in close relation with the auditory nerve it is possible that a nerve-route is thus provided by which the reflex movement of the eyes in the direction of the sound may be effected.

Disease of the nuclei of the eye-muscle-nerves, which is, in many cases at least, analogous with poliomyelitis, is called, after *Wernicke*, poliencephalitis superior or nuclear eye-muscle paralysis (opthalmoplegia externa nuclearis *sive* progressiva).

6. Trigeminal Nerve (par quintum).—If a line is drawn

through the substance of the pons, inclining, from the point of exit of the trigeminal nerve, a little spinewards towards the angle which the floor and roof of the fourth ventricle make with one another, a region, called on this account the convolutio trigemini, is met with, in which the root-fibres of the trigeminus converging from very various regions of the brain come in contact with certain of its nuclei of origin.

It will be easier to impress upon the memory the several bundles which converge towards this point if we bear in mind that they come from very different directions, from the spinal cord and from the cerebrum, from the side and from the middle line; and that they unite with other bundles which come from the sensory and motor nuclei of this region itself to form the two peripheral roots of the nerve (figs. 129 and 154, NVm and NVs; fig. 153, Nm and Ns).

The trigeminal nerve leaves the pons in two roots, the sensory much the larger (portio major), and the smaller motor root (portio minor). The two roots have completely different origins.

The point of exit from the pons lies somewhat in front of the region in which the converging roots collect (figs. 153 and 154). Hence the planes which we have chosen for our sections never show the trigeminal in its full extent as it traverses the pons. The sensory root, Rs (fig. 153), extends from the surface to the convolutio trigemini in a straight line; the motor root, Rm, in an arch convex brainwards. The origin of the several separate root-fibres will now be traced.

(1) By far the most important of all the origins of the trigeminal can be followed as far downwards as the second cervical nerve. We have learnt to recognise the crescentic bundle which lies on the outer side of the substantia gelatinosa, increasing in size as the sections are carried farther forwards and known as the **ascending root of the fifth** (racine bulbaire—in figs. 118 to 128, Va; in fig. 153, 1). It cannot be said with certainty whether the fibres of this root take origin in the substantia gelatinosa (which we do not at all understand from a histological point of view), or from the cells of the posterior horn (*Bechterew*); the latter view appears the more probable. Numbers of fibres are seen, in longitudinal sections, coming out of the centre of the posterior horn traversing the substantia gelatinosa and joining themselves to the trigeminal root. Since, however, remains of the substantia gelatinosa lie in the concavity of the ascending root, even as far forwards as the point at which it bends outwards, this substance must be in some way related to the trigeminus. We know no more than this.*

* See the discussion of this question by the *translator* on p. 185. The formation known as substantia gelatinosa Rolandi lies on the concave side of the brush of

294 CONSTITUENT PARTS OF THE TRIGEMINUS.

Longitudinal sections parallel to the floor of the fourth ventricle allow one to follow the ascending root of the trigeminus in its whole extent, they show that it turns into the sensory root exclusively (fig. 154).

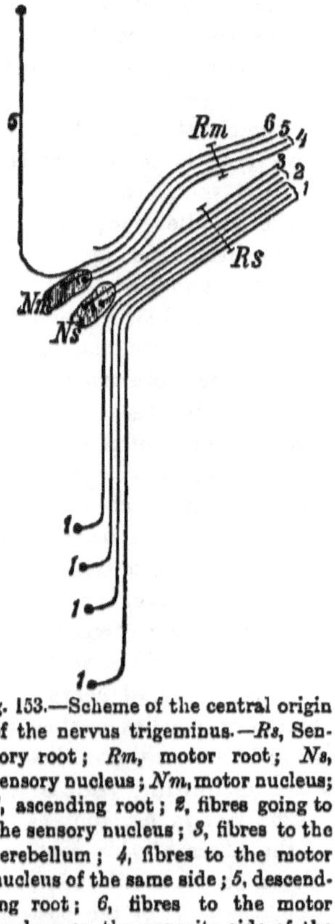

(2) The trigeminus appears to receive a **lateral addition** in the form of a bundle (fig. 153, *3*) which descends from the cerebellum on the side of the brachium conjunctivum; it joins the sensory root. *Bechterew* denies this cerebellar origin.

(3) A not inconsiderable number of fibres come to the trigeminus from the **middle line.** They extend more or less obliquely towards the plane underlying the ependyma, and occasionally traverse the posterior longitudinal bundle; they are partly visible in fig. 129, *Vx*. These fibres are of several kinds :—

(*a.*) Root-fibres arising out of the motor nucleus of the opposite side, and, perhaps, also out of the sensory nucleus.

(*b.*) Crural fibres which bring the trigeminal nerve into connection with the cortex cerebri by way of the raphe, and ought not, therefore, to be looked upon as root-fibres.

(*c.*) The so-called crossed descending root of *Meynert.* This observer says that the nerve receives an accession of fibres from the large darkly-pigmented cells of the locus cæruleus (substantia ferruginea) by means of a bundle which runs from this group of cells close beneath the floor of the ventricle median-

Fig. 153.—Scheme of the central origin of the nervus trigeminus.—*Rs*, Sensory root; *Rm*, motor root; *Ns*, sensory nucleus; *Nm*, motor nucleus; *1*, ascending root ; *2*, fibres going to the sensory nucleus ; *3*, fibres to the cerebellum ; *4*, fibres to the motor nucleus of the same side ; *5*, descending root ; *6*, fibres to the motor nucleus on the opposite side of the fourth ventricle.

fibres, into which all sensory nerves spread out as they enter the cerebro-spinal axis. The substantia gelatinosa is hardly found in the mid-brain, for the nerves proper to this part are purely motor. Nor is it found in the fore-brain, for the two sensory nerves which belong to this region, the olfactory and optic, have their gelatinous substance placed peripherally in the olfactory bulb and retina respectively.

wards to the raphe, and after crossing and piercing the posterior longitudinal bundle turns a little spinewards to join the sensory trigeminal root. Some isolated cells of the substantia ferruginea are to be found deep in the substance of the tegment and others in the roof of the ventricle.

(4) The trigeminus receives an important accession of thick fibres from the mid-brain by means of the **descending root** (radix descendens, anterior root, trophic root of *Merkel*), (figs. 130, 132 to 135, and 154, *Vd*, and 153, *5*).

The large round vesicular cells (45 to 60 μ in diameter) in which the fibres of the descending root originate do not form a compact group, but are either separate or united into little clumps which lie on the edge of the central grey matter, and may be followed as far as the plane of the anterior quadrigeminal bodies. In size and shape they closely resemble the cells of the substantia ferruginea, which latter show every transition, so far as pigmentation is concerned, to the neighbouring non-pigmented cells of this descending root.

The cross-section of the descending trigeminal root forms a long figure, slightly convex outwards in Man, but in other animals usually straight. It lies up against the posterior longitudinal bundle and the cells of the substantia ferruginea. The other cells which we have just been describing lie against its mesial border.

The farther we advance forwards the smaller is the number of fibres which appear in this root in cross-section. When the anterior corpora quadrigemina are reached the few distinct fibres which remain exhibit a tendency to turn towards the middle line in the roof of the aqueduct. That some of them reach it is shown by the exceptional occurrence in this region of one or more of the characteristic cells.

We may suppose that the descending root of this nerve which joins the portio major (or, as *Bechterew* believes, the portio minor) is thus to be accounted for. The large cells give origin to the thick root-fibres which run spinewards. These cells have also [presumably] a cerebral pole, from which a very much finer fibre extends in the roof of the aqueduct across the middle line, and thence to the great brain. According to this the absolute number of fibres which run spinewards in the descending root is not increased, but only their calibre.

(5) **Middle roots** of the trigeminal nerve :—

(*a.*) From the sensory trigeminal nucleus (accessory nucleus). This nucleus consists of a number of small irregularly-grouped masses of grey substance containing small scattered cells.

That part of the trigeminal trunk which extends from the ventral side of this nucleus into the transverse fibres of the pons receives, owing to its interweaving with the ascending root, which bends round in this situation, a peculiarly characteristic striation, clearly seen in fig. 129. The sensory nucleus presents a not inconsiderable sagittal extension (4 to 5 mm.), and may well be identified with the proper grey matter of the posterior horn rather than with the substantia gelatinosa.

(*b*.) The motor nucleus (upper trigeminal nucleus, noyau masticateur) can be easily distinguished from the sensory nucleus. It lies on the mesial side of the sensory root, and consists of a single round grey mass of large multipolar cells. Its sagittal extension is considerably less than that of the sensory nucleus. Here originates the principal part of the motor root. We may look upon it as the proximal end of that part of the anterior horn (including the lateral horn), which was separated from the central grey mass by the crossing of the pyramids (fig. 152).

[For morphological purposes it is somewhat important to distinguish between these horns, or rather columns, of cells; the nucleus of the fifth, like the nuclei of the seventh and eleventh, belongs to the lateral column, which gives origin to nerves supplying the muscles derived from the "lateral plates" (see figs. 156 and 157).]

Reviewing what we have said with regard to the *sensory root*, we find that its fibres may be classified as follows (fig. 153):—

    (1) Those of the ascending root, 1.
    (2) From the sensory nucleus, 2.
    (3) From the cerebellum, 3.
    (4) From the substantia ferruginea of the opposite side.

The *motor root* is formed of fibres—

    (1) Of the descending root, 5.
    (2) From the motor nucleus of the same side, 4.
    (3) From the motor nucleus of the opposite side, 6.

The **connections of the trigeminal nuclei with the higher brain** must be complicated in proportion to the width of distribution of the nerve and the intricacy of its primary connections.

We know nothing whatever about the central connections of the ascending root. The central paths connected with the descending root cross over the median line, as we have seen, in the region of the corpora quadrigemina anteriora, above the aqueduct of Sylvius. Reference has also been made to fibres which, starting from both sensory and motor nuclei, reach the crura cerebri *viâ* the raphe.

The cortical area for the muscles which the trigeminal nerve innervates occupies, perhaps, the lower third of the anterior central and the neighbouring portions of the middle and inferior frontal convolutions. A one-sided lesion of the cortex, especially when on the left side, paralyses the jaw-muscles on both sides (*Hirt*).

7. Facial Nerve (nervus communicans faciei, portio dura paris septimi).

—The nucleus of origin of the facial nerve lies very close to the spot at which the nerve takes its exit from the brain, but notwithstanding this the root-fibres have an extremely round-about course within the substance of the brain. Several times they turn in the wrong direction, and quickly leave it again before they have the good fortune to discover, after many wanderings, a way out of their prison. The way is made as narrow as possible owing to the pressing together of various structures.

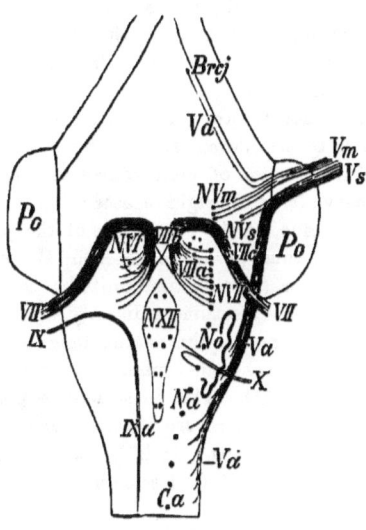

Fig. 154.—Schematic projection of the medulla oblongata.—*Po*, Pons; *Brcj*, brachium conjunctivum; *Va*, ascending, *Vd*, descending, *Vm*, motor, *Vs*, sensory trigeminal roots; *NVm*, motor, *NVs*, sensory trigeminal nucleus; *NVII*, facial nucleus; *VII a, b, c,* facial root; *VII*, point of exit of facial nerve; *NVI*, nucleus of abducens; *IXa*, ascending glossopharyngeal root; *IX*, its point of exit; *No*, nucleus olivaris; *X*. vagus (or glossopharyngeal) nerve, with the origin of certain fine fibres in the nucleus ambiguus, *Na*; *Ca*, anterior horn of the spinal cord; *Ca*, *Na*, *NVII*, *NVm*, column of motor nuclei; *NXII*, nucleus of hypoglossal nerve.

Only a single nucleus of origin of the facial nerve is known with certainty (anterior or inferior nucleus), *NVII* (figs. 125, 126, 127, and 154). It commences in the most distal part of the pontal region, and extends some 4 mm. farther forwards. It lies in the formatio reticularis, on the mesial side of the ascending trigeminal root, nearer to the trapezoid fibres than to the ventricular surface. This nucleus is very characteristic, and is not easily confused with other structures, such as the upper olive.

Large lightly-pigmented nerve-cells lie in a ground-substance, which stains darkly with carmine, and is broken up into little bits by irregularly-disposed medullated fibres (root-fibres of the facial). We must look upon this nucleus as a thickening of the grey substance in the formatio reticularis lateralis, and thus also as a continuation of the separated portion of the anterior horn. It is continued forwards by the motor nucleus of the fifth, although the two nuclei are not actually contiguous; spinewards its connection with the grey columns of the cord is kept up by means of the motor glossopharyngeal and vagus nuclei. The root-fibres come off from the nucleus either singly or in quite thin bundles. They converge in easy curves with a slight anterior deflection towards the part of the floor of the fourth ventricle, which lies on the dorsal side of the posterior longitudinal bundle ($VIIa$). This portion of their course is termed the nuclear crus of the facial nerve (crus of origin or ascending facial root).

Close to the middle line, on the dorsal side of the posterior longitudinal bundle, which is thus pushed away from the ependyma of the ventricle, the facial fibres (by this time united into a single compact bundle, oval in cross-section), assume a directly sagittal direction, and course forwards on either side the sulcus longitudinalis, distinctly lifting up the floor of the ventricle (enimentia teres) for about 5 mm., $VIIb$ (figs. 127, 128, and 154). This intermediate portion of the root of the facial continually grows in cross-section, owing to the accession of fibres from the facial nucleus. It is known as the intervening or middle portion (the ascending limb, fasciculus teres, *Stilling's* constant root of the trigeminus).

Suddenly the facial root leaves the course just described at a right angle, arches a short distance beneath the ependyma dorsally to the abducens nucleus, curls down its outer side into the tegmental region, and runs towards its point of exit almost in a straight line ventrally, laterally, and distally, $VIIc$ (issuing limb or root—fig. 126).

The double bending of the facial nerve is termed genu nervi facialis.

It is possible (fig. 127) to see in a single section the facial nucleus with its crus, the issuing limb on the outer side of the nucleus, together with the intermediate segment; and yet the three pieces of the root may appear to be in no way connected with one another.

Certain additions must still be made to what has been already said.

Fibres extend from the facial nucleus across the middle line to reach the root of the opposite side (fig. 153), probably they form part of the beautifully-curved bundle which advances towards the raphe, between the posterior longitudinal bundle and the intermediate portion of the root (figs. 127 and 128). We are bound to take for granted that fibres pass through the raphe to the pyramidal tract, by means of which they reach the opposite hemisphere, and are distributed to the lower part of the anterior central convolution (with the exception of its very lowest portion, *Exner*). Such a connection may be looked for in the bundle just mentioned, as well as in the fibræ arcuatæ, which are scattered through the formatio reticularis. Fibres are also supposed to join themselves to the issuing limb from the cells which lie near it (*Laura*).

Since the upper branches of the facial nerve, which contain fibres for the orbicularis oculi and frontalis muscles, are not (as a rule) affected in central disease of the facial nerve, a different source has been sought for them in the brain. For this purpose the abducens nucleus was hit upon, as we have already mentioned; but this "upper facial nucleus" or "nucleus abducens et facialis" has, as a matter of fact, no connection with the facial nerve. According to *Mendel's* researches, it is more probable that the fibres intended for the superior facial originate in the oculomotor nucleus of the same side and pass to the genu facialis in the posterior longitudinal bundle.

In animals the intervening segment is so short that it is reduced to an arch uniting the nuclear crus with the issuing crus, and containing the abducens nucleus in its concavity.

8. Auditory Nerve (nervus acusticus, portio mollis paris septimi).

—An inconvenience which is very commonly encountered in treating of the anatomy of the central nervous system makes itself felt in a peculiar degree when we come to deal with the auditory nerve, I refer to the constantly changing designations of the particular nuclei from which the nerve takes origin, as well as of its roots. The fault to which this complexity in nomenclature is attributable lies especially at the door of the various conceptions as to the proper designation of relative positions (front, back, upper, under). The comprehension of the central connections of the auditory nerve is also rendered difficult by the following circumstances of its anatomico-physiological relations :—

(1) We have to deal with two, or perhaps three, different nerves which make up the auditory trunk; to wit, (*a.*) the nervus cochleæ or proper nerve of hearing; (*b.*) the nervus vestibuli, intended for the semicircular canals, and having nothing to do with hearing; (*c.*) perhaps also the nervus intermedius Wrisbergi (portio intermedia), which joins the facial.

(2) Different physiological methods have yielded results which are difficult to harmonise—indeed are in some cases diametrically opposed.

Owing to the obscurity which reigns with regard to the central origin of the auditory nerve, we shall only attempt to give an account of such views as coincide most closely with the facts, without stopping to notice all the divergent opinions on the subject.

Two peripheral **roots** of the **auditory nerve** are generally acknowledged to exist, but even in this connection there is no uniformity of opinion.

These two roots are easily kept apart in one's mind, owing to the fact that they lie on either side the corpus restiforme, $Crst$ (fig. 155). All the fibres which reach the trunk of the auditory nerve on the outside of the restiform body constitute the lateral root, Rl. The fibres, on the other hand, which force their way through between the restiform body and the ascending root of the trigeminus belong to the mesial root, Rm. If a series of sections is prepared, in the manner which has been advocated, it will be seen that the lateral root is found nearer to the spinal cord than the mesial root (it first appears in a section a little behind the one represented in fig. 125); while the mesial root is still seen in sections farther forwards than the lateral root extends. The names commonly used to distinguish the two roots are thus explained—the mesial root is also known as the "deep," "superior," or "anterior" root; the lateral as the "superficial," "inferior," or "posterior" root. There is much agreement in the statement that the lateral root passes into the cochlear nerve, the mesial root into the vestibular nerve (*Forel, Onufrowicz, Flechsig, Bechterew*). According to this the real auditory functions belong to the lateral root, which might also be termed the radix cochlearis; while the mesial root (radix vestibularis) has other work to do, probably connected with the maintenance of equilibrium.

Sometimes the portion of the mesial root which lies farthest spinewards is counted to the posterior root, although a sharp division between this and the other part situate more brainwards is not possible.

Three grey masses are to be regarded as the **nuclei of origin** of the acoustic nerve.

[The fact ought not to be overlooked that in the ganglion spirale the cochlear ramus is connected with bipolar cells of moderate size.]

(1) The **chief nucleus** (central, inner, or posterior nucleus, nucleus of the posterior root, mesial part of the superior nucleus), $VIIIh$ (figs. 124, 125, 126). A symmetrical grey area is seen in the sections which show the most anterior points of exit of the hypoglossal nerve (fig. 123); it lies at first on the outer side of the chief nuclei of the glossopharyngeal and vagus nerves, but reaches to the raphe in sections farther forward, and assumes then a triangular form. Still farther brainwards it retires again from the middle line. It disappears in

PRIMARY CENTRES OF AUDITORY NERVE. 301

the region of the abducens nucleus. Only isolated small nerve-cells are found in this fairly extensive region. A small group of spindle-shaped cells is constantly found in the mesial angle of this triangular field where it attains to its greatest cross-section (nucleus funiculi teretis *seu* medialis). This group reaches beyond the limit of the nucleus, both on the cerebral and caudal side, but seems to have no direct relation to it, *Nft* (figs. 123–127).

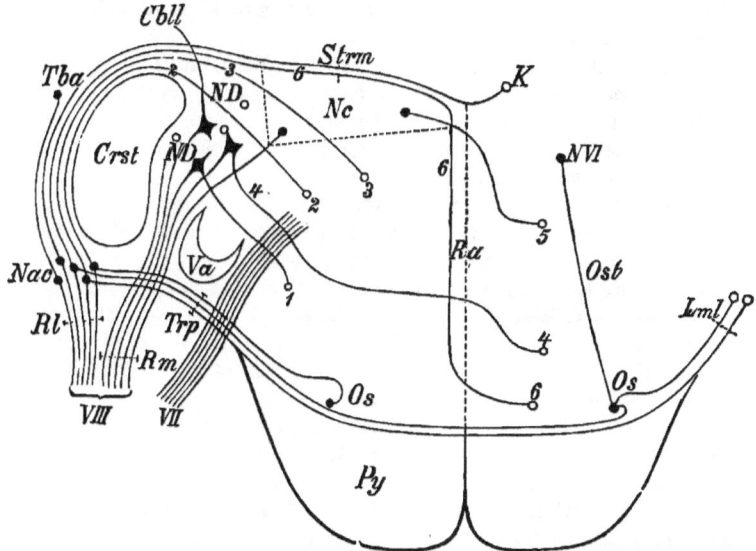

Fig. 155.—Scheme of the central auditory apparatus.—*VIII*, Peripheral root of auditory nerve; *Rl*, lateral, *Rm*, mesial divisions; *Nac*, nucleus accessorius; *ND*, large-celled nucleus; *Nc*, chief nucleus; *Tba*, tuberculum acusticum; *Trp*, corpus trapezoides; *Strm*, striæ medullares; *K*, conductor sonorus; *Os*, superior olive; *Lml*, lateral fillet; *Ost*, peduncle of the superior olive; *Crst*, corpus restiforme; *Va*, ascending root of the trigeminal; *Py*, pyramid; *Ra*, raphe; *NVI*, nucleus of the abducens; *VII*, facial root; *Cbll*, to cerebellum; *1-6*, central connections referred to in the text.

(2) The **large-celled nucleus**, *ND* (external acoustic nucleus, Deiters' nucleus, mesial nucleus of the anterior root, lateral part of the superior nucleus, inner segment of the restiform body), *VIIIa* (figs. 123–127).

On the mesial side of the corpus restiforme, where it first comes into existence, an area is met with consisting of nerve-bundles cut across and separated by very little intervening grey substance.

Roller has shown that the fibres which take origin in the cells present in this region pass directly into the auditory nerve and constitute, therefore, an ascending root. The farther we advance towards the proper acoustic region the more abundant does the whole space occupied by grey matter become, while the nucleus increases correspondingly. Especially in the most anterior sections (fig. 127), in which the ascending fibres are bending laterally and ventrally towards the point of exit of the nerve, numerous nerve-cells conspicuous for their size are scattered about. The cells of Deiters' nucleus are far more remarkable for their size in most animals than they are in Man.

(3) **Accessory nucleus,** *Nac* (anterior nucleus, lateral nucleus of the anterior root, lateral acoustic nucleus, auditory ganglion); the portion of this nucleus which lies between the two roots of the nerve is known as the nucleus inferior (or lateral nucleus of the posterior root), *VIIIac* (figs. 125, 126).

This nucleus lies partly outside the proper brain-substance, on the nerve-stem, like the spinal ganglia on the posterior roots. With the exception of the portion which is intercalated between the two roots, it lies within the lateral root and on its lateral side, and stretches up as far as the substance of the cerebellum. It is made up of small round cells closely packed together. At its proximal end the cells often exhibit a kind of capsule, reminding one of the arrangement found in the case of the spinal ganglia.

A mass, which is of no great size in Man, and is scarcely marked off distinctly from the accessory auditory nucleus, is known as the tuberculum acusticum, *Tba* (tuberculum laterale, superficial auditory nucleus). It also lies to the side of the lateral root and is reckoned as part of the central apparatus of hearing. In many animals, the rabbit for example, it shows a characteristic structure which places it on the same level as the front quadrigeminal body (*Onufrowicz*).

[In the mammalian embryo, and permanently in many lower vertebrates, *e.g.*, the frog, the accessory nucleus lies outside the brain-stem.]

In naming the three nuclei above described we have as far as possible avoided topographical expressions in the hope of preventing mistakes. The chief nucleus deserves to be so called on account of the large area which it occupies in cross-section. The large-celled nucleus is characterised by the form and size of its cells. The accessory nucleus may well receive this name, since in part it lies like an appendage outside the brain-stem proper.

We must now refer to the relation in which each of the two **roots** stands to the nuclei above named, and thereby to other portions of the brain.

1. The LATERAL ROOT.—By far the larger portion of the fibres of this

root arise from the accessory nucleus, which is therefore to be looked upon as the primary centre of hearing; [unless the accessory nucleus is considered as functionally, as well as structurally, homologous with a spinal ganglion]. It cannot, however, be denied that many fibres of the lateral root pass by the accessory nucleus, and either encircle or traverse the corpus restiforme in their course towards the large-celled nucleus (*Freud*). [There are several points of great difficulty in connection with this root, *e.g.*, its morphological relation to the ganglion spirale, the direction taken by the axis-cylinder processes of the cells of the large-celled nucleus with which its fibres join, and the manner in which these fibres join the cells.]

2. The MESIAL ROOT.—The larger part of this root comes out of Deiters' nucleus, nevertheless the root may have a variety of functions.

The fibres come partly through the ascending root from deeper and more distal planes, partly they spring out of the large nerve-cells which are especially numerous at this level. *Bechterew* says that the mesial root is connected with those grey masses (which are not well defined by the way) that lie in the lateral wall of the fourth ventricle on the dorsal side of the large-celled nucleus (Bechterew's nucleus, nucleus angularis, chief nucleus of the vestibular nerve).

A second portion of the mesial auditory root, that is to say, the part which lies on the spinal side, originates in the chief nucleus. It must, however, be allowed that very little more is known concerning its mode of origin than that the root-fibres leave the nucleus at its latero-ventral corner.

[The eighth nerve is a pure sensory nerve. It appears, however, to carry two distinct kinds of impressions. The organ of Corti is regarded as the organ for the analysis of sound. The cochlear nerve carries, we therefore suppose, the impulses generated by the impact of sound-vibrations of varying period. In the semicircular canals of the labyrinth originate impressions of orientation, while the hair cells of the utricle and saccule may take 'cognisance of the amplitude of sound-vibrations. Such is the allocation of function commonly ascribed to the several parts of the internal ear, although in neither case can it be said to be placed beyond the reach of doubt.

Turning to the auditory nerve, this, as already pointed out, consists distinctly of two separate parts—(1) the ramus lateralis *seu* cochlearis, which is the first to myelinate (when the fœtus is about 30 cm. in length); (2) the ramus medialis *seu* vestibularis, the fibres of which acquire their medullary sheaths somewhat later (when the fœtus is about 38 cm. in length).

Despite this complexity in its functions the auditory nerve is, like the olfactory and optic, limited in its distribution to a single organ of a single segment; we may, therefore, expect to find that, unlike the multisegmental fifth, ninth, and tenth nerves, it will have a defined connection with the central tube of grey matter within the cerebro-spinal axis. This is conspicuously the case, for the small-celled grey matter, which is the upward continuation of the posterior horn of the spinal cord, undergoes a great local development beneath the tuberculum acusticum. Into this grey matter the auditory nerve pours its impulses, and it is open to doubt whether a search for other nuclei of the auditory nerve is likely to prove profitable. The accessory nucleus is clearly a posterior root-ganglion. It hardly appears large enough to take the place of root-ganglion to the whole auditory nerve, nor is this by any means necessary, for just as we find bipolar cells on the fine filaments of commencing olfactory and optic nerves (in the olfactory bulb and retina respectively), so we find the cochlear nerve-fibres connected, before they leave the ear, with the bipolar cells of the ganglion spirale.

That sensory nerve-fibres join large multipolar cells appears unlikely to many anatomists. The *translator* is inclined to look upon the so-called large-celled nucleus of the auditory as homologous with the scattered cells of the posterior horn, and to suppose that these cells collect filaments from the sensory plexus and despatch medullated fibres to the cerebellum and elsewhere. It does not appear that the nucleus of Deiters atrophies after destruction of the ear (*Baginsky*).

It is well to remember that while the primary connections of the auditory nerve are likely to be simple, its secondary connections must certainly be extremely complex, for not only do sensations from the ear play an important part in orientation, but auditory impulses start innumerable protective movements, and also take part in the highest degree in intellectual life.]

If the **central connections** of the above-described nuclei are asked for the following may be given :—

(1) The CHIEF NUCLEUS.—Very little is yet known concerning its further connections. Out of the mesial angle of the triangle, plenty of fibres, not collected into bundles, extend through the posterior longitudinal bundles, and then through the raphe (*Freud*) to the tegmental region (*5*). Probably they form, therefore, the central connections of this nucleus.

Out of the cerebellum, according to *Edinger*, the nucleus receives fibres which probably come from the opposite flocculus. They run on the inner side of the corpus restiforme.

(2) The LARGE-CELLED NUCLEUS.—Its connection with the cerebellum is certain. The fibres which ascend from this nucleus to the cerebellum seem, in large part at least, to cross in the roof of the ventricle, ending

provisionally in the nuclei of the roof (in the nuclei emboliformis et globosus, according to *Flechsig*). Thence numerous bundles extend, directly or indirectly, into the brachium conjunctivum, in which they cross again, and disappear in the red nucleus of the tegment (*Flechsig*). That root-fibres of the auditory nerve merely traverse the nucleus to stream directly into the cerebellum is now denied with fair unanimity.

Strong fibres seem to originate in the large-celled nucleus, and to run in a ventro-mesial direction, crossing through the issuing limb of the facial nerve into the tegmental region, where they bend over and assume a longitudinal course (spinewards, perhaps, as well as brainwards), (*1*). Other thick fibres extend as arcuate fibres into the raphe (*4*) and the tegment of the opposite side; and in this way, perhaps, reach the great brain (fig. 127). The more dorsally-lying localities in which the auditory nerve originates, Bechterew's nuclei, as we have called them, are, according to *Flechsig*, united with one another by commissural fibres, which come out of the cerebellum in the brachia conjunctiva and bend round in arches in the posterior angle of their decussation. *Mendel* had already found that the auditory nerve sends a considerable bundle to take part in the formation of the brachium conjunctivum.

(3) The Accessory Nucleus.—The central connections of this nucleus seem to be very varied :—

(*a.*) Bundles which, as the striæ medullares, cross the floor of the fourth ventricle. A very considerable portion of the fibres which come out of the dorsal pole of the accessory nucleus, sling themselves round the corpus restiforme, and course towards the raphe, so close beneath the ependyma as to be visible from the surface. Most of these fibres turn down ventrally towards the pyramids (*6*) on the lateral surface of the raphe just before they reach the middle line, but they appear finally to cross over to the other side; according to *Meynert* they enter the pedunculus cerebelli as fibræ arcuatæ. Another set of fibres of the striæ medullares (K, fig. 155) cross one another in the most dorsal part of the raphe, and extend laterally and cerebralwards; their termination cannot be stated with certainty, often these fibres form a compact bundle, which lies beneath the ependyma of the ventricle, and is known as the conductor sonorus (fig. 15, K) [in German "*Klangstab*" which Bergmann Latinised as above]. Groups of middle-sized nerve-cells are found in the centre of the conductor sonorus. Fibres from the nucleus funiculi teretis join the striæ medullares.

(*b.*) Another portion of those fibres which serve as the central continuation of the accessory nucleus and encircle the outer side of the corpus restiforme, scatter on the mesial side of the latter, some of the diverging fibres of the bundle going to the large-celled nucleus, others to the chief nucleus. Most probably they only traverse these nuclei on their road to the tegmental region, where they join the posterior longitudinal bundles (*2*, *3*).

The bundles designated (*a.*) and (*b.*) used to be allocated to the lateral auditory root. Only lately have we learnt to consider at least the greater part of the fibres which border the corpus restiforme, as central connections of a primary auditory centre (nucleus accessorius).

(*c.*) It is supposed that numerous connecting fibres run from the accessory nucleus to the tuberculum acusticum; but in Man these are of no more importance than the tubercle itself.

(*d.*) A very important connection between the nucleus accessorius and other parts of the brain is established by means of the corpus trapezoides.

The bundles of fibres which constitute the **corpus trapezoides** (trapezoideum), a system especially conspicuous in animals, pass from the vicinity of the accessory nucleus towards the raphe, *Tr* (figs. 125 to 127). It is not hereby asserted, however, that all the fibres of the corpus trapezoides originate in the nucleus accessorius; some may come from the cerebellum or the corpus restiforme (*Kahler*). A small portion of the corpus trapezoides is in relation with the upper olive of the same side; the larger part of its fibres cross the middle line, and extend either into the upper olive or into the fillet of the opposite side. Large nerve-cells are scattered about through the substance of that part of the corpus trapezoides, which lies on the ventro-mesial side of the superior olive, lateral to the root of the abducens nucleus; they are certainly connected with its transverse fibres.

Commissural fibres, too, are supposed to be present in the corpus trapezoides, connecting the two accessory nuclei and, perhaps, also the tubercula acustica with one another (*Flechsig*).

The **superior olive** (nucleus olivaris superior, nucleus dentatus partis commissuralis, figs. 126, 127, 129, *Nos*, and fig. 155, *Os*) is insignificant in Man and many animals—*e.g.*, the horse; in other animals (carnivora, rodentia, and especially cetacea) it is well developed. It consists of a broad plate of grey substance folded from one to five or six times at most. The ground-substance of this structure scarcely stains at all with carmine. Its scattered round or fusiform nerve-cells (in the dog 40 μ in diameter) are surrounded by connective-tissue capsules.

In the dog the superior olive consists of two divisions, separated from one another by nerve-fibres. It is wrapped in by fibres on all sides.

The following are the connections of the upper olive as yet discovered:—

(1) With the nucleus accessorius of the opposite side, and to a certain extent of the same side by means of the corpus trapezoides.

(2) With the posterior corpus quadrigeminum through the lateral fillet (p. 258).

(3) With the abducent nucleus of the same side through the pedunculus olivæ superioris (fig. 155, *Ost*).

Other connections have been repeatedly put forward—*e.g.*, with the roof nuclei of the cerebellum, and with the lateral columns of the spinal cord (*Bechterew*).

Everything indicates that we have to look for the **cortical centre** of the sense of hearing in the temporal lobes, especially the gyrus temporalis superior, and partly also in the gyrus medius. In favour of this localisation we have, apart from the results of experimental investigations, the appearances presented in cases of word-deafness, in which disease, lesions of this region, especially on the left side, are almost always found. In the brains of deaf-mutes a very perceptible atrophy of the upper temporal convolution may be present, although the peripheral stem of the auditory nerve is intact.

Monakow has extirpated the temporal lobe in the rabbit, and found a consequent atrophy of the portion of the corona radiata which originates in it, and also of the corpus geniculatum mediale, as well as a portion of the zona reticulata of the optic thalamus. Thereby an unbroken chain between the peripheral nerves of hearing and the cortical acoustic field is completed, its links being radix cochlearis ; nucleus accessorius ; corpus trapezoides ; oliva superior ; lemniscus lateralis ; corpus quadrigeminum posterius ; ganglion geniculatum mediale ; lobus temporalis.

Spitzka's comparative observations also favour the existence of such an auditory route. In many cetacea he found an extremely marked development of the posterior auditory root, corpus trapezoides, posterior corpus quadrigeminum, and corpus geniculatum mediale.

The posterior corpus quadrigeminum and the mesial geniculate body seem, therefore, to occupy the same place in the central apparatus of hearing as the anterior quadrigeminal and lateral geniculate bodies in the apparatus of sight (*Baginski*).

We must not, however, forget that a less complicated route may lead directly from the nucleus accessorius to the great brain.

It may be noticed that neither of the acoustic nuclei has an undisputed claim to be considered the origin of the auditory nerve. In the case of the large-celled nucleus, and still more in the case of the "chief" nucleus, reasons have been repeatedly advanced for denying the possibility of any connection with the auditory nerve. Even the nucleus accessorius itself, to which it seems that the greatest importance as a centre of origin of the auditory nerve should be assigned, has been looked upon by *Huguenin* as the centre of a vasomotor nerve, the nervus intermedius Wrisbergi.

Neoplasms of various kinds originate in the nervus acusticus. According to *Virchow*, they are more frequent in this nerve than in any other. Chalky concretions have been repeatedly observed in the auditory stem.

9. Glossopharyngeal Nerve.—From the distal border of the pons spinewards we meet with a succession of nerve-roots, which at first come out through the side of the restiform body dorsally to the eminentia olivaris, and lower down in a line continuing spinewards in the same plane as far as the region of the sixth cervical nerve. These roots belong to the ninth, tenth, and eleventh pairs of cranial nerves.

Since the root-bundles of these nerves join one another it is impossible in the case of most of them to say, without preparation of their nerve-stems, to which of the three they belong, especially as their central origin also agrees in many points. The uppermost roots belong, without doubt, to the glossopharyngeal nerve; the lowest, especially when they come out from the spinal cord, to the spinal accessory.

The glossopharyngeal receives its fibres from three sources (fig. 124):

(1) The small-celled glossopharyngeal nucleus, nIX (upper part of the common accessorio-vago-glossopharyngeal nucleus, sensory or posterior glossopharyngeal nucleus).

This nucleus lies just below the ependyma of the ventricle, except where (farther brainwards) it is pushed more deeply into the medulla by the chief nucleus of the auditory nerve. Its small cells, which are for the most part spindle-shaped, form a compact rounded group, their long axes being generally placed in the direction of the issuing root-fibres. It is not improbable that the uppermost fibres which originate from this nucleus constitute the portio intermedia, and are continued by the chorda tympani into the lingual nerve (*Duval*). It is tempting to suppose that all the fibres which originate in this nucleus are concerned with the sense of taste.

(2) Large-celled glossopharyngeal nucleus, Na (motor or anterior glossopharyngeal nucleus, anterior column of origin of the mixed lateral system, nucleus ambiguus, nucleus lateralis medius). Large cells lie scattered about in the substantia reticularis grisea on the ventral side of the small-celled nucleus of the glossopharyngeal nerve. They are similar to the anterior horn cells of the spinal cord. From these cells fibres, not united into bundles, extend dorsally. Some of these fibres bend round in a latero-ventral direction in sharp curves, and join the glassopharyngeal nerve on its mesial side (*cf.* also fig.

151, *X*). Another set bend towards the middle line, just before they reach the floor of the fourth ventricle, cross in the raphe, and join the glossopharyngeal root of the opposite side. This large-celled nucleus of the glossopharyngeal nerve, which is to be looked upon, as already explained, as a remnant of the anterior horn cut off by the crossing of the pyramids, finds its serial continuation upwards in the facial nucleus, in which the grey masses are more compact, and ends above in the motor nucleus of the fifth. Fig. 154 shows this succession of motor nuclei. [The translator has put forward the theory that in the medulla oblongata the anterior horn is truly withdrawn to the mid-dorsal line, where it gives origin to the hypoglossal and sixth nerves, and that these nuclei (whether the nucleus ambiguus belongs to the same group as the antero-lateral nucleus of Clarke requires further elucidation), including the nucleus of origin of the spinal accessory nerve, are to be looked upon as the continuation upwards, not of the anterior, but of the lateral horn. It was with a view to obtaining light upon the problem of the segmentation of the head and the morphological value of the cranial nerves, as indicating the limits of its metamers—the vertebræ composing the skull was the aspect of the problem as introduced by Oken—that the translator hoped to find indications of segmentation in the position of the nuclei within the cerebro-spinal axis. It had for a long time been allowed that nerves are the most conservative organs of the body and longest perpetuate structural dispositions which other organs have discarded, but no attempt had, at the time, been made to look for evidences of metamerism in the nuclei from which the cranial nerves grow. The division of the grey matter in the spinal cord into anterior and lateral columns (horns) of large cells, as well as Clarke's column and posterior horn, is obvious; and the question which must be settled before attempting to homologise cranial and spinal segments is this—What is the meaning of the distinction between anterior and lateral horns? The answer to this question is given in the cervical region, where the two horns give origin to distinct nerves—the spinal accessory appropriating the lateral horn to itself. This is the key to the problem of the segmental disposition of the cranial nerves—for the seventh and motor part of the fifth, both in line of exit and in position of nuclei, are the successors of the spinal accessory in the occupation of the lateral vesicular column. The division of the cranial nerves into two lines was noticed by *Sir Charles Bell*, who supposed that the lateral group (in which he included the phrenic, the external respiratory of Bell, and the fourth) are thus disposed, on account of their participating in common in the function of respiration. Another change in the arrangement of the nerves also marks the medulla. The vagus nerve takes origin in the brain (few of the cranial nerves are what might be termed *pure* nerves) from a group of cells which is, as *Ross* pointed out, a local enlargement of Clarke's column at its upper extremity. Here, then, in the medulla we find that each segmental nerve is split up into four elements, anterior motor, lateral motor, visceral, and sensory, which, in the nerves of the spinal cord, are united into a single trunk (figs. 156 and 157). On this basis the *translator* grouped the cranial nerves for the purpose of making up the complement of each single metamer, bearing in mind, however, that, as already remarked, the cranial nerves are seldom pure nerves; even when fibres of one function greatly preponderate, any nerve may contain the vestiges of nerves of other function. To what extent associations and dissociations have thus occurred the comparative anatomist must decide. Thus much, however, may be accepted as a basis for morphological speculation: the cranial nerves, when called in in evidence of cranial segmentation, must no longer be arranged in linear series of equal segmental value, as supposed by *Balfour, Marshal, van Wijhe,* and others; but must be collected into groups, as in *Gegenbaur's* scheme, although on different lines.

Gaskell has taken up the *translator's* scheme, and seems to have carried it to a very fruitful termination, by showing that fibres from the visceral and the lateral columns of cells tend to travel in company to the splanchnic portions of the body, while the fibres from the cells of the anterior column go to the somatic muscles. As must already be abundantly apparent, even the most fundamental problems in the anatomy of the nervous system are not to be settled with scalpel and forceps, or with the microscope and series of sections; but physiological experiment and morphological speculation are alike called into requisition for the purpose of cutting paths from which to commence purely topographical explorations into the labyrinth.]

A certain harmony in the course of these root-fibres attracts attention. Facial, as well as glossopharyngeal roots (including the vagus) do not take the most direct road to their points of exit, but turn dorsally first. We may conclude from this disposition of the facial fibres that we are right in supposing that the motor glossopharyngeal fibres originate in the large-celled nucleus. The important connection of the nucleus with the root of the opposite side, suggests the view that the muscles which the nerve innervates (stylopharyngeus and constrictor muscles of the pharynx, for example) act bilaterally and simultaneously. There is a slight uncertainty about the innervation of some of the muscles attributed to this nerve.

(3) Ascending glossopharyngeal root, *IXa* (*Stilling's* fasciculus solitarius, ascending root of the mixed lateral system, ascending vagus root, "respirations-bündel" of *Krause*, figs. 121 to 124; and 151, *IXa.*).

RESPIRATORY BUNDLE. 311

The round cross-section of this tract is indistinctly seen on macroscopic investigation of the region which lies between the decussation of the pyramids and the descussation of the fillet; it is more distinct in sections carried through regions farther forward. It lies laterally to the small-celled accessorio-vago-glossopharyngeal nucleus. It is sharply circumscribed by the medullated fibres which encircle it.

The spinal origin of the ascending root of the glossopharyngeal nerve is not known with certainty; still it is not improbable that delicate bundles out of the posterior horn, ascending obliquely upwards and

 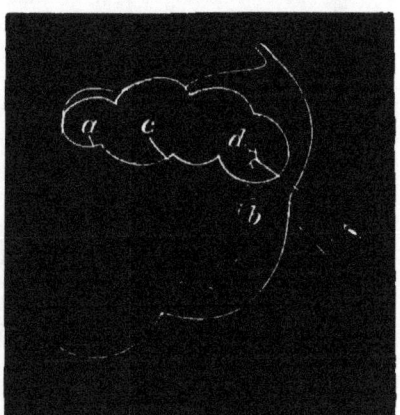

Figs. 156 and 157.—Diagrammatic section of the spinal cord and medulla, designed to show the relative positions of the centres in the central grey tube of the several roots of a spinal and segmental cranial nerve respectively; as also the grouping of the constituents of the roots.—*a*, Anterior horn; *b*, lateral horn; *c*, Clarke's column; *d*, posterior horn; *1*, anterior (somatic) motor root or nerve; *2*, lateral splanchnic (motor) root, which in the lower spinal cord joins the anterior root, in the upper spinal cord runs by itself as *2'*, the spinal accessory, while in the medulla it accompanies the posterior root; *3*, visceral root; *4*, the posterior or sensory root bearing a ganglion. The spinal roots unite into a common trunk. With the exception of those of the trigeminus. The cranial roots do not form permanent associations.

inwards, combine to form this bundle on the border of the central grey substance. In transverse sections just below the pons (fig. 124) the ascending root, which in lower sections lay laterally (and dorsally) with regard to the vagus and glossopharyngeal nerves, turns suddenly from the longitudinal into the horizontal direction, and passes outwards as a thick compact bundle through the ascending root of the trigeminus to its point of exit by the side of the corpus restiforme.

Thus it forms the most proximal root-bundle of the IX, X, XI group, so that its interpretation as glossopharyngeal root is certainly correct. It is not impossible that during its longitudinal course a few fibres from this bundle join the vagus nerve; since, however, by far the largest part of it bends outwards into the glossopharyngeal root, the name we have given to it is justified. Small masses of grey substance which lie on this bundle are called by *Roller* the "glossopharyngeal flock." The anatomical similarity between the ascending trigeminal and ascending glossopharyngeal roots must be pointed out (*cf.* fig. 151). One might suppose that the latter serves to convey the sensory impressions of common sensation which come from the glossopharyngeal area.

The three separate functions, motility, taste, and general sensation, which the glossopharyngeal nerve subserves, may without hesitation be assigned to the three separate origins of the nerve.

It is not yet possible to make assertions with regard to the cerebral connectors of the glossopharyngeal nuclei. Perhaps they are to be looked for in the fibræ arcuatæ and the raphe.

10. Vagus Nerve (pneumo-gastric nerve, nerve of the lungs and stomach).—There is very little to add to what has already been told about this nerve in connection with the glossopharyngeal.

The vagus nerve gets its fibres from the same source as the last-described nerve, with the exception of the ascending glossopharyngeal root from which the vagus derives few fibres, if any (figs. 122 and 123).

(1) The small-celled (sensory) nucleus of the vagus provides the sensory fibres. At its margin and scattered about outside larger darkly-pigmented cells are found.

(2) The large-celled (motor) vagal nucleus gives it its motor fibres.

The mode of origin from each nucleus corresponds to what we already know with regard to the glossopharyngeal nerve.

11. Accessory Nerve (accessorius Willisii, nervus recurrens, spinal accessory nerve).—It is customary to describe two different modes of origin for the accessory nerve. The proximal part of the root-bundle comes out as a continuation of the vagus origin, between the olive and the corpus restiforme, accessorius vagi (*seu* cerebralis), *XI* (fig. 11). The distal portion of the nerve, accessorius spinalis, arises by a row of root-filaments from the surface of the lateral column of the cord on the outer side of the posterior roots, from the level of the first cervical nerve down to the fifth or sixth (and, exceptionally, the seventh). The accessorius vagi has exactly the same origin as the vagus itself, from which (on the surface of the brain) it cannot be separated. Farther down the roots unite for the time being with the accessory trunk, but in their extra-cranial course join them-

ORIGIN OF SPINAL ACCESSORY.

selves definitely with the vagus; so that it is best to speak of them as the most distal vagal roots, and to restrict the term accessory to the spinal roots which are purely motor.

In many sections taken from the upper part of the cervical cord a strong curved bundle convex dorsally is seen to enter the lateral column (at a spot the relation of which to the point of entrance of the posterior root varies), to traverse the lateral column and pass into the grey matter in the region of its processus reticularis (figs. 118 and 153). At the level of the decussation of the pyramids it is often difficult to distinguish the roots of the accessory nerve from those fibres of the lateral column, which are making their way obliquely towards the middle line.

In the grey substance of the spinal cord the root-fibres of the accessory nerve either course straight ventrally to the nerve-cells which lie on the border of the anterior horn, or else they reach these cells after first running longitudinally for a certain distance within the grey substance, v (figs. 158 and 159). The nerve-cells just mentioned are, therefore, considered to constitute the proper accessorius nucleus, $n\ 1$ and $n\ 2$.

 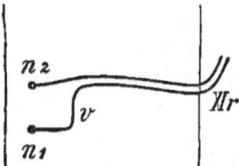

Fig. 158.—Diagram showing the disposition of the nervus accessorius Willisii in cross-section.—n, Cells of origin; v, respiratory bundle; XIr, issuing root; rp, posterior, ra, anterior spinal roots.

Fig. 159.—Scheme showing the disposition of the n. accessorius Willisii in longitudinal section.—n_1, n_2, Cells of origin; v, ascending segment of the root (Krause's respiratory bundle); XIr, issuing root.

The account just given (after *Roller*) corresponds best with the facts, but other very divergent views as to the origin of the accessorius are held.

The cells of the lateral horn, as well as those of the processus reticularis, have been claimed as the accessorius nucleus. That fibres which come from the other side of the medulla join the root-bundles is most probable. *Roller* allows the accessorius a further accession in fibres which arise in the lateral column, while *Darkschewitsch* describes fibres which reach the accessorius from the nucleus of Burdach's column.

Just as little uniformity reigns amongst the accounts given with regard to the level in the cord from which the accessory fibres arise. Many think they arise from the lateral horn throughout the whole length of the cord (*Krause*, *Clarke*), others only as far as the fifth cervical (*Huguenin*).

Dees agrees in the main with *Roller*, but makes more exact statements as to the position of the nucleus accessorius. This group of cells lies, he says, in the middle of the anterior horn on the cerebral side of the first cervical nerve; at the fourth cervical nerve it has moved back to the lateral border of the anterior horn; and from there to the sixth cervical it remains at the base of the lateral horn. The longitudinal segment, *v*, which many of the root-bundles present, lies, according to Dees, in the angle between the anterior and posterior horns; perhaps, therefore, it corresponds with Krause's respiratory bundle. The root-bundles must bend longitudinally after their origin in the nuclei and course brainwards.

12. **Hypoglossal Nerve.**—The most important origin of the hypoglossal nerve is from a grey area on the ventral side of the central canal; farther brainwards it lies in the floor of the fourth ventricle alongside the sulcus longitudinalis; the spinal portion is, at the decussation of the pyramids, the only portion of the anterior horn which remains attached to the central grey matter. It is characterised by large multipolar nerve-cells just like ordinary anterior horn-cells. This grey column, which can be followed as it lies up against the raphe as far brainwards as the striæ medullares, is termed the large-celled nucleus of the hypoglossal nerve, $NXII$ (chief nucleus, *Stilling's* classical hypoglossal nucleus, figs. 120 to 123, and 154). The coarse hypoglossal fibres show many twistings and curvings inside this nucleus. United into thick bundles they extend thence to its point of exit on the outer side of the pyramids. The most distal fibres are directed brainwards in a remarkable degree, so that they do not in our sections show their whole length (fig. 120). The lower olives are traversed by many hypoglossal fibres which come into no anatomical connection with them; the fibres suffer thereby, however, a number of distortions from their otherwise rather straight course, both in sagittal and in frontal planes.

The mesial angle of the large-celled hypoglossal nucleus is occupied by a rounded column of small cells of unknown meaning; we have already met with it in the chief auditory nucleus as the nucleus medialis or nucleus funiculi teretis.

A second origin is from the small-celled hypoglossal nucleus of

Roller. By this is meant a not distinctly circumscribed round clump of small nerve-cells which lies close up against the ventral side of the large-celled nucleus; in the sections farther brainwards it surrounds the root-bundles of the hypoglossus. Perhaps hypoglossal fibres end here.

It is possible that the large multipolar cells which are seen in the neighbourhood of the hypoglossal roots also give an accession of fibres to them (*Duval, Koch*). *Laura* regards the nucleus ambiguus as an accessory nucleus of this nerve.

Some of the root-fibres when just to the ventral side of the nucleus bend towards the middle line joining the substantial bundle which originates in the vagus nucleus; probably these are the fibres which are going to the hypoglossal nerve of the opposite side.

Commissural fibres between the two nuclei seem to be present, as well as fibres which take part in the formation of the posterior longitudinal bundle. Various connections with other parts of the brain appear to be established by means of the fibres of the medullary layer which, lying dorsal to the hypoglossal nuclei, gives the white colour to the floor of the ventricle, *m* (fig. 123). Many of these fibres bend laterally and unite into a strong column which traverses the vagus nucleus and ends in an unknown manner. *Koch*, who looks upon this stratum as made up chiefly of fibres connecting the different cells of the hypoglossal nucleus to one another, speaks, consequently, of fibræ propriæ nuclei hypoglossi; he thinks, however, that commissural fibres also go out from them.

According to general supposition the hypoglossal nucleus is connected with the great brain by the well-known route *viâ* the longitudinal fibres of the raphe and the pyramidal tract.

Roller thinks that the chief nucleus may have other relations besides its connection with the hypoglossus.

A process analogous with the poliomyelitis of the spinal cord may cause destruction of the motor nuclei of the medulla oblongata and the rest of the brain-stem as far as the third ventricle. Ophthalmoplegia nuclearis has been already mentioned; a disease to which the cells of origin of the hypoglossal nerve first fall victims, then those of the facial nerve, as well as the vagus and glossopharyngeal (chiefly their motor nuclei), and, exceptionally, also the cells of the motor root of the trigeminal nerve, is known as glosso-labio-pharyngeal paralysis (progressive bulbar paralysis, polioencephalitis inferior).

C. THE CEREBELLUM.

1. Central Nuclei.—We have already discovered that the cerebellum presents a peripheral grey layer or cortex, as well as certain internal masses of grey matter, and various columns of fibres which take part in the formation of its medullary substance.

Reserving the description of the finer histological details for treatment later on, we must now examine the central grey masses of the cerebellum. Neither the corpus dentatum with its appendages (the nuclei emboliformis and globosus) nor the nucleus of the roof actually reach the surface of the ventricle; but they come very close to it, being separated therefrom by a thin white layer only.

(1) The **corpus dentatum** is a purse-shaped many-plaited sheet of grey matter surrounding a mass of white substance, distinguished by the numerous large veins which it contains (nucleus medullaris corporis dentati). The opening into the bag is directed forwards and medianwards. The thickness of the grey band is from 0·3 to 0·5 mm.

It contains nerve-cells, not very closely packed together, of from 20 to 30 μ in long-diameter, and a varying amount of pigment. Most of the cells are so arranged that a single process is directed into the medullary centre, while two or three processes, which divide dichotomously, are directed towards the outer medullary substance of the cerebellum. Numerous medullated fibres not united into distinct bundles traverse the grey matter of the nucleus from without inwards, whilst other fibres usually of considerable calibre run through the grey matter itself in a direction parallel with the surface of the sheet. A fairly close network of finer fibres also occupies the whole thickness of the grey matter.

The cells of the corpus dentatum develop very early in the human fœtus, being distinctly recognisable between the sixth and seventh months of intra-uterine life.

(2) The **nucleus of the roof,** Nt (figs. 17, 160), is to be looked upon as the central nucleus of the vermis. It has a not well-defined triangular or oval form, and is about 6 mm. in its sagittal diameter. Only a thin layer of medullary substance separates it from the ventricle. It reaches upwards for about a half or two-thirds of the thickness of the vermis. It is least well defined at the back. In the median plane it almost reaches the nucleus of the roof of the opposite side. Large vesicular ganglion-cells (40 to 90 μ in diam.) containing a great deal of yellow-brown pigment are found in it as well as numerous nerve-fibres, many of which (united into coarse bundles) extend transversely across to the nucleus of the opposite side, Dt (decussation of the nucleus of the roof). Remarkably thick axis-cylinders (5 μ in diam.) are met with also as well as quantities of granules.

(3 and 4) **Nuclei emboliformis et globosus** are only separated pieces of the corpus dentatum, which they resemble in structure.

The same central grey masses are found in animals, but the corpus dentatum is never so much plicated as in Man; even in monkeys it is relatively a broader and less folded sheet, while in lower mammals it is merely a diffuse grey mass. In birds, in correspondence with the considerable reduction in the size of the cerebellar hemispheres, only a single roof-nucleus is present; it is covered with a thin sheet of medullary substance, and bulges out on either side into the dorsal prolongation of the fourth ventricle which is a characteristic feature of the cerebellum in birds.

2. The Medullary Substance of the Cerebellum.—Three
mighty columns of fibres, as well as certain other tracts of less dimensions, converge from either side to form the medullary mass of the cerebellum.

The origin of the **corpus restiforme** has already been explained (p. 260); we have also described (p. 262) the way in which the spinal constituents of the corpus restiforme turn into the cerebellum, the fibres of the two sides crossing one another apparently in the "anterior commissure and decussation," and how the portion of the restiform body derived from the olive loses itself in a plexus of fibres which envelopes the corpus dentatum in a kind of coat, the "fleece" of *Stilling*.

The portion of the corpus restiforme which ends, without crossing, in the cortex of the vermis, may be inferred, from the results of *Monakow's* investigations, to come from the lateral cerebellar tract. *Vejas* denies that any part of the restiform body crosses in the cerebellum.

Many fibres of the restiform body probably reach the cortex cerebelli. Owing to the directly anterior course of the corpus restiforme the fibres intended for the back of the cerebellum must bend off at an acute angle ("neck of the cerebellar peduncle").

The fibres which enter the cerebellum by the **pedunculi pontis** are disposed in thin plates which split off from the main mass as the several branches and twigs of the "arbor vitæ cerebelli." It seems as if the whole of the cortex, both of the hemispheres and of the vermis, is plentifully supplied with pontine fibres. A crossing of these fibres in the vermis is not proved. Possibly the capsule of the corpus dentatum receives fibres from the pons. A more detailed account of the disposition of the fibres of the pons has been already given (p. 254), and it has been already pointed out that they probably provide for a crossed connection between the cerebrum and cerebellum.

The third connection of the cerebellum extends brainwards, the **brachium conjunctivum** (superior cerebellar peduncle, crus cerebelli ascendens, processus cerebelli ad corpora quadrigemina *seu* ad cerebrum, brachium copulativum).

Almost all the fibres from the centrum medullare corporis dentati pass out of the hilum into the brachium conjunctivum, of which they constitute the most important components. They are termed its intra-ciliary constituents, as coming from the "corpus ciliare." The brachium conjunctivum also contains extra-ciliary fibres, derived from the fleece, as well as a few, perhaps, from the cortex cerebelli. When its fibres are first collected into a bundle the brachium lies on the mesial side of the corpus restiforme (*cf.* fig. 129). Just below the ependyma at the lateral angle of the ventricle lies a bundle which can be easily unravelled; it joins the brachium conjunctivum and runs brainwards with it as far as the locus cœruleus, with which it is always connected. Its presence is to be associated with the fact that we can always find within the brachium conjunctivum (especially when we make our sections in the long axis of this column) a number of spindle-shaped cells of as much as 90 mm. in diameter and containing dark-brown pigment. These fusiform cells are laid with their long axes in the direction of the fibres. The bundle is known as the lateral longitudinal bundle of the roof of the ventricle. At their spinal end these fibres seem to turn, just in front of the striæ acusticæ, on the dorsal side of the corpus restiforme, outwards towards the stalk of the flocculus. Concerning the connections of the brachium conjunctivum with the auditory nerve see p. 305.

As the brachia conjunctiva coming out of the substance of the cerebellum converge towards the corpora quadrigemina they are covered by the inferior (lateral) fillets which come up from the outer side. They further show a tendency, as we have seen (figs. 127 to 131), to draw ventralwards and towards the middle line, and between the posterior and anterior quadrigeminal bodies they begin to cross. The decussation is at its height beneath the centre of the anterior quadrigeminal bodies (decussation of the brachia conjunctiva, Wernekinck's commissure, tegmental decussation). The greater part of the brachia certainly cross here, but it has been already stated that they contain fibres which do not take part in the decussation (*Arnold, Mendel*). It has also been supposed that in the posterior angle of the decussation are situate fibres commissural between the two cerebellar hemispheres, connecting together (that is to say) the two nuclei of origin of the auditory nerves, so that the decussation would be analogous to the chiasma nervorum opticorum, and might hence be called the chiasma nervorum acusticorum (*Meynert*).

After their crossing, the brachia conjunctiva extend, as round columns ([once called] the white nuclei of the tegment), a short distance farther brainwards; soon, however, they swell out owing to the intercalation of small pigmented nerve-cells into a mass, also round in cross-section (figs. 131, 132), which in the fresh state is light brown in colour, the red nucleus of the tegment (nucleus ruber tegmenti, olive supérieure of *Luys*). We still need a more detailed account of the histology of the red nucleus. The fibres which pass out of it are collected into small bundles while still within its substance, giving to it a peculiar striped or punctate appearance, *Ntg* (fig. 135).

It is not possible to give a really satisfactory account of the fate of the fibres which leave the red nucleus. Most probably these fibres lose themselves in the ventral part of the optic thalamus, as described by *Forel*. Some of them go, perhaps, to the cortex cerebri (*Meynert*), and it is just as possible that the red nucleus and the nucleus lenticularis are connected together (*Wernicke*), (*cf.* fig. 168).

Other connections of the cerebellum, besides the three peduncles, also exist. Cerebellar roots of several cranial nerves have been described, but in no case are they certainly proved. Possibly the sensory root of the fifth receives an accession from the medullary centre of the cerebellum (*cf.* p. 294). The fibres which have been indicated as the cerebellar root of the auditory nerve are, probably, only secondary connections between the cerebellum and the large-celled nucleus of the nerve (p. 305 and fig. 157). Some of the fibres, probably, reach the roof nucleus of the opposite side. On either side of the middle line a thin tract of fibres, frenulum veli medullaris anterioris, passes out of the region of the corpora quadrigemina into the cerebellum in the velum medullare anterius, beneath the lingula.

Several divisions of the **medullary centre** of the cerebellar hemispheres may be described, (1) the medullary centre of the corpus dentatum; (2) the "fleece" or plexus of fibres which stands in intimate relation with the corpus dentatum, enclosing it in fact; (3) *Stilling* has described certain different sets of fibres in the remaining greater mass of the white matter which are very difficult to distinguish; (4) a layer, 0·2 to 0·5 mm. in thickness, which lies in contact with the inner layer of the cortex of the cerebellum, following its contour, the "garland-like" bundle which connects the different lobules together, *g* (fig. 160); (5) in sagittal sections numerous bundles of fibres are seen cut across in front (brainwards) of the corpus dentatum, they belong to the great (anterior) cerebellar commissure; other bundles are cut across on the dorsal side of the corpus dentatum, they constitute the dorsal cerebellar commissure.

The **medullary centre of the vermis** (fig. 160) is often called the corpus trapezoides, a name which should be avoided, since it belongs to a different structure. A sagittal section through this mass shows the nucleus of the roof, *Nt*. This nucleus lies on a layer of sagittal fibres (median sagittal basal-bundle, Bs), which can be followed brainwards into the fibres of the velum medullare anterius, Vma. On either side it is joined by the lateral longitudinal bundles of the roof of the ventricle. Single dark pigmented nerve-cells are also found between the fibres of these bundles.

The **large anterior commissure,** DC, is met with on the cerebral side of the nucleus of the roof, separated from it by a layer of fibres, some 0·2 mm. broad. It is at least 0·4 mm. distant from the cortex. Above the anterior half of the nucleus of the roof the layer which was at first only 0·2 mm. in thickness increases to 1 mm.; from here it extends, always diminishing in size, far into the vertical branch, Rv, of the arbor vitæ, in which it ends in a point. Above the nucleus of the roof an indistinctly curved continuation of the anterior commissure, dc, consisting of separate bundles, cut transversely, can be followed close under the cortex, as far as the commencement of the horizontal branch, Rh. Especially in its most strongly-developed part the anterior commissure is split by fibres which come out of the anterior border of the nucleus of the roof into bundles which are fusiform when cut across with the long axis of the spindle placed sagittally. Frontal sections (fig. 131) show that not a few fibres of the anterior commissure, which lie on the dorsal side of the nucleus of the roof, descend in the median line between the nuclei of the two sides, cross one another here, and then, apparently, assume a sagittal course. It appears, therefore, that decussating and commissural fibres are intermingled in the anterior commissure; and, hence, it is often termed the decussation-commissure ("Kreuzungscommissur").

The **commissure of the roof nuclei,** Dt, consists of a second set of fibres, independent of the anterior commissure. We have already mentioned that numerous round bundles run from one side to the other within the substance of the nuclei of the roof; they are most numerous, perhaps, in their anterior portions. These rounded bundles of fibres also form the dorsal boundary of the nuclei of the roof and are arranged in a gently-sloping line, so that the last of the bundles are found in the horizontal branch. The latter portion of the roof commissure, dt, is nothing more than the middle portion of the dorsal cerebellar commissure.

Behind (distalwards to) the nucleus of the roof no transverse fibres united into bundles are found in the medullary centre; some such are found, however, far back in the horizontal branch where it breaks up into a number of smaller branches (posterior cerebellar commissure). Longitudinal fibres are found almost exclusively in the medullary branches of the arbor vitæ, so that they lie in the plane of sections cut at right angles to the convolutions. Sections must be strictly sagittal when the vermis is cut, while for the hemispheres they should diverge to the sides posteriorly. These longitudinal fibres lie in the centre of each medullary branch directed towards the general medullary centre; on the other hand, the garland-like tracts of fibres above mentioned lie up against the cortex. In all places where the medullary branches divide dichotomously, or where lateral branches come off from them at right angles, a thickening of the medullary substance due to an increase in its connective-tissue is visible. In stained sections these spots are coloured more deeply on account of this preponderance of connective-tissue.

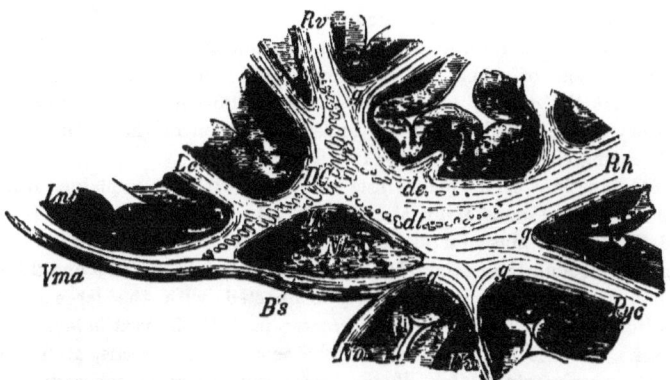

Fig. 160.—Sagittal section through the cerebellum some millimeters to one side of the middle line. *Magn.* 5.—*Vma*, Velum medullare anterius; *Lng*, lingula; *Lc*, central lobule; *Rv*, ventral medullary ramus; *Rh*, horizontal *ditto*; *Pyc*, pyramis cerebelli; *Uv*, uvula; *No*, nodulus; *Bs*, sagittal basal tract; *DC*, anterior great commissural decussation; *dc*, its posterior prolongation; *Nt*, nucleus of the roof; *Dt*, decussation of the nucleus of the roof; *dt*, its posterior prolongation; *g*, garland-like fasciculi.

The following **connections** of the cerebellum are almost certainly established; others, not sufficiently determined as yet, may also exist:—

(1) With the spinal cord and the after-brain by means of the corpus restiforme.

(*a.*) With the lateral cerebellar tract, and so with Clarke's column and the posterior roots of the same side.

(*b.*) With the nuclei of the posterior columns of the same and the opposite side, and therefore, indirectly with the posterior roots of both sides.

(*c.*) With the olive of the opposite side.

(2) Only slight connections with the mid-brain through (as is supposed) the frenulum veli medullaris anterioris.

21

(3) With the 'tween and fore-brains—(*a.*) Through the pedunculus pontis, and by means of the contra-lateral pes pedunculi cerebri with the cerebral hemisphere of the opposite side. (A portion of this connection has been described as the frontal pontine tract.)

(*b.*) Through the brachium conjunctivum with the red nucleus of the opposite side, and, thence, with the optic thalamus.

(*c.*) Indirectly with the nucleus lenticularis through the opposite olive and the central tegmental tract.

(4) With certain cerebral nerves; undoubtedly with the auditory, that is to say, with one of the nuclei of origin of the radix vestibularis; probably also with the trigeminus.

It appears as if a direct connection between the cerebellum and the anterior roots of spinal nerves is wanting. One may suppose, however, that the cerebellum is affected by stimuli coming from various sensory regions, and is capable under the action of these stimuli of influencing the release of motor impulses.

The functions of the cerebellum can be still further explained from the anatomical data given above. Of all sensory impressions it is chiefly those of muscular sensations which are conducted to the cerebellum through the nuclei of the posterior columns (pp. 199 and 260). Further, this organ is intimately connected with the large-celled auditory nucleus from which the greater part of the vestibular nerve arises. We are bound to regard the semicircular canals, after the exact experiments of *Golz, Mach, Breuer* and others, as the organs of the sense of equilibration; the sensations which they set up are transferred directly to the cerebellum for further elaboration.

Impressions of muscle-sense and of equilibration (as well as visceral sensations, which may be conducted to the cerebellum through the lateral cerebellar tracts) do not, so fully as other sensations, help to make up intellectual life; they continually, however, exert an influence on the threshold of consciousness and so modify the movements of the body without needing the intervention of the cortex of the great brain. These sensations find a meeting place in the cerebellum from which they direct our movements; the necessary force for the production of a co-ordinated movement is probably measured out to each single muscle-contraction from this centre. It can hardly be supposed that influences on motility which come out from the cerebellum enter the cortico-muscular pyramidal tracts on their way through the pons, despite the close interweaving of the two systems of fibres, for we are bound to believe that the fibres descending from the cortex to the spinal cord go through the pons without interruption. The physiological connection which the cerebellum brings about between certain sensory impressions and motor impulses may,

ELEMENTS CONTAINED IN CORTEX CEREBELLI.

therefore, take place through the cortex cerebri; or, as is more probable, through other parts of the great brain (fig. 140, C^5).

3. Cortex Cerebelli.—The boundary between the cortex and medullary substance of the cerebellum is nowhere quite sharply marked; at the summits of the lobules it is quite obscured; it is more distinct at the bottom of the fissures (fig. 161).

The bodies described as "granules" (*cf.* p. 127 and fig. 56) are found scattered about everywhere between the bundles of white fibres, although sometimes they are arranged in rows. Towards the surface they are more closely packed together, and constitute the **granular layer** ("rust-coloured" layer, since it is marked out macroscopically by its yellow-brown colour). The layer of granules is thinnest at the bottom of the fissure, thickest at the summits of the convolutions. In France these bodies are termed myélocytes.

Fig. 161. — Cross-section through a convolution of the cerebellum. *Carmine staining. Magn.* 15.

The granules are not disposed regularly throughout the layer; they always constitute rounded groups where they are most numerous. Here and there amongst the granules a very few indubitable nerve-cells are seen, fusiform or round in shape, pigmented, attaining to a maximum of 30 μ in diameter. Often these cells are quite wanting, and their number varies very much in different individuals. A large number of small cells which do not stain with hæmatoxylin ("eosin-cells," as *Denissenko* calls them) are also found in the granule layer; presumably they are nerve-cells.[*]

The medullated nerves of the central white substance give up their parallel arrangement as soon as they enter the closer layers of the granules to form a neat network throughout the whole breadth of the layer. Moreover, the spaces between the groups of granules is filled up with a network of closely-felted fibres, in addition to a small amount of neuroglia. The felt-work consists, as is proved, of indubitable

---

[*] The *translator* has failed to find these cells, as figured and described by *Denissenko*, in the cerebellum of any animal, after the most careful search with the methods recommended by him. Clumps of grey matter, made up for the most part of non-medullated fibres, lie amongst the granules. They may easily be mistaken for groups of cells.

connective-tissue fibrils, as well as probably non-medullated fibres and the processes of the granules. [The most conspicuous of the fibres from the arbor vitæ pass, without branching, through the granule-layer, diverging with great regularity towards the bases of the cells of Purkinje.]

The layer of the cortex cerebelli which follows next is distinguished by its peculiar large nerve-cells. They form a sheet one-cell thick, which invests the granular layer (figs. 161 and 163). The second or middle layer is, therefore, usually termed the **layer of large cells.**

The cells just mentioned, named, after their discoverer, the cells of Purkinje, have a round, somewhat flattened, shape, like a lens or a melon-seed.

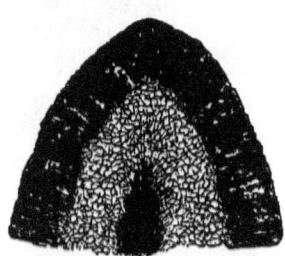

Fig. 162.—Cross-section through a lobule of the cerebellum. *Weigert's staining. Magn.* 15.¦

The transverse diameter of these cells attains to about 30 μ, their longitudinal diameter about 38 μ; it is not possible, however, to fix exactly the boundary between the cell and its peripheral process, so that the longitudinal diameter is usually stated to be somewhat greater than we have given it. Their thickness varies from 20 to 30 μ.

Purkinje's cells contain a large round nucleus with distinct nucleoli. Neither nucleus nor nucleoli possess processes as described by *Denissenko*. An exceedingly delicate cell-membrane, which is described as being continued on to the cell-processes, is not yet proved for certain, but its existence is not improbable.

The cell-body exhibits a striation which surrounds the nucleus as with a sling, and extends towards the peripheral process. It may be pointed out that these cells have not, as so many large cells have (*e.g.*, those of the spinal cord, cortex cerebri, and optic thalamus), any pigment granules, or, at any rate, but very few such granules—a fact of physiological importance.

From the pole of the cell, which is turned towards the granule-layer, originates the so-called central process, broad at its base, and rapidly growing thinner, which, on account of its fineness, is soon lost amongst the granules (occasionally, two such processes are present).*

Only in very fortunate preparations or after the use of the sublimate

---

* [By staining with Weigert's hæmatoxylin, the central process can be followed back into a distinct medullated nerve-fibre. A well-prepared section stained in this way leaves very little doubt as to the connection of each cell of *Purkinje* with an unbranched fibre of the medullary centre.]

method can one follow the process farther down. In copper preparations it tears off very easily owing to its delicacy. Hence opinions vary as to its ultimate fate. *Koschewnikoff*, *Schwalbe*, and *Beevor* believe that the process goes undivided into the axis-cylinder of a medullated nerve; *Denissenko*, contrary to all other observers, asserts that it is surrounded with a medullary sheath, soon after issuing from the cell. It has only rarely been described as dividing. In recent times it has been *Golgi* who has chiefly asserted that numerous branches leave the central process.*

Golgi says that these lateral branches are very thin, and show a tendency to turn backwards towards the surface of the cerebellum; and that the proper axis-cylinder process, instead of dividing dichotomously like the other processes, keeps its independence, and can be followed without diminution in thickness into the medullary substance. It is, therefore, certain only that Purkinje's cells are connected with the fibres of the medullary centre by their central processes, most probably it is the radially running fibres of the centre with which they are connected. How this happens or whether the granules of the granular layer play any part in this connection cannot as yet be said with certainty.

A thick peripheral process, which is directed towards the surface, originates from the pole of each of Purkinje's cells. It belongs, however, altogether to the layer which comes next on the outer side, the molecular layer, with which it will, therefore, be described.

The granules of the granular layer extend to a certain extent into the large-celled layer, and even into the molecular layer. The outermost of these granules are considerably larger than those which occur in the deeper parts of the granular layer. A not inconsiderable tract of medullated fibres, which appears to envelope the granular layer, and extends both on the inner and outer sides of the cells of Purkinje, stretches parallel to the surface in the direction of the longitudinal axis of the convolutions. A good number of connective-tissue fibres, bearing in the same direction, are seen amongst these nerve-fibres. Other connective-tissue fibres surround the cells of Purkinje in a fairly close network.

On the whole, the large-celled layer is a very loose one [in hardened preparations it appears, owing to the shrinking of the cells of Purkinje, more open than is natural]; so that sections of the cerebellum are prone to break across through this layer, and into it small effusions of blood are apt to be poured.

* [The caution already given as to the deductions to be drawn from the use of Golgi's method should be borne in mind. It effects a precipitation of solid particles in the lymph paths surrounding the fibres.]

It should be noticed that at the bottom of the fissures the cells of Purkinje stand far apart while they are closely packed together at the apices of the convolutions. The breadth of the granular layer is proportional to the number of the large cells.

It is tempting to associate this proportional relation with the development of the fissures and convolutions, but such an interdependence is not to be found. Rather is it the case that the number of Purkinje's cells is directly proportional to the extent of free surface exposed, since each cell has to provide for an equal segment of the cortical surface. Since, on the convexity, the superficial area is greater than it is in the concavity, the number of Purkinje's cells varies accordingly. The number of granules in the granular layer depends, as already mentioned, upon the number of large nerve-cells, and certainly varies as these cells vary, although as yet the physiological connection between the two is not cleared up.

The most external or **molecular layer** (finely granular or grey layer) covers the whole of the surface of the cerebellum to a uniform depth of 0·4 mm. In it the peripheral (protoplasmic) processes of Purkinje's cells are distributed (fig. 163). Each process from the peripheral pole of the cells consists, as a rule, of a short thick trunk directed straight outwards towards the surface, which soon divides into two similar chief branches disposed horizontally. From the chief branches fairly strong branches come off again at right angles and run towards the surface. All the thicker processes which originate from these branches (the case is different with the finest terminal twigs) run either parallel to the surface or else vertically to it. In the two middle fourths of the molecular layer they are almost exclusively parallel to the surface.

A single peripheral process as just pictured is only to be seen distinctly on the convexity of the convolutions. The nearer we approach the bottom of the furrows the closer does the point of division of the single stem approach to the cell, until at last at the bottom of the fissure two horizontal processes come off separately from the cell (fig. 164).

The thick branches (apart from the fine twigs which they give off directly) gradually dissolve into a network of excessively delicate fibres which extends as far as the free surface, and is best exhibited in its marvellous richness by Golgi's method of precipitation of silver or corrosive sublimate (fig. 164). [The ultimate twigs may be still better shown by the method of staining, first with carmine alum and then with Weigert's hæmatoxylin. They can hardly be said to taper, for the greater part of the branching is accomplished in the deeper strata of the molecular layer, and when once a terminal process is constituted it runs towards the surface, where it tends to fall back again like spray from a fountain, maintaining a uniform diameter for a considerable distance. In the shark and other animals in which numerous horizontal limbs come off from the cell, these give rise at once to radial branches which do not subsequently divide, but traverse the whole thickness of the molecular layer with a gently undulating course.]

If sections of the cerebellum are made at right angles to the surface, but in the direction of the convolutions, not, as in the preced-

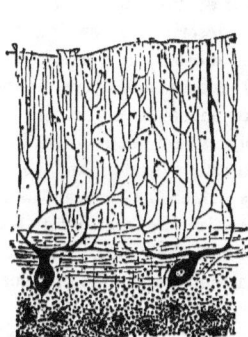

Fig. 163.—Vertical section of the cortex from the lateral surface of a cerebellar convolution. *Carmine staining. Magn.* 90.

Fig. 164.—A Purkinje's cell exposed by a section vertical to the surface and at right angles to the long axis of a convolution. *Golgi's staining. Magn.* 120.

ing case, at right angles to them, a different picture is seen. The lateral extension of the peripheral branches is absent; their ramifications occupy a segment of the molecular layer, not broader than the cell is thick. Hence it follows that the peripheral processes of Purkinje's cells are disposed in two dimensions only, like the stem and branches of an espalier fruit tree, not like the ramifications on all sides of a free-standing tree; a circumstance not without physiological importance.

Medullated fibres rise out of the granular and large-celled layer into the molecular layer, and pass either directly towards the surface or in other directions, but they are only to be seen in the inner half of this layer and even there in small numbers (fig. 163).

Occasionally medullated fibres are seen running parallel to the surface beneath the pia mater or in the middle of the molecular layer (*Beevor*).

Fig. 165.—A Purkinje's cell exposed by a section vertical to the surface and parallel with the long axis of a convolution. *Golgi's staining*.

Various cell-elements are scattered about through the molecular layer, viz., (1) the already-mentioned large granules (only in the deepest layers); (2) smaller nuclei, apparently free; (3) connective-tissue cells; (4) small cells, which are in all probability nerve-cells. One of the most important questions (which as yet has hardly been answered) is as to the ultimate fate of the finest twigs which arise from the branching of the peripheral processes of Purkinje's cells. Not seldom they have been described as ending freely on the surface. Undoubtedly a portion of the terminal twigs bend round and turn inwards again at the surface or more deeply in the molecular layer. Probably when deep down in the layer they collect into axis-cylinders and so form the medullated fibres which have been already remarked as occurring in the molecular layer; or, on the other hand, they may, still in a non-medullated condition, enter the plexus in the granular layer.

These views must, however, be regarded as hypothetical and intended to help us out of the embarrassment in which our inability to discover such a termination of these twigs as will satisfy the physiological necessities of the case places us; the opinion here expressed is not by any means founded on observation.

It is worthy of remark that not only are coarse anastomosing branches between the cells of Purkinje wanting, but that even the very finest processes of the cells fail to unite with one another, no proper nerve network, in the strict sense of the word, is present in the molecular layer. [Many preparations, especially of the cerebella of lower vertebrates, show brushes of fine non-medullated fibres passing from the granular to the molecular layer between the cells of Purkinje.]

The arrangement of the connective-tissue of the molecular layer deserves especial mention.

Between the proper vascular pia mater and the cortex cerebelli lies a delicate membrane (basal membrane), from which connective-tissue fibres with pyramidal bases start off into the substance of the cortex (radial fibres). It was first described by *Bergmann*. On account of their delicacy these fibres cannot (except in new-born animals) be followed far into the cortex. Their existence is best proved in cases of pathological (? inflammatory) changes in the cortex cerebelli to which the more delicate tissue-elements fall a prey, while the coarser connective-tissue skeleton remains intact. It is, then, possible to convince oneself that these radial fibres traverse the molecular layer as far as the layer of large cells, running parallel to one another without dividing (*cf.* p. 335 and fig. 167).

In the deeper strata of the molecular layer are found the connective fibres which have been already described as running parallel to the surface. [Indeed, the molecular layer shows most distinctly two sets of striæ which cross one another at right angles, viz., the radially-disposed processes of Purkinje's cells and a much closer system of tangential fibres. If a process of Purkinje is teased out it shows as it were an arrangement of horizontal pegs or thorns on either side, the adhering fragments of tangential filaments. Along the median line in the shark's cerebellum, and in other cases, the molecular layer is destitute of processes of Purkinje's cells, and the strata of tangential filaments are very distinct. Where these cells are absent, and the molecular and granular layers come into contact, the tangential fibres appear to pass into the granular layer. Nothing in the molecular layer is more conspicuous than the rectangular arrangement of its striation.] Nuclei are but seldom observed in these different, so-called, connective-tissue fibres.

The unimportant spaces which lie between the several elements of the molecular layer, including the blood-vessels, are occupied by finely granular neurogleia.

In all parts of the cerebellum the cortex exhibits the structure just described; no local differentiæ are yet known; we may well conclude from this circumstance that over the whole of the cerebellum the function is the same.

A striking similarity of structure is exhibited by the cortex cerebelli throughout the whole vertebrate series. It is not to be denied that a certain harmony between the size of the animal and the diameter of the cells of the cerebellum obtains amongst mammals. This parallelism affects the cells of Purkinje in the first place, but the granules also to a certain extent.

Apart from these variations in size, the cortex cerebelli presents in mammals a remarkable uniformity in structure; but the complexity of the branching system of Purkinje's cells is nowhere else so great as it is in Man; this character is most conspicuous if the human cerebellum is compared with that of small mammals, especially rodents.

The connective-tissue is more compact in the cerebellum of many animals than it is in Man. In the cat the basal membrane with its radial fibres can be easily seen, and the fibres can be followed for a long distance into the molecular layer.

The structure of the cerebellum of birds closely coincides with that seen in mammals. *Tenchini* and *Staurenghi* assert that the large-celled layer is very strongly developed in the eagle. Wider differences are met with when we come to other classes. In reptiles, amphibia, and fishes the large-celled layer is especially broad, owing chiefly to the numerous medullated fibres lying parallel to the surface. In consequence, Purkinje's cells are not arranged in a single sheet, but lie many cells thick. Moreover, the cells just named do not present in the lower three classes of vertebrates the round form characteristic of birds and mammals; they rather tend to assume a much folded fusiform or triangular shape. Their peripheral processes ramify in a somewhat different manner to those of birds and mammals. After branching a few times they run directly towards the surface, giving off only a few lateral branches which cannot be followed farther. Near the surface the connective-tissue grows so soft and loose that this part of the cortex often looks like a delicate lace-work.

A further peculiarity of the cerebellum of many lower animals consists in the reduction of the central medullary substance to a minimum; sometimes it is completely wanting in places, all the medullated fibres lying in the granular layer. [In plagiostome fishes the vermis is occupied by a large central mass of granules which surrounds the diverticulum of the ventricle, while the medullated fibres lie between this granular layer and the cells of Purkinje; not beneath the former as in mammals. The same difference in stratification is seen in the hemispheres. The so-called restiform bodies of these animals (in reality they are parts of the cerebellum, perhaps the homologues of the atrophied mammalian tæniæ) show a still greater alteration in the arrangement of the layers; for the fibres lie on the surface under the pia mater, while the molecular layer is placed beneath the ependyma of the ventricle. Reptiles exhibit a somewhat similar formation as far as the massing of the granular layer in the vermis is concerned; but in these animals the white fibres line the ventricular diverticulum, forming a compact medullary layer beneath the lateral parts of the cerebellum, and diverging rapidly into the granular formation of the vermis. The number of Purkinje's cells is distinctly proportional to the quantity of granules, so that where the latter are thickly massed the large cells are unable to find room without sheering over one another.]

The **histological development** of the cerebellum is fairly well worked out. In Man it consists at first of a quantity of round granules (gleia-granules) in which, about the middle period of embryonal life, a band free from granules makes its appearance, lying parallel to the surface, between it and the granular layer [but leaving a layer of granules beneath the pia mater. This layer is a conspicuous object in sections of cerebella from new-born animals]. This band is the developing molecular layer which even now presents a considerable likeness to the molecular layer in the adult. At the same time, or even a little earlier, the medullary centre, at first, of course, formed of non-medullated fibres only, advances towards the surface. At the end of the sixth month the Purkinje's cells can be usually, but not always, recognised along the inner border of the molecular layer. At birth nerve-cells are usually very visible, although their processes are as yet but slightly branched.

Although the breadth of the molecular layer slowly increases, that of the outer granular layer remains, even up to birth, about the same; only then does it begin to dwindle, and, finally, at varying periods of development, disappear.

At birth the outer granular layer can be divided into two nearly equally broad strata. The superficial granules are for the most part employed in the construction of the basal membrane, while the deeper ones move, by and by, into the substance of the molecular layer.

It has been already mentioned that the nerve-cells of the corpus dentatum cerebelli are among the earliest to be developed. At the sixth month they are marked out by their striking development, but hitherto we have been able to make no use of this circumstance for the explanation of their function.

Lastly, we must notice certain small grey clumps which, if great care is used, can be found in many cerebella in the midst of the medullary substance. Usually they are very small, scarcely visible, or at any rate not larger than a grain of millet; but, exceptionally, they may attain to a diameter of 1 cm. They contain, besides a network of fibres, irregular club-shaped nerve-cells, very similar to Purkinje's cells; also granules like those of the granular layer and a close capillary network (fig. 166). *Pfleger* first called attention to these small heterotopic collections of cortical substance.

[The cerebellum certainly offers the most favourable opportunity for studying the structure of cortex, and determining what is the plan of formation of this tissue, for although, at first sight, its cortex appears unlike that of the cerebrum, there is no sufficient reason for thinking that the two are fundamentally dissimilar. In the cortex of all parts of the brain are found large nerve-cells and granules in varying proportion. The nerve-cells provide for the nutrition of descending fibres. The granules are in all probability small nerve-cells borne by the non-medullated processes, into which ascending fibres break up on reaching the grey matter. Such is the *translator's* theory, based upon the study of the cerebellum. The medullary substance of this organ contains more fibres than are needed to supply one to each cell of Purkinje. Amongst the granules are clumps of matrix containing non-medullated fibres. Non-medullated fibres may be seen to break through into the molecular layer between Purkinje's cells, while the ultimate processes of these cells curve backwards towards the granular layer. There is no reason to suppose that the cells of Purkinje differ from the large cells of the cortex cerebri in not giving rise to efferent fibres. On the contrary, pathological observations point to such a connection, since they are not

Fig. 166.—Section through a small heterotopia in the medullary substance of the cerebellum. *Magn.* 40.—*a*, Medullary substance; *b*, the grey patch.

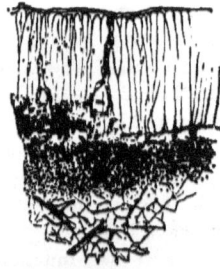

Fig. 167.—Vertical section of the cortex cerebelli from a case of encephalitis in which the connective-tissue elements of the cortex were not destroyed. The radial fibres of the molecular layer are distinct, so too are the holes out of which Purkinje-cells have fallen. Some medullated fibres remain in the connective-tissue framework between the granular layer and the medullary substance. *Weigert's stain. Magn.* 60.

affected by ascending degenerations. Far as such a theory may be from giving a complete explanation of cortex formation, our conception of this tissue is simplified by looking upon it as a field in which impressions received along the ascending fibres and their fibrils to which the granules belong, are collected after their distribution through the cortex by the processes of the large cells for transmission downwards to the grey matter which surrounds the central canal.]

4. Blood Vessels of the Cerebellum.—In Man the cerebellum is chiefly supplied with blood from the vertebral artery.

Three arteries reach it on either side (*cf.* fig. 180).

(1) Arteria cerebelli inferior posterior, which comes off as a rule from the uppermost portion of the vertebral artery, but sometimes from the commencement of the basilar artery; (2) arteria cerebelli inferior anterior from the basilar; and (3) arteria cerebelli superior from the front of the basilar shortly before it splits into the two posterior cerebral arteries. The superior cerebellar artery is very constant; whereas the other two are often wanting, though usually on one side only. The anterior inferior cerebellar artery has the smallest caliber of the three. All three leave the main artery at right angles. Within the pia mater the vessels divide repeatedly, and only small delicate branches enter the substance of the cerebellum. From the anterior inferior cerebellar artery a larger branch comes off, however, and is directed towards the corpus dentatum, through the hilum of which it enters the medullary substance, arteria corporis dentati. The larger veins in the interior of the medullary substance of the corpus dentatum havé been already mentioned.

The capillary network of the cerebellar cortex shows certain peculiarities corresponding with the stratification of the nerve-elements.

The arteries and veins enter the molecular layer vertically, and continue in this direction as far as the cells of Purkinje. There the capillary vessels form an abundant close network with oval meshes, the long axes of the ovals also arranged radially as it appears. The upper layer of the cortex contains no capillary network (*Oegg*). In the granular layer we find a capillary network of rather narrow mesh. On passing on into the medullary substance, the meshes of the network become wider, and are elongated in the direction of the fibres. Larger vessels, both arteries and veins, attract attention in the zone of the cells of Purkinje; they run almost parallel to the surface, and are devoted to the nourishment of the large cells. At birth the cerebellum contains comparatively few, but wide, vessels; already, however, the peculiarities characteristic of their distribution are recognisable.

5. Pathological Changes in the Cerebellum.—On the whole the anatomical changes in this organ due to disease are similar to those which occur in the rest of the brain. In this place we shall only point out such as are characteristic.

ATROPHY of the cerebellum has been often described; a distinction must be made, however, between a strikingly diminutive, but otherwise normal, cerebellum and a cerebellum in which the falling off in size is associated with sclerosis of its tissue.

CONGENITAL ATROPHY belongs to the first class only, the latter class of atrophies are, of course, acquired.

Where one hemisphere of the cerebellum is atrophic the opposite cerebral hemisphere is usually also diminished in size. Atrophy of the opposite olive is almost always associated therewith.

EMBOLISM of a cerebellar artery is very rare. Owing to the three arteries of the cerebellum coming off at right angles from a much more considerable trunk, it will be easily understood that the embolus, as a rule, floats away in the basilar artery, and is not stopped until it reaches the posterior cerebral artery.

Large APOPLEXIES of the cerebellar substance are much more rare, too, than in other parts of the brain. This is due to the fact that in the cerebellum hardly any besides the very smallest arteries occur; the only artery of somewhat larger size, arteria corporis dentati, is consequently the commonest source of extended bleeding into the cerebellum.

CAPILLARY HÆMORRHAGES are sometimes met with especially in the cortex of the cerebellum. In these cases one can usually see that the little effusion of blood has spread out horizontally in the layer of Purkinje's cells where it found the least resistance.

Amongst TUMOURS found in the cerebellum the first place is occupied by tubercle. It occurs much more often in the cerebellum proportionally to its size, than in any other part of the brain. Often several masses of tubercle are present at the same time; usually they originate in the pia mater and are sharply marked off from the surrounding brain-substance. They may be so large that a whole hemisphere, or even more, of the cerebellum is changed into a tuberculous mass. Glioma and carcinoma also belong to the more frequently occurring tumours. Passing over certain other new formations (such as fibromas, sarcomas, &c.) we will notice some which are interesting on account of their rarity, *e.g.*, dermoid cysts (*Clairat, Irvine, Heimpel*), osteomata (*Ebstein*) and echinococci which have penetrated from the fourth ventricle.

INFLAMMATORY PROCESSES often affect the cerebellum and its meninges. Purulent meningitis may be of traumatic origin, but it is usually secondary to disease of other spots on the surface of the brain or on surrounding bones (*e.g.*, temporal bone).

Inflammation of the substance of the cerebellum produces local softening of the tissue—cysts and abscesses—which are relatively very common in this organ. Sometimes a whole hemisphere is turned into a cyst or an abscess-sac. Diverticula of the fourth ventricle may develop into cysts within the substance of the cerebellum; sometimes they still communicate with the cavity of the ventricle.

In consequence of circumscribed chronic encephalitis the nervous constituents of the cerebellar cortex and underlying white matter may come to grief almost completely, only the connective-tissue framework remaining intact; the preparation looking as if it had been successfully macerated. Such specimens show us the arrangement of the connective-tissue in the cerebellum more plainly than any others (fig. 167). Isolated intact nerve-fibres are, however, to be found in the nuclear layer and in the central medullary substance after all the nerve-cells have disappeared (*Hess*), the spaces in which the cells of Purkinje used to lie being still recognisable.

Secondary degeneration may be followed far into the medullary substance after destruction of portions of the cerebellar cortex (*Borgherini*); especially when the lesion of the cortex is quite a small one. A number of well-preserved fibres may also be found coming out of the diseased piece of cortex amongst the fibres of this degenerated bundle. From this it follows that of the fibres connected with the cerebellum some must have their trophic centres within this organ, others must have their trophic centres elsewhere.

There is very little to be said about pathological changes in the **pedunculi cerebelli** and the **pons.**

In atrophy of one of the hemispheres of the cerebellum, degeneration of the corresponding brachium pontis and the same side of the pons is especially noticeable. Changes in the brachia conjunctiva are also found after extirpation of one of the hemispheres, and the existence of uncrossed fibres can then be demonstrated (*Marchi*).

Small aneurysms are common in the arteries of the pons, and hence apoplexies of this region are not rare. Patches of softening and tumours (especially tubercle) have been repeatedly seen. Aneurysm of the basilar artery must have a lasting effect owing to its pressure on the pons.

The pons is a favourite spot for sclerotic lesions in disseminated sclerosis.

The pathology of the brachia pontis is the same as that of the pons. Independent isolated disease of the other peduncles of the cerebellum is rare.

D. THE CEREBRUM.

In treating of the great brain we will first describe the central grey masses which it contains, their intimate structure and connections with other parts of the brain; then we will attempt to unravel as far as possible the separate tracts which occupy its medullary substance, and only when this has been done shall we devote a more detailed consideration to the minute structure of the cortex cerebri.

1. THE GANGLIA OF THE GREAT BRAIN.

(1) **Thalamus Opticus.**—On both free surfaces of the thalamus, mesial as well as dorsal, a superficial stratum covering the grey matter is to be recognised apart from the ependyma of the ventricle. On the

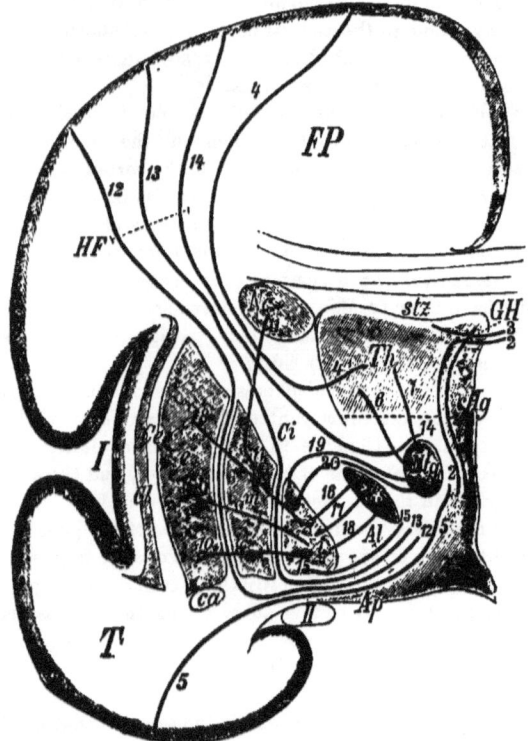

Fig. 168.—Diagrammatic frontal section through the great brain.—*FP*, Frontal parietal lobe; *T*, temporal lobe; *I*, island of Reil; *II*, tractus opticus; *HF*, fibres from the tegment; *Nc*, nucleus caudatus; *Th*, thalamus opticus; *GH*, ganglion habenulæ; *cHg*, central grey matter; *Ce*, capsula externa; *Ntg*, nucleus tegmenti; *Cs*, corpus subthalamicum; *i, m, e*, the three segments of the nucleus lenticularis; *Ci*, capsula interna; *Cl*, claustrum; *ca*, anterior commissure; *Ap*, ansa peduncularis; *Al*, ansa lenticularis; *Stz*, stratum zonale; *1* to *20*, tracts of fibres referred to in the text.

mesial surface this layer is formed of grey matter, the central grey substance of the ventricle. On the dorsal surface it consists of tracts of white fibres, stratum zonale.

The central grey matter lining the thalamus is a continuation of the grey matter which surrounds the aqueduct and terminates in front in the lining of the infundibulum. Basewards it forms the floor of the third ventricle, or grey commissure of the floor, in which the chiasma nervorum opticorum is embedded (fig. 16).

The central grey lining is not everywhere distinguishable from the true thalamic substance. It consists of a ground-substance similar to that found in the masses of grey substance, and containing nerve-cells and fibres, the further connections of which are not known.

The middle commissure (commissura mollis) is a part of the central grey lining and contains a considerable number of nerve-fibres, not united into bundles, some of which run laterally into the thalamus, others bend over so as to course in various directions within the central grey lining parallel to the wall of the ventricle. A few of the fibres are continued into the inferior (inner) peduncle of the thalamus (*Fritsch, Holländer*).

At the very back of the tænia medullaris thalami lies the ganglion habenulæ (figs. 13, 16, 20, and 168, *GH*), a group of small nerve-cells not well defined in Man. Fibres from the pedunculus conarii and others coming out of the tænia medullaris enter this ganglion. Other sets of fibres of the two nerve-bundles just named probably merely traverse or pass over the ganglion.

A large column of fibres, usually visible to the naked eye, which leaves the ganglion habenulæ and passes towards the base of the brain with a slight convexity outwards deserves especial notice. It runs between the central grey lining and the substance of the thalamus itself and appears to end, as seen in frontal sections, on the mesial side of the red nucleus (fig. 168, *1;* fasciculus retroflexus, Meynert's bundle). According to *Forel* and *Gudden* this bundle extends to a group of nerve-cells (ganglion interpedunculare) which lies in the hinder part of the substantia perforata posterior, and is very distinct in certain animals—rodents and bats, for example. The cells corresponding to this ganglion are disposed in a diffuse manner just in front of the commencement of the pons, in the middle line of the most basal part of the tegmental region.

A part of the pedunculus conarii crosses over Meynert's bundle and joins it on its outer side. · The two then extend basewards together, *2*.

The pedunculi conarii have little to do with the pineal gland, which is only a rudimentary organ in Man. They constitute the crossed portions of Meynert's bundles, but it cannot be said for certain how they arise; it may be asserted, however, that they get many of their fibres from the opposite thalamus and ganglion habenulæ, *3*.

Darkschewitsch has proved that the peduncles of the pineal gland contain, in addition, fibres which form a crossed connection between the tractus opticus and the oculomotor nucleus (*cf.* p. 289).

The **stratum zonale** forms a coat not quite one millimeter thick on the upper surface of the thalamus; it is composed of medullated fibres running for the most part in a sagittal direction. The following sets of fibres take part in its formation.

(1) Fibres from the lateral root of the optic tract which, passing superficially to the corpus geniculatum laterale, spread out over the thalamus (p. 281).

(2) Fibres which extend out of the occipital lobes, and, perhaps, the temporal lobes also, in the sagittal medullary strata, coursing forwards to reach the surface of the pulvinar.

(3) Fibres out of the peduncle of the thalamus to be described later on.

(4) The fibres to that part of the tænia thalami called pedunculus conarii.

The lateral boundary of the grey mass of the thalamus is not everywhere sharply defined. Quantities of fibres stream into the thalamus on this side, so that grey and white matter are mixed up (stratum reticulatum). In animals the lateral surface of the stratum reticulatum is noticeable for its richness in medullated fibres (lamina medullaris externa).

So far as the bundles of fibres which stream into the thalamus originate in the cortex cerebri, they take part in forming the **corona radiata of the thalamus,** of which the following are the principal constituents:—

(1) Fibres from the frontal lobe which pass to the thalamus in a sagittal direction between the nucleus caudatus and nucleus lenticularis in the anterior portion of the internal capsule (fig. 168, *4*). A bundle, very inconsiderable in size in Man, comes from the cortex of the lobus olfactorius (*cf.* p. 275). It separates from the tract of fibres destined for the anterior commissure, and extending farther backwards enters the front part of the optic thalamus.

(2) Fibres from the parietal lobe which, piercing the posterior part of the internal capsule in thin bundles, sink into the lateral surface of the thalamus.

(3) Great bundles of fibres from the occipital lobe, and to a certain extent the temporal lobe also, which pass forwards in the sagittal medullary stratum to the thalamus (optic radiations of Gratiolet, posterior peduncle of the thalamus). A few of these fibres enter the stratum zonale in the manner already described.

(4) The inferior peduncle (stalk) of the thalamus, which comes from the temporal lobe. It will be described shortly.

We have already mentioned the three nuclei of the thalamus (of which the anterior or upper is the smallest), the columna fornicis passing in the anterior part of the central grey lining of the ventricle to the front of the thalamus, and the bundle of Vicq d'Azyr (pp. 64 and 74). Owing to this latter bundle turning a little outwards as it descends it diverges from the columna fornicis, leaving room for a sagittal tract of fibres which is not very well defined in this place (the inferior peduncle of the thalamus) to pass between the two, b (fig. 168, see also fig. 139).

By this peduncle fibres are conducted from the temporal lobe, and perhaps also from the globus pallidus, beneath the nucleus lenticularis, to the base of the thalamus. A part of these fibres reaches the surface of the thalamus and helps to form the stratum zonale; by *Wernicke* it is termed the inner peduncle of the thalamus. Another part extends in the sagittal direction forwards on the outer side of the fornix, as has just been mentioned. The term "inferior" peduncle, and still more the term "internal" peduncle, is used very loosely by many authors.

If the tractus opticus is dissected away from the base of the brain the crus cerebri is exposed as it disappears in the substance of the hemisphere. The structures on the base of the brain which invest the crus as it enters the hemisphere—sling themselves around it, as one might say—are termed collectively the ansa peduncularis, Ap (substantia innominata, fig. 168). The inferior stalk of the thalamus is an important constituent of the ansa peduncularis.

We shall see later on that all the bundles which enter into the formation of the ansa peduncularis have in their disposition certain points in common. They all agree in passing medianwards out of the region which lies on the ventral side of the nucleus lenticularis, and, therefore, as we learn from frontal sections (figs. 138, 139, 168) in arching round the ventral part of the internal capsule, as the direct continuation of the now covered-up pes pedunculi cerebri is called. They closely invest this structure before they separate to pass in various directions beneath the thalamus.

It has been already mentioned that fibres out of the lateral root of the optic tract, passing under the corpus geniculatum laterale, stream in a brush into the pulvinar. Connections of the thalamus with the anterior quadrigeminal and lateral geniculate bodies exist for certain; with the nuclei lenticularis et caudatus they are probable. It is likely that a part of the posterior commissure is also connected with the thalamus. Lastly, the thalamus is united with the tegment and the spinal cord in manifold ways not yet clearly understood. The fibres of the laminæ medullares seem especially to establish these connections, of which the most important and best known are with the red nucleus, *6, 7* (fig. 168; *cf.* p. 318), and the fillet (*cf.* p. 257). By many (*Meynert, Wernicke,* &c.) the posterior commissure is looked upon as the commencement of the crossed tegmental connection of the thalamus.

The thalamus is thus in connection with almost all parts of the cortex cerebri; with frontal, parietal, and occipital lobes through the internal capsule; with the temporal lobe through the ansa peduncularis; with the spinal cord and with the tegmental region of the medulla oblongata through the mesial fillet and posterior commissure; lastly, with the cerebellum by the red nucleus and brachium conjunctivum. Many other connections besides those enumerated above certainly exist.

With regard to the **minute structure** of the thalamus, it may be said that the nucleus externus is very rich in white fibres, hence its light colour. The nerve-cells of the thalamus are for the most part fairly large and strongly pigmented.

(2) **Nuclei Lenticularis et Caudatus.**—The nucleus caudatus and the external segment of the nucleus lenticularis, which are united to one another in a number of ways, may be considered as modified portions of the cortex cerebri (*see also* p. 37).

A thickening is formed in the floor of the anterior cerebral vesicle which constitutes the rudiment of the above-named grey masses; even in the completely developed brain the putamen is continuous with the grey covering of the substantia perforata, which is undoubtedly homologous with the rest of the cortex.

Apart from these genetical connections with the cortex, *Wernicke* has proved conclusively that fibres which are the homologues of the corona radiata come off from the nucleus caudatus and putamen, and enter, for the most part, the two inner segments of the nucleus lenticularis (globus pallidus), making use of it as an intermediate station, *8, 9, 10, 11* (fig. 168).

The fibres coming out of the putamen collect on its mesial border into distinct coarse bundles which, traversing the lamina medullaris nuclei lenticularis lateralis, reach the globus pallidus. Corresponding bundles from the nucleus caudatus, crossing the anterior division of the internal capsule, extend both to the lateral lamella and also to the second segment of the nucleus lenticularis, where they agree with the fibres from the putamen in assuming a mesial direction (*11*); it is these fibres which give a radial striation to the globus pallidus when seen in frontal section.

The two laminæ medullares, to which a third plate of white fibres is sometimes added owing to the splitting of the inner segment of the nucleus lenticularis into two, consist, as far as we can judge, of fibres which come, for the most part, out of the nucleus caudatus and putamen, but course basewards, not taking part in producing the radial striation of the globus pallidus.

According to *Edinger*, other bundles of fibres which come from the cortex of the parietal lobe, and receive their medullary sheaths at an earlier period than the rest of the fibres of the great brain, also take part in the formation of the laminæ medullares (*12, 13*). These fibres are called tegmental by *Edinger*, but he reckons amongst them still other fibres (*14*) which do not enter the nucleus lenticularis, but pass spinewards beneath the thalamus and above the red nucleus to join the fillet (*cf.* p. 257).

Lastly, fibres are found which come out of the grey substance of the middle segment of the nucleus lenticularis and bend round into the internal lamina medullaris; for the sake of simplicity they are omitted from fig. 168.

All the fibres which run basewards in the medullary laminæ turn medianwards beneath the globus pallidus, from which they receive additions (*15*). In this way they form the ansa lenticularis, *Al* (better ansa nuclei lenticularis), another constituent of the ansa peduncularis.

The internal capsule lies mesio-dorsally to the nucleus lenticularis, separating it from the nucleus caudatus and the thalamus, or rather from the regio subthalamica (stratum intermedium).

The ansa nuclei lenticularis as it traverses the most mesial and basal parts of the internal capsule comes into this regio subthalamica; farther back it lies beneath the red nucleus on the base of the brain, near the middle line; beyond this it cannot be followed with any certainty. Since the posterior longitudinal bundle increases quickly in the neighbourhood of the ansa lenticularis, *Wernicke* is of opinion that it is connected with it by means of fibres ascending in the raphe. The fibres of the posterior longitudinal bundle exceed those of the ansa peduncularis in caliber, and hence such a connection is only possible on the supposition that nerve-cells are intercalated between the two sets of fibres. *Bechterew* and *Flechsig* are of opinion that there is no connection between the ansa peduncularis and posterior longitudinal bundle. By these authors the ansa peduncularis is considered as prolonged down through the central tegmental tract as far as the inferior olive (*cf.* p. 262). It was on this ground that we described the nucleus lenticularis as connected with the inferior olive of the same side and, by this means, with the cerebellar hemisphere of the opposite side.

When it is considered that the posterior longitudinal bundle is medullated much earlier than the ansa lenticularis, it will be understood that a direct connection between the two is impossible.

Edinger supposes that a considerable portion of the ansa peduncularis enters the corpus subthalamicum.

Still other fibres enter the regio subthalamica from the nucleus lenticularis, all, it is unnecessary to add, traversing the internal capsule.

We have already seen the corpus subthalamicum, *Cs* (nucleus of Luys), on the dorsal side of the internal capsule in this region, and more in the mid-brain, above the pes pedunculi cerebri, the substantia nigra Soemmeringi. Fibres from the internal segment of the nucleus lenticularis enter both ganglionic masses, *16, 17, 18*. They are delicate fibres which can be seen in frontal sections coming from the dorso-mesial surface of the nucleus lenticularis and traversing the internal capsule. The uppermost (most dorsal, *19, 20*) of these fibres do not enter the corpus subthalamicum itself, but collect, after passing the internal capsule, into a compact bundle which constitutes the dorsal portion of the capsule of the corpus subthalamicum and enters the red nucleus according to *Wernicke* (tegmental bundle from the nucleus lenticularis).

The ventral part of the capsule of the corpus subthalamicum is, according to *Kahler*, formed of fibres which originate in the nucleus lenticularis and join in part the ansa lenticularis.

The ansa nuclei lenticularis and the inferior peduncle of the optic thalamus together make up the ansa peduncularis (*cf.* p. 339). A system of fibres is supposed to push its way between these two systems (posterior medullary lamina of the tegment), which *Meynert* regards as passing over into the posterior longitudinal bundle.

The external capsule, if it sends any fibres into the nucleus lenticularis, sends to it but inconsiderable tracts; hence in hardened preparations it is very easy to peel off the external capsule from the nucleus lenticularis; often this occurs as the result of hæmorrhage in this region.

All the tracts which turn spinewards after leaving the nucleus lenticularis reach the tegment. Although not proved it must be regarded as most probable that the nucleus lenticularis is connected with the crusta. Fibres which, coming out of the laminæ medullares, as well as fibres from the nucleus itself, and enter the internal capsule, in which they mingle with peduncular fibres, may perhaps be looked upon as establishing this connection.

With the exception of the fibres already described as passing through the nucleus lenticularis very little is yet settled as to the connections of the nucleus caudatus. We have some right to believe, however, that fibres pass from the nucleus caudatus directly to the crusta *viâ* the internal capsule, and by this route reach the region of the pons (forming part of the frontal pontine tract, p. 254).

Very extensive connections of the nuclei lenticularis et caudatus with the cortex cerebri are described by *Meynert*. He thinks that fibres from the frontal and parietal lobes reach the nucleus caudatus by way of the internal capsule. Such connections with the cortex have, however, been most emphatically set aside by *Wernicke* and other recent writers on the subject, at any rate as far as concerns the outer segment of the nucleus lenticularis and the nucleus caudatus. The fibres seen by *Meynert* are regarded as at the utmost only fibres which are passing through the nucleus lenticularis, not ending in it. Only in animals can it be proved that the bundles of fibres in question by no means all pass through the nucleus lenticularis (*Kowalewski*); further, it is admissable to believe that the great grey masses of the putamen and nucleus caudatus are connected with other parts of the cortex cerebri by means of "association fibres" just in the same way as the other several portions of the cortex are united with one another.

More detailed investigations into the **minute structure** of the ganglionic masses just described, to which in this connection may be added the nucleus subthalamicum and substantia nigra, are still desirable.

(*a.*) **Nucleus Caudatus**—In that part of its head which lies upon the internal capsule fibres that have streamed in from behind and below can be followed far towards its upper surface. Most of its nerve-cells are small, round, or fusiform.

(*b.*) **Nucleus Lenticularis.**—Not only is the outer tegment of this nucleus like the nucleus caudatus in colour, but it also agrees with it in minute structure. The bundles of nerve-fibres which collect towards the lamina medullaris externa have already found mention. The lighter colour of the globus pallidus depends upon a difference in quality in the ground-substance, although it would be difficult to define this difference histologically. It chiefly depends, however, upon the yellow pigment which its moderate-sized nerve-cells contain, as well as upon the number of medullated fibres which traverse the two inner segments of the nucleus lenticularis.

(*c.*) **Corpus subthalamicum,** mentioned first by *Luys* and described in greater detail by *Forel.* Its greatest thickness amounts to 3 to 4 mm., its breadth to 10 to 13 mm., its sagittal diameter to 7 to 8 mm. Its shape is that of a lens, lying on the pes pedunculi cerebri.

This body is characterised histologically by a close network of the very finest medullated fibres, amongst which coarse fibres are almost wholly wanting. Multipolar nerve-cells of moderate size containing light brown pigment are scattered about its substance. Only a few regions in the central nervous system are distinguished by so close a capillary network as the corpus subthalamicum; it possesses this latter character in most animals—the dog, amongst others.

(*d.*) **Substantia Nigra Soemmeringi.**—This substance contains spindle-shaped nerve-cells of moderate size, containing in Man little lumps of dark-brown pigment. From one-third to one-half of the cell is usually filled up with this pigment, which does not appear until extra-uterine life. The cells of the locus cœruleus are distinguished from those of the substantia nigra by their round vesicular form and greater diameter. Pigment is never present in the cells of the substantia Soemmeringi in animals.

2. THE MEDULLARY CENTRES OF THE GREAT BRAIN.

The greatest extent of this mass of medullary substance, so considerable in size in Man, is seen in a section carried through the centrum semiovale Vieussenii on a level with the corpus callosum. It is made up of three systems of fibres:—

(1) Fibres which extend from the cortex cerebri to the ganglionic masses of the 'tween-brain, or deeper down to the mid-brain, hind-brain, after-brain, and spinal cord: the corona radiata.

(2) Fibres which connect identical areas in the two hemispheres: commissural fibres.

(3) Fibres, some shorter, some longer, which bring different spots in the cortex of the same hemisphere into functional association; collectively we shall term them "association fibres" [fibræ propriæ].

So few medullated fibres are found in the human cerebrum at birth that it looks grey and gelatinous. Between the second and third week after birth the pyramidal tract begins to myelinate. In sagittal sections it is easy to see how this tract extends from the internal capsule towards the two central convolutions, beneath which it forks, ansa Rolandica (*Parrot*). After the first month the occipital lobes begin to whiten, after the fifth month the frontal lobes; but the myelination of the fibres of the great brain is not completed until after the ninth month of extra-uterine life (*Parrot*).

(1) **Corona Radiata.**—The fibres of the corona radiata, considered as a whole, converge like an open fan towards the internal capsule—a crown of rays may be shown in dissected preparations, to rest upon the 'tween-brain. The region next above the internal capsule,

where the fibres coming from various places meet, the stalk of the fan, is termed the pedunculus coronæ radiatæ.

The following more important parts of the corona radiata may be distinguished:—

(*a.*) From the anterior part of the frontal lobe, the frontal pontine tract and the anterior peduncle of the thalamus.

(*b.*) From the central convolutions and neighbouring areas, the pyramidal tract and perhaps also *Edinger's* tegmental system of fibres, as well as bundles for the thalamus.

(*c.*) From the posterior parts of the parietal lobe and also from the occipital lobe, fibres for the thalamus (running chiefly in its posterior peduncle), as well as fibres for the external geniculate and anterior quadrigeminal bodies, and also for the posterior part of the hinder segment of the internal capsule (sensory tracts of the sagittal medullary stratum).

(*d.*) From the temporal lobe, fibres for the thalamus, a part of which run in its inferior peduncle, while others join the sagittal medullary stratum. Of the last division the majority are probably not meant for the thalamus, but for the back of the internal capsule and so for the crusta. Fibres seem also to extend from the temporal lobe to the internal geniculate body.

Beside these most important constituents of the corona radiata, it contains others which are not as yet sufficiently determined. The fibres which pass from the nucleus caudatus and the putamen into the globus pallidus, some of the medullated fibres on the basis of the tractus olfactorius, as well as part of the fornix, must be considered as equivalents of the corona radiata.

To what we have already said about the anatomical disposition of the **fornix** (p. 73), something must still be added here.

The fornix contains many fibres which originate in the region of the cornu Ammonis and seem to end in the corpus mammillare, being, therefore, analogous to the fibres of the corona radiata. A small portion of the fornix which streams on to the septum pellucidum [præcommissural fibres of Huxley] ought to be reckoned as fibres of association, since the septum pellucidum belongs to the cortex.

Each **corpus mammillare** is, according to *Gudden*, divided into two separate ganglia—a mesial one containing small cells and a lateral one containing large nerve-cells. A large part of the columna fornicis (called its radix) pushes itself between the two ganglia, partly to enter their substance, partly to form their capsule.*

[* The *translator* has submitted a theory which he thinks deserves mention, although anything like a discussion of the evidence upon which it is based would be out of place in a text-book. He looks upon the fornix as containing the con-

Out of the mesial ganglion arises the bundle of Vicq d'Azyr (ascending crus of the fornix of *Meynert*) which first ascends directly and then bends more anteriorly to terminate in the tuberculum anterius of the thalamus. If, in the corpus mammillare, a simple turning over of the radix columnæ fornicis into the bundle of Vicq d'Azyr is not probable, an undeniable relationship between these two tracts nevertheless exists. A smaller bundle extending backwards to the tegment arises in the mesial ganglion.

The lateral ganglion also sends a bundle of fibres backwards to the tegment (pedunculus corporis mammillaris, *Meynert's* tegmental bundle of the corpus mammillare). In the rabbit it lies quite superficially on the inner border of the crusta; in Man it is situate more deeply. It is pierced by fibres of the oculomotor nerve, *Pcm* (fig. 135).

Sometimes a bundle, about 1 mm. broad, is seen stretching, quite superficially, from the corpus mammillare over the tuber cinereum, and disappearing beneath the chiasma some 4 or 5 mm. from the mesial border of the crusta (*Lenhossék*). This bundle, stria alba tuberis, turns outwards beneath the tractus opticus, to the fornix of which it ought to be regarded as a detached fasciculus.

Lenhossék describes yet another bundle of fibres arising in the medullary covering of the corpus mammillare. It passes more deeply through the tuber cinereum in a sagittal direction forwards, and spreads out into the substantia perforata.

(2) **Commissural Fibres of Cerebrum.**—By means of the corpus callosum and anterior commissure, a connection is effected between

> tinuation of the olfactory tract, chiefly on the following grounds :—(1) There is a great similarity of minute structure between the olfactory bulb and the sheath of grey matter (fascia dentata), which invests the margin of the cortex where it is folded over in the hippocampus. (2) The fascia dentata is continued up beneath the fimbria (posterior pillar of the fornix) for a distance varying apparently with the acuteness of the animal's sense of smell. (3) The fornix is of very much smaller cross-section in the anosmatic marine mammalia than it is in animals with a moderate sense of smell. (4) There is reason to think that, in mammals, the cerebral hemisphere twists over upon itself during its growth (*cf.* Appendix A) in such a way that the olfactory tract would necessarily be folded, like the fornix, around the peduncles. (5) In reptiles, and other low vertebrates in which this rotation has not occurred, the olfactory tract is connected with the grey matter lining the third ventricle which seems to be homologous with the front of the thalamus. (6) Unless the olfactory tract has such a connection with the central grey tube, the corpus mammillare, and front of the thalamus having the same relation to the olfactory tract as the corpus geniculatum and back of the thalamus have to the optic tract, the first nerve differs from all other nerves in being immediately connected with the cortex cerebri; in all other cases sensory nerves find their primary centres in the central grey tube.]

identical spots on the cortex of the two hemispheres; in Man, at any rate, it looks as if each individual area on the general surface of the great brain was without exception united to its corresponding contra-lateral area. It is not, however, certain that the commissural system for the several regions of the cortex is everywhere equally well-developed. [*Sherrington* finds that secondary degenerations in the corpus callosum do not extend from the injured area directly to the homologous spot on the opposite hemisphere, but show a tendency to spread out.]

(*a.*) **Corpus Callosum.**—From that part of the corpus callosum which can be exhibited by simply drawing aside the two hemispheres and exposing the bottom of the great longitudinal fissure, the fibres stream into the two hemispheres. In this free portion of the corpus callosum the fibres run horizontally, but soon after entering the hemispheres they diverge upwards and downwards to the parts of the cortex for which they are destined. Since the cerebral hemisphere exceeds the corpus callosum in length (*cf.* fig. 29), it is clear that all the fibres of the latter cannot continue in the same frontal plane; but both at its front and back the fibres must form arches as they curve round to the front of the frontal and the back of the occipital lobe respectively. In front, at the genu corporis callosi, the fibres destined for the frontal lobes of the two hemispheres form the forceps anterior. The streaming out of the fibres of the rostrum corporis callosi into the convolutions of the two sides, may be termed with *Henle* the white commissure of the floor (commissura baseos alba).

Since the splenium is only the rolled up posterior edge of the corpus callosum, chiefly the fibres for the back parts of the hemisphere come off from it. They run off on either side as a strong white column in the same kind of curve as those for the anterior parts of the brain forming the forceps posterior. The external capsule receives a considerable number of fibres from the corpus callosum; on their way to it they have to cross fibres intended for the internal capsule (*cf.* figs. 21, 138, 139). The great bundles of fibres which come off from the back of the corpus callosum, form, as they curve backwards, the outer wall of the lateral ventricle in its posterior and inferior horns (tapetum).

It may be regarded as certain that the corpus callosum provides for the whole of the surface of the cerebrum, with the exception of the inferior and anterior portions of the temporal lobes and the olfactory lobes (tractus olfactorii).

Since no single fibre can be isolated in its whole course from a certain spot on the cortex in one hemisphere to the corresponding spot on the opposite hemisphere, it will be readily apprehended that from time to time a voice is raised in favour of a different meaning for this part of the brain. In particular is it asserted by many people [especially *Hamilton*] that the corpus callosum represents a great crossed communication between the cortex and the opposite internal and external capsules.

Complete or partial deficiency of the corpus callosum has been repeatedly observed in Man. In lower mammals the corpus callosum is but weakly developed; in monotremes and edentata as well as the submammalian vertebrates it is almost completely wanting.

[*Osborn* has shown that the corpus callosum is present in all vertebrates. It is, however, the commissure of the cortex *par excellence*, and when this formation is rudimentary the corpus callosum is equally undeveloped.]

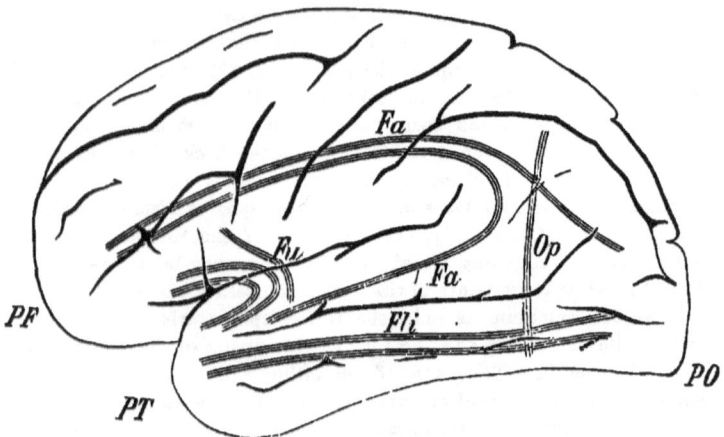

Fig. 169.—Diagram of the associating tracts of the cortex cerebri.—*PF*, Frontal pole; *PT*, temporal pole; *PO*, occipital pole; *Fa*, fasciculus arcuatus; *Fu*, fasciculus uncinatus; *Fli*, fasciculus longitudinalis inferior; *Op*, fasciculus occipitalis perpendicularis.

(*b.*) The **anterior commissure,** which is an accessory to the corpus callosum for the connection of the cortex of the olfactory lobes and parts of the temporal lobes, has already been described in detail.

(3) **Fibres Connecting together Different Areas on the same Hemisphere.**—It is necessary to distinguish between short fibres which connect together neighbouring cortical regions and long or considerable tracts which unite together parts of the cortex situate some distance away from one another. Collectively the two sets of fibres are spoken of as belonging to association-systems, since they are

regarded as systems for bringing into functional connection distant parts of the brain and so providing the mechanism for concerted actions. It would be more convenient if commissural fibres, as well as association fibres, were included in a single group, so that all the homodesmotic fibres of the cerebral cortex which have an analogous function were classed together.

The short fibres which unite neighbouring convolutions are to be seen in properly prepared dissections, arching beneath the cortex at the bottom of the fissures; *Arnold's* fibræ arcuatæ *seu* propriæ.

Amongst the long association-bundles (fig. 169) which may be demonstrated by defibering are reckoned :—

(*a.*) Fasciculus uncinatus, *Fu*, at the entrance of the Sylvian fossa, extending from the inferior frontal convolution to the gyrus uncinatus and the apex of the temporal lobe.

(*b.*) The fasciculus longitudinalis inferior, *Fli*, which is of all the long association-bundles the most easily demonstrated, runs as a broad tract from the anterior part of the temporal to the apex of the occipital lobe.

(*c.*) The fasciculus arcuatus, *Fa* (*seu* longitudinalis superior), consists of sagittally-disposed fibres, beneath the inferior and middle frontal convolutions, running partly towards the occipital lobe and partly arching round towards the apex of the temporal lobe; it is not easy, however, to make a good preparation of this tract.

(*d.*) The cingulum is an arched tract which lies in the medullary substance of the convolution of the same name. For the greater part of its course the cingulum lies up against the corpus callosum along the line of junction of its body and its radiating fibres. In frontal sections of the brains of animals it can, as a rule, be recognised by its circular cross-section.

(*e.*) The perpendicular occipital fasciculus of *Wernicke*, *Op* (*cf.* fig. 21, *Fov*), descends from the upper angle of the inferior parietal lobule vertically to the lobulus fusiformis.

On the outer side of the cingulum we come upon the place where the fibres of the corpus callosum and the internal capsule meet one another at right angles and interlace. It is immensely difficult in this region to distinguish individual tracts of fibres; farther outwards these two sets of fibres are more in accord as to their course.

The fibres of the external capsule have a fan-shaped, downwardly-converging course, corresponding to the disposition of the convolutions of the island of Reil. They seem to belong exclusively to the cortex of the island and to have nothing to do with the lateral segment of the nucleus lenticularis. Some of the fibres extend, as already mentioned, towards the corpus callosum.

3. CORTEX CEREBRI.

The wall of the anterior cerebral vesicles, both primary and secondary, but especially the latter, develops into the grey substance of the cortex cerebri. Certain parts of the developed wall of the anterior cerebral vesicle do not seem to enter into the formation of the superficial layer of the great brain; nor do they either in date of development or in histological characters agree with the cortex cerebri in the stricter sense of the term. Further, embryological observations are necessary to justify the conception that they really are homologous with the cortex. We have already made the acquaintance of some of these portions of the brain which do not look at the first glance as if they ought to be accredited to the cortex—viz., the grey substance of the tractus olfactorius, the nucleus caudatus, and the putamen.

If we cut, no matter where, at right angles into the surface of the hemisphere, the cortex (in the limited sense of the word) appears as a dark bordering band. The breadth of this band not only varies in different individuals, but in the same brain it depends upon the locality. It varies from 1·5 mm. to 4 mm., and is usually thicker at the apex of the convolutions than it is in the fissures. Its maximum breadth is attained at the upper part of the central convolutions and in the lobus paracentralis, its minimum near the occipital pole. In old age, with advancing atrophy of the brain, the diminishing thickness of the cortex is very noticeable.

Even in microscopical observation of fresh brains a stratification of the cortex parallel to the surface is evident, owing to the different colour of its layers. The differences in colour of the several layers are not equally distinct in all brains or in all parts of the cortex of the same brain.

Kölliker distinguished an outer white, middle grey, and inner yellowish-red layer. The narrowest of the three is the white layer on the surface, the other two are about equally broad. Between the two inner layers lies a not well-defined white band, while another band, or even two bands, sometimes occupy the middle of the inner layer; these are known as Baillarger's stripes (outer and inner). Hence the reader will gather that *Baillarger* distinguished six layers in the cortex.

This stratification is most likely to be seen in the superior frontal or anterior central convolutions. In the neighbourhood of the calcarine fissure, extending a little into the surrounding convolutions, especially the cuneus, Baillarger's outer stripe although narrow is sharply marked and easily seen in all brains (fig. 27). It has received the name in this locality of Vicq d'Azyr's stripe owing to its descrip-

GENNARI'S STRIPE.

tion for the first time by this anatomist. Before him, however (on February 2, 1776), *Gennari* saw this band traversing the cortex, "lineola albidior admodum eleganter," and described it very exactly, considering the topographical knowledge of the time. He figured it

Fig. 170. Fig. 171. Fig. 172.

Fig. 173. Fig. 174.

Fig. 170.—Vertical section through the human cortex cerebri where it covers the posterior part of the middle frontal convolution. *Carmine staining.* *Magn.* 20.
Fig. 171.—Cortex of the lobulus paracentralis.
Fig. 172.—Cortex of the cuneus in the fissura calcarina.
Fig. 173.—Cortex of the gyrus cinguli.—*Ccll*, Corpus callosum ; *Jgr*, indusium griseum ; *Stlm*, stria longitudinalis medialis.
Fig. 174.—Cortex of the subiculum cornu Ammonis at its most projecting part.

as well as the other stripes of Baillarger. It would, therefore, be only just to re-christen Vicq d'Azyr's stripe at any rate as Gennari's stripe (lineola albida Gennari).

Just as with the naked eye the appearance of the cortex is not the same in all regions, so we find that (unlike the cortex cerebelli) its microscopic structure varies in its several parts. An exact account of all the local differences in minute structure in the cortex cerebri is not yet available, although the more important details with regard to certain regions have been described.

We will commence our description with a section taken from the posterior end of the middle frontal convolution, and subsequently point out the more important features by which other regions are distinguished. The stratification visible with the naked eye in the cortex of a fresh brain is produced by the arrangement of the various tissue-elements of which it is composed, not, it is true, in definite strata, but still in layers with a certain amount of regularity.

The layers of the cortex cerebri are usually classified according to the form, size, and distribution of their nerve-cells. We shall, therefore, study first a carmine-coloured preparation. The direction in which the section is cut should be such that the nerve-fibres which stream into the cortex are divided, as far as possible, in the direction of their course. This direction can be determined by breaking a small piece of hardened brain including a portion of medullary substance. The plane of cleavage corresponds to the direction of the bundles of fibres, as can be seen by the characteristic radial striation.

Immediately beneath the pia mater which limits it towards the epicerebral space, a layer 0·25 mm. thick is met with (neurogleia layer, stratum moleculare, ependyma-formation), in which small ganglion-cells are scattered about irregularly in an apparently homogeneous ground-substance (fig. 170). On its outer border is to be seen a narrow stratum (10 to 30 μ in thickness) consisting exclusively of a close connective-tissue feltwork containing many Deiters' cells. It gives to the preparation when slightly magnified a dark contour.

The second layer (of small pyramidal cells, outer nerve-cell layer) is about as thick as the molecular layer from which it is sharply marked off. It contains a great number of nerve-cells, not more than 10 μ in height, closely packed together. They are for the most part pyramidal in shape, the apices of the pyramids being directed towards the surface.

The third layer, 1 mm. thick (layer of large pyramids, formation of the cornu Ammonis, middle nerve-cell layer), is not well marked off from the preceding one.

The pyramidal cells become more widely separated from one another,

PYRAMIDAL CELLS.

and larger in size the farther we descend from the surface, so that the largest (as much as 40 μ in breadth) are to be looked for deepest down. These large cells afford us the best opportunity of studying the peculiarities of pyramidal cells (figs. 55 and 175).

The pyramid may be imagined as evolved from a fusiform cell. The spindle-cell must be supposed to be placed radially to the surface; it gives off two terminal processes, the outer one of which is very gradually derived from the cell-body, being formed indeed by a tapering off of its substance; it can be followed a long way towards the surface. The other process originates more suddenly from the cell with which it is connected by a conical base; it turns directly towards the medullary substance, but cannot, as a rule, be followed far. Besides these two chief processes the cell gives off a number of secondary processes (from 4 to 12). The largest and most regularly disposed of these come off from the deepest part of the cell-body which gains, therefore, a considerable girth. The cell acquires in this way the shape of a cone or pyramid with its point directed outwards.

The processes are named in terms applicable to this pyramidal shape. The first of the chief processes which runs towards the periphery is named the apical process, the other one, which passes deeply, the basal process. The accessory processes which come off from

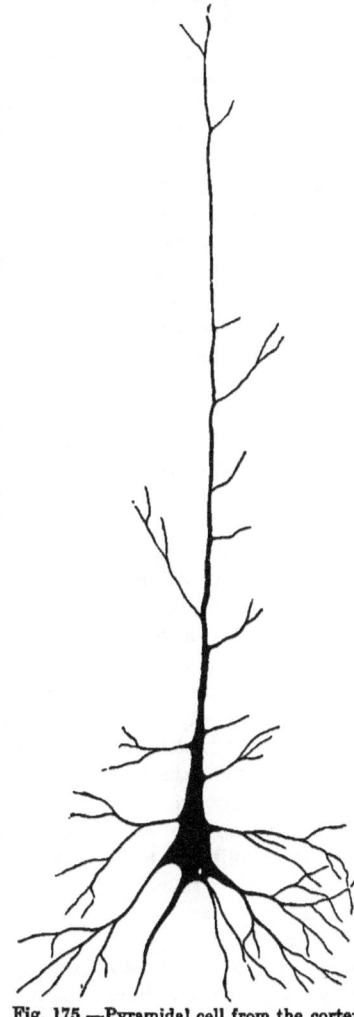

Fig. 175.—Pyramidal cell from the cortex cerebri. *Mercury preparation.* *Magn.* 200.

the circumference of the base are named lateral basal processes, all the rest are simply lateral processes.

The apical process can be followed, sometimes as far as the layer of small pyramids, but hardly ever into the molecular layer. On its way from the cell it gives off a variable number of side branches, which have somewhat thickened bases and lie at right angles to the apical process, dissolving farther on into the finest possible network. The chief process thus becomes gradually thinner and more delicate, and finally takes part, as may be supposed, in the formation of the network just named; that this is the way in which it ends must, however, be considered as by no means settled.

The middle basal process, which is often very difficult to find, is supposed to be continued directly into a medullated fibre, and hence is termed the axis-cylinder process. It either possesses no side branches or but very few, and does not decrease in thickness in its course. The axis-cylinder process is supposed, in exceptional cases, to come off from the side of the cell. Even under the most favourable conditions it is only rarely possible to prove that it passes directly into a medullated fibre.

The accessory processes are distinguished from the chief processes by the manner of their division, which is dichotomous; only after continued forking do they end in the network.

The protoplasm of the pyramidal cells is finely granular; sometimes a delicate striation can be recognised. A little clump of light-yellow pigment is always found in the cell, usually nearer to its base than apex. The nucleus is round or oval, or imitates on a diminished scale the pyramidal form of the cell. Cells with round and cells with pyramidal nuclei occur side by side. It is not yet ascertained whether this difference in the form of the nucleus is associated with a difference of function or is only a result of hardening.

The nucleolus is conspicuous owing to its high refraction.

Owing to the manner in which the apical process arises it is impossible to determine the length of the cell; the division between its body and the apical process being an arbitrary one.

The nerve-cells of the cortex cerebri, especially the pyramidal cells, are, on various grounds, considered as concerned in carrying out psychical functions. It is very probable that the largest pyramids have a psychomotor *rôle*.

It will be easily understood that special attention has been directed towards any possible alterations in their structure which might lead to a disturbance of their normal function. The investigation suffers from the circumstance that as yet our knowledge of the normal structure of a nerve-cell is so incomplete that only the wider departures from this normal structure can be recognised as such. It seems, however, important (for the reason just given) to look a little more closely at these elements.

A pericellular space of varying breadth surrounds the larger pyramidal cells. It often contains from one to five lymphoid cells [leucocytes].

The nerve-fibres come up from the medullary centre in close bundles easily seen in carmine preparations; they pass outwards towards the surface in regular order to lose themselves one after another in the neighbourhood of the third layer. The spaces between the groups of nerve-fibres are occupied by nerve-cells arranged in columns.

Not a few connective-tissue cells (spider cells) with many processes are found throughout the whole thickness of cortex; they are seen most distinctly in the zone of the large pyramids.

The fourth layer of the cortex (layer of small irregular nerve-cells, granule formation, mixed nerve-cell layer) is about 0·3 mm. wide in the section under description. The spaces between the radiating bundles, which now contain more fibres, are occupied by small cells still arranged in columns. These cells are about 8 to 12 μ in diameter, round, angular, or irregular in shape, and indubitably nervous. Very little can be alleged with certainty as to the number and further course of their processes. It may be noticed that similar cells are to be found scattered irregularly through all the layers of the cortex. Not a few large pyramids, and small pyramids, too, are met with amongst the small polygonal cells.

In the fifth layer the bundles of medullated fibres coming from the centre claim the largest share of space. The small irregular cells become rapidly sparser, but the fourth layer is not marked off sharply from the fifth. In the fifth layer cells of moderate size make their appearance, presenting every gradation of form from spindles to pyramids (layer of fusiform cells, claustral formation). Since, for the most part, they correspond in direction with the medullated fibres, a single process, similar to the apical process of the pyramid cells, is seen as a rule; especially at the bottom of the fissures, however, it often happens that these cells lie parallel with the surface. In this situation the layer is narrow and sharply-marked off from the medullary substance; whereas at the apex of the convolutions the cells spread into the medullary substance, from which the cortex is not, therefore, clearly limited.

The carmine method teaches us little with regard to the arrangement of the FIBRES IN THE CORTEX; one of Weigert's methods must be used for this purpose. It is chiefly in pathological cases that we desire to obtain exact information as to their relative abundance.

Still it often happens that, despite the most careful preparation (pieces of the brain being put to harden a few hours after death), sections stained according to Weigert's method do not give such complete information with regard to the arrangement of the nerve-network in the upper layers of the cortex as we desire.

Friedmann has devised the following more reliable modification of Weigert's method:—

Small pieces of the brain as fresh as possible are put into a fixing medium similar to those given on p. 6. It should consist of

Osmic acid 2 per cent. solution, . . .	2 parts.
Chromic acid 1 per cent. solution, . . .	7 parts.
Acetic acid, ...	0·2 to 0·5 parts.

In this mixture the pieces lie for about a day (five hours to one day according to circumstances); after a short washing in water they are hardened further in strong alcohol, and then in a few days embedded in celloidin. The sections, when cut, are placed for two and a-half to three hours in alum-hæmatoxylin in an incubator at 30° to 45° C. Differentiation is usually effected, as in Weigert's method, in potassic ferridcyanide in seven to fifteen minutes. If they remain for a longer time in the differentiating fluid some of the fibres are discoloured, but the rest come out so much more distinctly that it is often worth while to leave the sections in the ferridcyanide if they are afterwards to be examined with high powers. *Exner* discovered the great wealth of medullated fibres in the cortex, especially in its uppermost layer, by means of his osmic acid method (p. 12), which can be employed when permanent preparations are not needed.

Sections stained as above described (Weigert and Freidmann's or Exner's methods) exhibit, just beneath the pia mater, a border of connective-tissue, *a* (fig. 176), destitute of nervous elements. Beneath this, corresponding to the outer half of the stratum moleculare (1) follows a layer (*b*), almost entirely occupied by medullated fibres. Most of these fibres are thin, but some coarse ones occur amongst them; they run parallel with the surface, tangential to the arc which the convolution forms (tangential border zone, zonal fibres). In the inner part (*c*) of the molecular layer is found a moderately close network of fine medullated fibres crossing one another at various angles.

A similar network (*d*) occupies the layer of small pyramids (*2*).

In the layer of large pyramidal cells (*3*) the fibres are arranged radially; they are collected into bundles which are more definite in its deeper parts. In the middle of this layer a region (*f*) is found, which appears in preparations made according to Weigert's method as a dark band, the number of interlacing fibres being very great, and the feltwork close. This layer corresponds to the outer stripe of Baillarger.

In the area marked *g*, which corresponds partly to the fourth and partly to the fifth layer of the cortex, the radial fibres are not only close-set and conspicuous, but the bundles contain more thick fibres than they do farther outwards.

In the middle of the fourth layer the network of fibres again becomes closer, making a second dark band, *h*, which corresponds to Baillarger's inner stripe; it is narrower and less marked than the band, *f*.

We have already seen in carmine preparations that the fibres which radiate outwards from the medullary centre occupy the greater part of layer 5.

The layer marked *i*, therefore, corresponds to the deepest part of the fourth, and the whole of the fifth layer.

In our description of the cortex cerebri we have followed *Meynert*, in recognising five layers as the typical arrangement. Certain of the layers may be again divided, as we have seen. The fourth and fifth layers are not always clearly distinguishable, and, consequently, many anatomists rank them as one (the layer of small nerve-cells, or inner nerve-cell layer).

Schwalbe recognises two chief zones, of which the inner comprises the bundles of radiating fibres, while in the outer, which is of about the same thickness, the bundles quickly fall to pieces. Baillarger's outer stripe, *f* (fig. 176), about forms the boundary between the two, which falls, therefore, in the middle of the third layer, not, as often stated, in the line between the second and third layers.

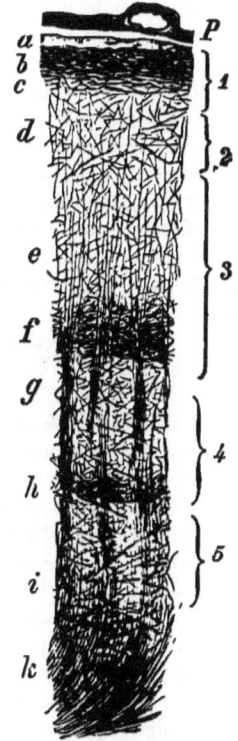

Fig. 176.—Cortex cerebri (frontal lobe). *Weigert's colouring. Magn.* 50.—*P*, Pia mater; *1-5*, Meynert's five layers; *a*, layer of superficial connective-tissue; *b*, layer of tangential medullated fibres; *c*, deeper part of the molecular layer; *d*, network of fibres in the layer of small pyramids; *e*, outer part of the third layer; *f*, external stripe of Baillarger; *g*, fibre-network of the third and fourth layers; *h*, internal stripe of Baillarger; *i*, deepest part of the fourth and fifth layers; *k*, medullary centre.

[Although the time has not yet arrived for making an authoritative statement as to the meaning of the intimate structure of the cortex, it is to many minds impossible to confide to the memory a collection of facts without first arranging them in a sequence according to some generalisation, however hypothetical.

The grey matter which covers the surface of the cerebral hemispheres is a field for the association of afferent sensory impulses. In it they are placed in communication with efferent roads along which they travel either immediately or at some future time; or to speak more correctly, the efferent impulse is not the unchanged afferent impulse directed into a descending path, but the product of afferent impulses just received, combined with impulses liberated from their resting places in the tissue of the brain.

The tissue in which the combination and reflection occurs is a plexus with cells for its nodal points. Of the cells some are connected with long processes or fibres. They appear to form three classes only— (1) cells belonging to afferent or sensory fibres, the so-called granules; (2) the cortical terminals of efferent motor fibres, the pyramids; (3) the trophic cells of the associating fibres or fibræ propriæ of the cortex, fusiform cells (fifth layer). The third class of cells was recognised as belonging to the fibræ propriæ by *Meynert* on account of their disposition beneath the sulci with their long axes parallel to the surface. The very large number of fusiform cells in the deepest layer of the cortex of the porpoise brain (in which the axial fibres must bear an unusually small proportion relatively to the associating fibres) appears to the *translator* strongly to confirm Meynert's hypothesis.

It seems probable, therefore, that in the cortex afferent fibres, after subdivision, are distributed to the plexus through the granules; efferent (radial) fibres take origin in pyramids; tangential fibres are supplied from either end of fusiform cells (*cf.* p. 331).

The plexus is supported by neurogleia cells (connective-tissue cells of epithelial origin). They are distinguished by their small darkly staining nuclei, scanty cell bodies, and numerous non-tapering processes.]

The cortex does not present exactly the same structure in all parts. In some parts the difference affects the number and size of the elements only; in other places the deviation from the type-form which we have described is due to an arrest of development (tractus olfactorius, septum pellucidum), or to a striking alteration in arrangement (cortex of cornu Ammonis).

There can be no doubt that the differences in structure are associated with differences in function. Anatomical considerations, therefore, lead one to the conclusion that all regions of the cortex are not functionally equivalent.

Nowhere do we find sudden breaks in structure, but one formation passes gradually into another.

If we move forward from the part of the brain which we have just been describing, namely, the frontal lobe near to the central fissure, we encounter no real change in structure, although as we approach the **frontal pole** the large pyramids become smaller. In the anterior central convolution some of these cells are of remarkable size.

The higher we mount towards the great longitudinal fissure on the central convolutions the larger do we find these cells to be; they reach their maximum size (65 μ) in the **lobulus paracentralis**, where they deserve the name of "giant pyramids" given to them by *Betz*. The third layer also increases in thickness, *pari passu*, with the increase in size of the cells. In the **posterior central convolution** large cells are only found near the longitudinal fissure and in the margin of the convolution where it adjoins the fissure of Rolando.

Some points must be referred to in regard to the giant pyramids. They are usually plump in form, not distinctly pyramidal; in size they greatly overtop the other cells, transitional cells being hardly present; they are usually arranged in small groups or nests of two to five, many of them being embedded in the layer of small irregular cells. According to *Betz*, axis-cylinders of striking thickness are found beneath the giant cells, and he supposes that one runs out from the base of each of them. *Bevan Lewis* thought that the large pyramids were arranged in larger groups, corresponding to the regions marked out by *Ferrier* as motor areas.

[The minute structure of the area of the cortex, stimulation of which produces movements of the muscles, is not sufficiently different from the structure of the rest of the cortex to allow us to believe that there is any such distinction in function between the two as would justify the appelations "motor" and "sensory" areas. The histologist, as far as he is able to draw conclusions as to function from his observations of structure, is bound to say that the grey matter all over the great brain contains elements suitable for receiving the terminations of afferent, and forming the starting points (trophic centres) of efferent fibres. Small cells (granules) and full-bodied cells (pyramids) are everywhere present. The differences between the several regions depend upon the number and distribution of the former rather than upon the size of the latter. The pyramids found in the limb- and trunk-areas are larger than those occurring elsewhere. If it is true that, other things being equal, the size of a cell varies as the length of the fibre over the nutrition of which it presides, the differences in size of the pyramids of the cortex may be explained by remembering that in the area just mentioned the pyramids are the starting points of long fibres which traverse the spinal cord; while the fibres which go to the muscles of the face, eye, ear, &c., are much shorter. At the present moment it is a question under dispute whether the epileptic contractions which follow electric stimulation of the occipital cortex are produced by impulses originating in this part of the cortex and passing thence straight to the cord, or by impulses transferred to the cord *viâ* the "motor area."

In the occipital cortex, especially of the cat and some other animals, very large pyramids occur at intervals. They strongly suggest that afferent visual impulses originate direct cortical messages for the limbs.]

As would be anticipated, the part of the **occipital cortex** which is distinguishable as such by the naked eye, on account of Gennari's stripe, shows certain peculiarities of structure (fig. 172). The molecular layer is thinner (0·15 to 0·20 mm.). The layer of small pyramids is disposed according to the ordinary type. In the third layer, which is 0·8 mm. thick, the pyramids do not increase regularly in size from the surface downwards; but, on the other hand, in the deepest part of the layer, pyramids of remarkable size are found singly (Meynert's solitary cells) or in groups. About at the level of these cells, or a little superficially to them, the interweaving of nerve-fibres is remarkably close, giving rise to the appearance of Gennari's stripe which is homologous with Baillarger's outer stripe. The fourth layer is very strongly developed in this region, being much wider than usual (0·6 mm.) and interrupted by a layer poor in cells. It contains the same kind of cells as elsewhere, except that some of the above described solitary cells are found in its intermediate band. (In the picture this layer is made too light.) The fifth layer is narrow and somewhat ill-defined.

Meynert says that the layer of irregular cells contains two intermediate layers. He, therefore, distinguished eight layers (eight-layered type).

The **gyrus fornicatus** does not throughout its whole course represent the real border of the cortex, but leads up to it both in the portion (gyrus cinguli) which lies above the corpus callosum and farther on in the gyrus hippocampi (subiculum cornu Ammonis)

The cortex of the **gyrus cinguli** is about 3 mm. broad on its peripheral side (fig. 173), but it narrows as it approaches the corpus callosum to about 1 mm.; finally, where it rests upon the corpus callosum, the sulcus corporis callosi (or as it has also been termed ventriculus corporis callosi) intervening, the cortex mantle appears to be sharply cut off. As a matter of fact it is continued medianwards over the surface of the corpus callosum as a very thin layer (20 to 30 μ

thick) the indusium griseum corporis callosi, *Jgr*, which rises into a ridge 0·3 to 1·0 mm. high, *Stlm* (stria longitudinalis medialis, nervus Lancisii). The most lateral part of the indusium, usually a little thickened, is designated the ligamentum tectum (striæ longitudinales externæ *seu* laterales).

The cortex of the gyrus cinguli presents nothing characteristic in its first two layers. The third layer contains for about its outer half only small pyramids; in its inner half pyramids almost all of the same medium size (about 25 to 30 μ). They lie, for the most part, at the bottom of the third layer where it adjoins the fourth; so that between them and the small pyramids is situate an intervening layer, with few cells, distinctly striated by the traversing apical processes of the large pyramids (stratum radiatum). Next comes the layer of irregular cells, not distinctly arranged in columns; and, lastly, an inconspicuous fifth layer. The narrowing of the cortex occurs principally at the expense of the third layer, the larger pyramids becoming more and more scarce and finally disappearing. By the time the fibres of the corpus callosum break through it, the second and fourth layers of the cortex are already fused together.

Single nerve-cells of small size can be found occasionally in the ligamentum tectum. They are met with in larger numbers in the stria longitudinalis medialis. In this band a deeper layer of grey substance containing small irregular nerve-cells and a peripheral layer rich in medullated nerve-fibres may be distinguished; the stria owes its white colour to the latter.

Thus we see that we have to look for the real edge of the cortex-mantle in the stria longitudinalis medialis, with the fascia dentata (outer arcuate convolution). In front, the stria longitudinalis medialis passes over into the pedunculus corporis callosi, which ascends from the basis cerebri; behind, it passes not only into the fascia dentata cornu Ammonis, but also into the white layer which we have already met with under the name substantia reticularis Arnoldi.

In many respects, but especially as regards its third layer, the cortex of the subiculum cornu Ammonis has an obvious likeness to that of the gyrus cinguli. We will discuss the subiculum in connection with the cornu Ammonis itself to which it leads up.

The **parietal lobe** or region lying behind the posterior central convolution is characterised by a thickly-packed layer of small pyramidal cells, which is intercalated between the third and fourth layers (*Bevan Lewis*). These cells must not be confounded with the small variously-shaped cells of the deeper layers.

The cortex of the **convolutions of the island of Reil** does not depart from the ordinary type (*Herbert C. Major*). *Meynert* is of opinion that the claustrum ought to be included with the cortex, since it contains fusiform cells, which correspond in form and size with the cells of the usual fifth layer ("claustral formation"); they are disposed parallel with the surface. Between the claustrum and cortex proper lies a layer of white fibres (lamina fossæ Sylvii), which is thin beneath the fissures, broad under the summits of the convolutions.

In the **nucleus amygdaleus,** which lies beneath the uncus, the same cells are to be found as in the cortex; but it is difficult to define their typical arrangement. [The nucleus amygdaleus is directly continuous with the claustrum.]

We are still in need of more detailed descriptions of the rest of the cortex. Only three regions in which the structure is obviously peculiar can be described.

(1) We have already given an account of the atrophied cortex of the **tractus olfactorius** (p. 266).

(2) The **septum pellucidum** is that part of the wall of the cerebral vesicle which was cut off from the general surface by the development of the corpus callosum. It is of scarcely any importance functionally, and in structure is also exceedingly rudimentary.

The part of the lamina septi, which looks into the ventriculus septi pellucidi, corresponds with the free surface of the cortex. It is not like the real ventricle-walls covered with epithelium, but by a distinct, although thin, superficial layer rich in medullated nerve-fibres. Next to this comes a grey layer containing a good many nerve-cells. Nearer to the fifth ventricle the cells are distinctly pyramidal, and provided with an apical process directed medianwards—i.e., towards the surface, corresponding to the free surface of the rest of the cortex. In the deeper layer the cells are irregular. Towards the lateral ventricle the septum presents a layer of medullated fibres, covered with the usual ventricular ependyma. The spaces in the septum for vessels are most of them remarkably wide. Often the lamina septi is not so well developed as this, and the distinction of the several layers is then very difficult.

The **cortex cornu Ammonis.** We have already explained (p. 81, cf. fig. 29) how as we advance into the lateral ventricle through the sulcus hippocampi we come across the following structures, each placed longitudinally:—Subiculum cornu Ammonis (gyrus hippocampi), fascia dentata, fimbria, and then the proper cornu Ammonis, lastly the unimportant eminentia collateralis Meckelii.

[While above the corpus callosum the margin of the cortex (where it is pierced by the corpus callosum and by crura cerebri coming up to the great brain from the brain-stem) is thick, blunt, and uniform (except for its prolongation as the indusium) the continuation of the margin where it bounds the "porta" on the lower side is peculiarly disposed. Almost as soon as the margin of the cortex passes over the corpus callosum, it becomes considerably thinner and folds upon itself, first outwards and then inwards again. It is this thin folded edge of the cortex which constitutes the cornu Ammonis (hippocampus). The extreme edge of the mantle is again thickened into a ridge and received into a special sheath of grey matter, the fascia dentata. The internal white lining of the hemisphere where it meets the external grey covering is thickened into a ridge, the fimbria.]

The part of the gyrus hippocampi, which adjoins the gyrus occipito-temporalis lateralis shows a cortex structure deviating but little from the ordinary type. But as soon as the convexity of the gyrus hippocampi is reached, changes make their appearance, which become more conspicuous as the fascia dentata is approached, and are the precursors of the proper cornu Ammonis (figs. 174, 177, and 178).

The molecular layer is very much broader; the increase in thickness being due chiefly to a great development of the peripheral medullary zone (lamina medullaris externa, *Lme*, fig. 177). The thickening is not uniform, the superficial layer of medullated fibres

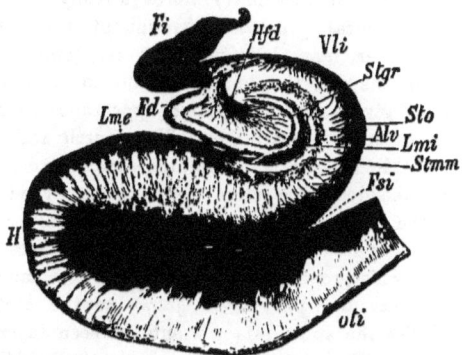

Fig. 177.—Transverse section through the cornu Ammonis. *Pal's staining. Magn.* 4. —*H*, Gyrus hippocampi; *oti*, fissura occipito-temporalis inferior; *Fsi*, fissura subiculi interna; *Lme*, lamina medullaris externa; *Fd*, fascia dentata; *Fi*, fimbria; *Vli*, descending horn of lateral ventricle; *Stmm*, stratum medullare medium; *Lmi*, lamina medullaris interna; *Alv*, alveus; *Sto*, stratum oriens; *Stgr*, stratum granulosum; *Hfd*, hilum fasciæ dentatæ.

forms a succession of prominent ridges projecting into the cortex, with valleys between (fig. 174). This varying thickness of the medullary covering is the cause of the easily recognisable reticulated white marking of the region in many fresh brains. The superficial fibres run "tangentially" as in other parts of the cortex, and are, therefore, exposed in length in transverse sections. The chief mass of the medullated fibres, upon which the considerable thickening of the layer depends, is disposed horizontally from before backwards.

From the superficial medullary layer, especially from its eminences, a rain of medullated fibres pours into the more deeply-lying medullary substance.

The nerve-cells are not arranged in a symmetrical sheet in the layer of small pyramids; rather do they form a chain of hills, each resting with its base upon the deeper part of the cortex, while its apex is received into one of the valleys of the superficial layer. From this description it follows that a real molecular layer hardly exists, for it is almost entirely occupied by longitudinal medullated fibres.

Cells of the smaller sort, transitional between the second and third layers, are almost entirely wanting in the third layer, just as in the gyrus cinguli. It contains hardly any but large pyramids with long apical processes. The largest pyramids in the deepest part of the stratum are about 40 μ n length.

The radial bundles of medullary fibres already mentioned run parallel with the apical processes throughout the whole layer. Besides these, however, many longitudinal fibres, some coarse, others fine, which run in the direction of the convolution, are cut across in this layer. They give a curious spotted appearance to the section, visible even in carmine preparations. The fourth and fifth layers are fused into a thin sheet which contains almost exclusively small irregular cells embedded in a close network of nerve-fibres which twist about in the most varied directions, but lie (especially as the medullary centre is approached) longitudinally for the most part.

The cornu Ammonis proper may be considered to commence at the place where the vascular pia mater grows to the cortex of the subiculum. Whilst the subiculum presents, as seen in cross-section, an arch of cortex with its convexity towards the middle line, the cornu Ammonis is joined to it as an arch with its convexity directed outwards into the inferior horn of the lateral ventricle, *Vli*.

In the cornu Ammonis, in preparations made according to Weigert's method, a three-fold layer of medullary fibres is shown.

The lamina medullaris externa splits into two; one of these layers is simply the superficial medullary layer of the cortex remarkably thickened, *Lmi* (nuclear layer, lamina medullaris involuta). Its fibres run in the plane of the section when cut transversely. The other layer derived from the lamina medullaris externa of the subiculum is also rich in medullated fibres, *Stmm* (stratum medullare medium). It lies parallel to the nuclear layer, but its nerve-fibres run for the most part obliquely or longitudinally.

The third layer of the cornu Ammonis, *Alv* (alveus), covers the surface which is directed towards the inferior cornu of the ventricle. It is the continuation of the central white matter of the subiculum thinned out to cover the cornu Ammonis.

The alveus proper consists of bundles of fibres closely packed together and interwoven in a complicated manner. On its deep surface, towards the cortex of the cornu Ammonis that is to say, the alveus is resolved into a layer of fibres not united into bundles, but running principally in arches parallel to the curvature of the cornu Ammonis (stratum oriens of *Meynert*).

In the transition from the subiculum to the cornu Ammonis the nerve-cells behave as follows :—The hills of small pyramids become fewer and lower, and at last these cells of the second layer disappear; as the smaller cells are lost the large pyramids retire to the deepest stratum only of the third layer [where they constitute a uniform sheet, little more than one cell thick]; the fourth and fifth layers of cells disappear almost completely.

The following layers are now to be distinguished in the cornu Ammonis (fig. 178) :—

(1) Nuclear layer, *Lmi*—this layer is separated from the fascia dentata for a short distance by the fold of pia mater, further on they fuse together. Fusiform cells are to be found scattered amongst the nerve-fibres.

(2) The stratum moleculare, *Stm*, reaches to the stratum medullare medium, and is constituted like the layer of the same name in the typical cortex.

(3) Stratum lacunosum, *Stl* (stratum reticulare *seu* medullare medium), may be supposed from its position to correspond with the layer of small pyramids. Its tissue is singularly loose; a good number of capillary vessels, made more conspicuous by the large spaces in which they lie, make an obvious network. The behaviour of the numerous medullary fibres in this region has already been described. A few small irregular nerve-cells are also found in this layer.

(4) Stratum radiatum, *Str*. The apical processes of the large pyramids of the next layer produce a marked striation of this stratum, which is the more distinct owing to the almost complete absence of cells.

(5) Stratum cellularum pyramidalium, *Stp*, is made up of large pyramids of almost uniform size (40 μ) in close order.

FASCIA DENTATA.

(6) Stratum oriens, *Sto;* scattered fusiform cells, representatives of the cells of the fifth layer of the cortex, lie amongst the medullated fibres.

(7) Alveus, *Alv.*

(8) Towards the ventricle the alveus is covered by a rather thick ependyma, *E,* with the usual form of epithelium.

On to the convexity of the cornu Ammonis two terminal structures are fixed (fig. 177), the one, the fimbria, composed entirely of medullated fibres with thick connective-tissue septa; the other, the fascia dentata, made up for the most part of grey matter.

Fimbria, *Fi* (bandelette de la Voûte), is in immediate connection with the alveus proper; it consists of thick bundles of longitudinal fibres [and passes into the posterior pillar of the fornix.]

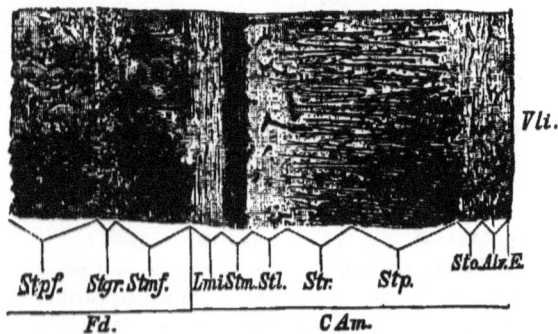

Fig. 178.—Cortex of the cornu Ammonis and a part of the fascia dentata. *Magn.* 20. —*CAm,* Cornu Ammonis ; *Fd,* fascia dentata ; *Vli,* inferior horn of the lateral ventricle; *E,* ependyma; *Alv,* alveus; *Sto,* stratum oriens; *Stp,* stratum cellul. pyramid. ; *Str,* stratum radiatum ; *Stl,* stratum lacunosum; *Stm,* stratum moleculare ; *Lmi,* lamina medullaris interna ; *Stmf,* stratum moleculare fasciæ dentatæ ; *Stgr,* stratum granulosum ; *Stpf,* stratum of pyramids beneath the fascia dentata.

Fascia dentata, *Fd* (corps godronné), represents the real edge of the cortex [from which it is so sharply defined, and differs so much in structure, that it may be supposed to be a kind of grey matter not elsewhere present in the cortex, but added to its margin in this situation]. It squeezes itself into the concavity of the cornu Ammonis, with which in some places it grows together, as described above.

Here we find two kinds of nerve-cells :—(1) A narrow layer parallel with the surface of the fascia dentata, *Stgr* (stratum granulosum *seu* corporum nervorum arctorum), made up of closely-packed cells of a rounded-angular or pyramidal shape ; as a rule, the nuclei are surrounded by so little protoplasm that they might be considered as "granules." Hardly any ground-substance is left between these cells.

The arch formed by this layer as seen in transverse section is open towards the fimbria at the "hilum."

(2) The second kind of nerve-cells which we meet with in the fascia dentata, correspond to the large pyramids of the cornu Ammonis; they are dispersed with an irregular stratification throughout the whole of the space enclosed by the stratum granulosum.

The fascia dentata shows the following layers, therefore:—

(1) A distinct superficial sheet of medullated fibres (stratum marginale), the continuation of the nuclear layer, but far thinner than the latter. This layer is not distinct in carmine preparations, especially when but slightly magnified; it is, therefore, not represented in fig. 178; it should be looked for at the spot where the lamina medullaris interna and the cornu Ammonis grow together.

(2) Stratum moleculare, *Stmf.*

(3) Stratum granulosum, *Stgr* (see above).

(4) The nucleus fasciæ dentatæ, *Stpf* (layer of pyramidal cells). The cells, as well as the arched fibres of the stratum oriens, enter through the hilum and scatter in all directions. [They are continuous with the sheet of pyramidal cells of the cornu Ammonis rather than with the fascia dentata.]

Farther forwards the fimbria grows progressively smaller; on the other hand, the fascia dentata enlarges and sinks at last into the uncus.

As soon as the digitations proper of the cornu Ammonis appear we have to deal with an undulating curved sheet of cortex, covered on its surface by the medullary layer known as the alveus. The fascia dentata, which is always recognised by the quite characteristic stratum granulosum, pushes itself on the under side of the cornu Ammonis into the front part of the fissura hippocampi. Behind, the fascia dentata becomes progressively thinner, and finally disappears.*

The type-structure of the cortex mantle, as pictured above, can be recognised with slight modifications through the whole of the mammalian series. The relative number of nerve-cells, their size, and the thickness of the layers which contain them are variable. On the

---

* The attempt has always been made, as in the account given above, to homologise the several layers of the fascia dentata with the superficial strata of the cortex in other situations. To the *translator* this appears impossible, for it is so unlike other parts of the cortex in structure as to be clearly, as its anatomical disposition also indicates, a different formation into which the margin of the cortex is received as into a scabbard. Since the fascia dentata varies in development directly as the size of the olfactory bulb; and since its principal constituents, the granules, are indistinguishable from those of the bulb, it is not impossible that it is a part of the olfactory bulb which is adherent to and caught up by the margin of the mantle.

whole, the size of the nerve-cells varies as the size of the animal. In Man, the molecular layer is relatively thin; the cortex showing a greater wealth of nerve-cells proportional to the increased dignity of this organ (*Meynert*).

In lower mammals the cortex cells are distinguished from those of Man by a difference in internal structure. This difference in constitution is exhibited in their behaviour towards hardening fluids in a way which is not noticed with other parts of the central nervous system. If, for example, small pieces of the cortex of some rodent animal are prepared by hardening in potassic bichromate, it is found on examination that in place of many of the pyramidal cells the sections exhibit rounded spaces communicating with the characteristic radiating channels. Cell-nuclei surrounded by irregular indistinctly defined finely granular protoplasm lie in these spaces. The difference from the human type depends upon *post-mortem* changes, indicating a difference in the chemical constitution of the cells. The cell has come to grief although we were in a position to place the brain in the hardening fluid while absolutely fresh, whereas, although we must always wait until disintegrative changes have set in, before we can harden the human brain such appearances are much more rare.

Similar spaces are seen in imperfectly hardened human brains and in various pathological conditions.

In the lower classes of vertebrates the cortex departs farther from the human type; a description of the differences would be out of place in this book.

The cortex cerebri of the human embryo exhibits numerous round nuclei (also called "gleia-nuclei") which represent the rudiments of the later developed cell-elements.

The nuclei are arranged in successive layers (*Lubimoff* counted six layers in a five months' fœtus), which give the appearance in section of a series of bands, lighter and darker according to the quantity of nuclei. In the deeper layers the nuclei are arranged in columns, between which spaces are left for the passage of the, as yet, non-medullated fibres.

The first pyramidal cells are to be seen in the sixth month of intrauterine life (*Vignal*). At birth they are very numerous and well formed in the deeper layers (*Lemos*) (*S. Fuchs* says in the upper layers also); but still no medullated fibres are present.

According to the investigations of *S. Fuchs*, the first medullated fibres make their appearance in the radiating bundles of the posterior central convolution during the second month of life. Medullated fibres are found in the superficial tangential layers in the fifth month; although the other layers of the cortex are joined to this, the formation in them of medullated fibres seems to be completed not earlier than the seventh or eighth year. It may be asserted that, as a general rule, the fibres first myelinated are those which are subsequently the thickest.

Two structures which are met with attached to the great brain may be now described, the conarium and hypophysis.

(1) **Conarium** (glandula pinealis, epiphysis cerebri). Its connection with the brain is chiefly effected by a bilateral white column of fibres, pedunculus conarii. Its connections with the posterior commissure (and so with the oculomotor nucleus and the central visual apparatus) are especially important. It has been proved that the pineal gland is a vestigial structure representing an unpaired eye. In many saurians, especially in Iguana tuberculata (*Wiedersheim*), an organ which corresponds in structure with an eye is found lying beneath a thin plate of pigment-free membrane in the parietal region. It is connected by means of a tract of nerve-fibres with the epiphysis.

[The fact that in certain reptiles the pineal body resembles an eye in structure was noticed by *v. Graaf*, but credit is due to *Spencer* for appreciating the immense value of this discovery, and systematically examining the pineal body and parietal foramen in all animals in which these structures show traces of their former importance. *Spencer* discovered that, although there is no reason to suppose that it is now functional in any animal, the pineal eye when best developed exhibits a complicated arrangement of bacillar and nervous elements. Nothing about this prehistoric cyclopian organ is more interesting than the fact that the rods are directed inwards towards the centre of the eyeball as in invertebrates. *Gaskell* has shown that the size, structure, and connections of the ganglion habenulæ in the lamprey indicate that it is the proper ganglion (central grey matter) of the pineal eye.

The nearest extinct ancestors of the saurians in which the pineal eye may be supposed to have been functional, seems to have been the labyrinthodonts which flourished during the time when the coal measures were being formed.]

The pineal gland receives an enveloping capsule from the pia mater which sends vascular sepiments into the organ.

In sections it is seen to consist of a rather close meshwork of connective-tissue trabeculæ. Numerous cells, seldom larger than 20 μ, are found in the alveoli. According to *Bizzozero* two sorts of cells are to be distinguished—one of rounded form with two or three rapidly tapering processes dividing into numerous little branches; the other fusiform, with sharper and more irregular contour; these latter contain yellow or red-yellow pigment-granules, and their processes are more distinct and larger, and end in a fine network.

Many of the cells of the pineal body, however, show no recognisable processes.

Quantities of nerve-fibres pass through the organ (*Darkschewitsch*), so that its intimate connection with the nervous system is beyond doubt, even if it is impossible to recognise the nervous character of all its cells.

[In the pineal glands of young persons, as in the cerebral part of the pituitary body, little can be recognised under the microscope, but a granular basis with fairly numerous easily stained nuclei, and occasional, irregular, usually coiled yellowish fibres.]

In the adult pineal bodies, concretions of phosphate and carbonate of lime (brain sand, acervulus) are often found. They are small stratified bodies arranged in nodules, resembling mulberries in form; the nodules may be as large as hemp-seeds. Rod-like and club-shaped or branched pieces of calcified connective-tissue are also to be found in the conarium. These concretions seem to be quite wanting in animals, although in the horse they may be replaced by very fine granules of phosphate of lime (*Faivre*).

(2) **Hypophysis** (glandula pituitaria colatorium) is a body about as large as a bean, somewhat less in its sagittal than in its frontal diameter. It is connected with the rest of the brain by the infundibulum.

A sagittal section shows that the apparently single body [is invested with a thick capsule of dura mater, and] is composed of two distinct divisions, often separated by a small lymph space; the anterior lobe (epithelial portion, hypophysis proper) is larger than the posterior lobe (cerebral portion, lobus infundibuli).

The ANTERIOR LOBE is composed of alveoli, surrounded by connective-tissue membranes, and containing two different kinds of cells, the larger of which stain more strongly with hæmatoxylin than the other or smaller kind (*Flesch*). [The epithelial character of the cells is very evident. Their round vesicular nuclei, firm homogeneous yellowish cell-bodies, often vacuolated or containing droplets of fat, and destitute of cell-wall, their polyhedral shape and grouping in alveoli, all tell of their origin, and recall to the histologist other organs made up of apparently functionless epithelial cells, such as the cortex of the suprarenal capsule, corpora lutea, &c.] The anterior lobe is formed by an involution of the mucous membrane of the mouth, and is, therefore, homologous with the buccal glands.

The POSTERIOR LOBE must be looked upon as a veritable part of the brain. In it are found bundles of fibres crossing one another in every direction; their histological nature is still doubtful. Besides numbers of small cells, scattered large pigmented cells are met with, which may be looked upon as deficiently developed nerve-cells.

BLOOD-VESSELS OF THE GREAT BRAIN.

We will only briefly mention here the manner of division of the finer blood-vessels within the cerebrum. The course of the larger vessels and their arrangement, especially on the basis cerebri, will be learnt later on.

It must be accepted as a law for the cerebrum, as for other parts of the central nervous system, that the richer any region is in nerve-cells the closer is the capillary network which supplies it. We still need more exact accounts of the course of the vessels in many parts of the brain.

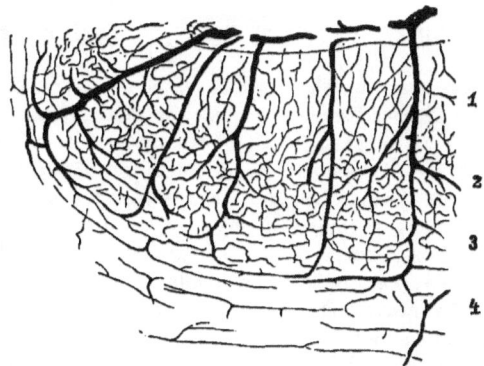

Fig. 179.—Injected cortex cerebri of the dog. *Magn.* 25.—*1*, The few-celled layer; *2*, region of pyramids; *3*, the internal (deeper) stratum of the cortex; *4*, white substance.

The application of the law just formulated may be observed in the cortex cerebri (fig. 179). Arteries and veins descend vertically from the pia mater. The larger ones give off relatively wide lateral branches, and traverse the cortex to reach the medullary centre. The smaller ones are used up in the cortex.

In the cortex at least three kinds of capillary network can be distinguished—

(1) In the molecular layer it is comparatively wide-meshed, *1*.

(2) A very close network in the neighbourhood of the pyramidal cells, *2*, which becomes a little looser in

(3) The deepest cortical layers, *3*.

The capillary network of the medullary substance beneath the cortex is very open, the meshes being placed, as a rule, with their long axes parallel to the surface.

The corpus geniculatum laterale, the corpus subthalamicum, and the nuclei of the nerves are distinguished from other grey masses by their richness in capillary vessels.

Many anatomical relations are not clearly appreciated, except in injected preparations; for example, the division of the corpus mammillare into two, in the dog.

The stronger vessels respect the median plane, even in the corpus callosum; but the capillary network extends across the middle line; only in a few parts of the central nervous system, however, are the median anastomoses between the capillaries abundant.

PATHOLOGICAL CHANGES IN THE GREAT BRAIN.

Those life-processes, which, in contra-distinction from reflex actions, occur in sight of consciousness, require for their normal course the integrity of a portion, at any rate, of the cortex. In all diseases in which consciousness is dimmed and the intellect disturbed, for more than a very short time, pathological changes may be looked for in this tissue; but, as already explained, many of these changes escape our notice on account of our ignorance of the limits of variation of the structural appearances in health of many of the nervous elements, particularly the most important of all—the nerve-cells. We shall, therefore, be often obliged to record a negative result in cases in which we were justified in expecting to find demonstrable alterations in the tissues. The distinctions between functional diseases are, however, constantly fading away, and their number, as classified, consequently diminishing. This statement is as true of the whole nervous system as it is of the cortex cerebri.

In this place we can only mention the most important pathological accidents to the cortex cerebri which have as yet been described.

The following changes occur in *senile atrophy* according to *Kostjurin*:—

(1) Pigmentary and fatty degeneration of many nerve-cells, perhaps vacuole-formation also.

(2) Diminution in the number of the nerve-fibres in all layers of the cortex.

(3) Atheroma of blood-vessels as well as overgrowth of the connective-tissue in their walls even to obliteration.

(4) Diminished limitation of the connective-tissue.

(5) Amyloid bodies in the periphery of the cortex.

The cortex behaves somewhat similarly in other slow atrophic processes—*e.g.*, chronic simple lunacy.

DEMENTIA PARALYTICA gives us a different result. While in the condition last named the atrophy is simple and primary, we have to deal now with sclerotic atrophy. The essential process is a diffuse, primary sclerosis of the cortex which leads to atrophy. It makes itself felt in the frontal lobe first. The sclerosis is preceded by a condition of irritation which seems to justify the expression, peri-encephalitis chronica. The risk in using this name is that one might be misled into supposing that the centre of interest of the disease lies in the meninges, whereas the brain coverings only play a secondary *rôle*.

In very acute cases in which we are able to recognise an early stage in the diseased processes, we are struck with the quantity of lymphoid bodies which surround the blood-vessels throughout the whole brain. These leucocytes probably migrate out of the blood, and pass through the ground-tissue of the cortex as "wandering cells," before being changed into stellate cells. Perhaps also the stellate or "spider" cells normally present in the cortex provide further material by proliferation. It is in the overgrowth of these cells belonging to the connective-tissue, that we have to look for the cause of the sclerosis. As soon as these new cells occupy so much space as to surround and press upon the normal nerve-tissue cells, the latter atrophy. The result of this process is seen in old-standing cases, not only in the degenerated nerve-cells (especially fatty-pigmentous degeneration, sclerosis of the cells, and enlargement of the pericellular spaces), but also in the remarkable diminution in the quantity of medullated fibres (*Tuczek*). This disappearance of medullated fibres advances from the periphery inwards; so that, as a rule, the outermost layer of tangential fibres is most affected, whereas in senile atrophy the decrease in the number of fibres affects all the layers equally. According to *Tuczek*, the convolutions most constantly and distinctly affected are those on the orbital surface of the frontal lobe, especially on the side of the great longitudinal fissure; next those of the island of Reil, and the left inferior frontal convolution. The other frontal convolutions, the gyrus fornicatus and the superior temporal convolutions, are likewise often diseased. All other parts of the cortex are, it is supposed, affected in a less degree only, or not at all (occipital lobe). A decrease in the number of fibres may be met with in other conditions besides dementia paralytica and senile atrophy, long-standing epilepsy for example (*Zacher*). The changes in the structure of the cortex cerebri to be seen in dementia paralytica assume many forms however; hence the descriptions of pathologists differ widely from one another.

A series of cases has been published in which spaces, or veritable CYSTS, have been found within the cortex. They were often found in the brains of those who had died of dementia paralytica. The cause of these hollow spaces is, perhaps, not identical in all cases. Very often they originate in connection with the vessels, being enlargements of the perivascular or adventitial spaces that is to say. A circumscribed parenchymatous inflammation may also be the cause of the decrease in tissue. In the latter case the cortex cerebri appears from the outside to be unevenly atrophied; on making sections it is seen to be cavernous, the spaces being often occupied, especially in their peripheral part, by a loose connective-tissue network, through which not a few well-preserved nerve-fibres course (*J. Hess*). Single bundles

Fig. 180.—Encephalitic cysts of the cortex cerebri with secondary degeneration of the white substance. *Weigert's colouring. Magn.* 4.

of degenerated fibres are seen deep down in the medullary centre in situations corresponding to the foci of cortical disease (fig. 180).

EFFUSION OF BLOOD into the great brain is very common; capillary effusions are met with in the cortex. Several kinds of disease of the intracerebral vessels lead to rupture. Alterations in the muscular coat (fatty and granular degeneration, &c.) are especially important. Rupture is also due to alterations of the intima, such as atheroma; the atheromatous patches, becoming detached, block vessels further on, and as the result of the embolism the vessel wall is ruptured.

Very often, but not always, miliary aneurysms are found in the tissue surrounding large cerebral apoplexies. Every effusion, whether large or small, spreads out in the direction in which it encounters least resistance. All apoplectic lesions are transformed eventually into apoplectic cysts or cicatrices.

Patches of DISSEMINATED SCLEROSIS may be found anywhere in the brain; but they are not so common or so extensive elsewhere as they are in the walls of the lateral ventricles. Sometimes the brownish gelatinous degeneration is seen to surround the whole ependyma of the ventricle so far as it rests on white substance. *Gowers* terms very small patches of sclerosis which are arranged in the deepest layer of the cortex MILIARY SCLEROSIS.

A very peculiar form of SCLEROSIS is LIMITED TO THE CORNU AMMONIS in which this structure becomes as hard as cartilage and much shrivelled. This form of sclerosis is almost restricted to epileptics; it is present in more than half of all the cases of epilepsy (*Pfleger*); it may be uni- or bilateral.

A DIFFUSE SCLEROSIS of the brain due, in part at any rate, to overgrowth of the interstitial connective-tissue may be almost equally distributed over the two hemispheres. As the result of this process the brain becomes as hard as leather and almost like cartilage in appearance. It is a rare form of sclerosis; occurring most often, perhaps, in idiot children.

INFLAMMATORY PROCESSES in the brain may have different causes. That form of encephalitis which leads to the formation of abscesses is usually the result of external injuries or else is due to the spread of purulent inflammation from surrounding structures (caries of the temporal bone especially). Metastasis of abscesses from distant organs, and especially from gangrenous disease of the lung, is not rare. In pyæmia, especially when puerperal (*Rokitansky*), numerous little metastatic abscesses from the size of a grain of hemp up to that of a bean are occasionally formed. Embolic and thrombotic softenings are classed with the inflammatory processes in the brain. The various kinds of bodies which may lead to embolism of the arteries of the brain have been already enumerated (p. 147).

In certain acute infectious diseases, *e.g.*, splenic fever and variola hæmorrhagica, very numerous effusions of blood may make their appearance in the brain-substance and in the pia mater. They are probably caused by accumulation of the infective medium in small heaps in the vessels, resulting in embolism. The large vessels which lie in the pia mater outside the brain-substance proper, at the base, and elsewhere, are also subject to embolism, in the production of which other causes come into play. The emboli for the larger vessels come, as a rule, from the left side of the heart or from the aorta.

The thrombus is usually a patch of atheromatous or of syphilitic inflammatory substance formed in the intima. In regions in the brain in which, owing to occlusion of the blood-vessels, necrosis of nervous tissue has ensued, very numerous capillary hæmorrhages often occur (red softening); later on the blood-pigment is taken up by innumerable fat granule-cells (yellow softening). Sometimes no colouring of the lesion by blood-pigment is noticeable (white softening). Softened patches are often found in the cortex as the result of meningitis tuberculosa; they lie beneath much diseased spots in the pia mater and often stretch into the medullary substance.

Tumours of the brain are very common; sometimes they originate in the membranes, sometimes in the brain-substance. Peculiar to nervous-tissue is the form of tumour termed glioma, in the formation of which nerve-cells, as well as other tissue elements, probably take part (*Fleischl, Klebs*). Gummata and solitary tubercles are very frequent, so too are sarcomata of the most varied kinds; melano-sarcomata are not rarely observed; but neither form of sarcoma, perhaps, ever has a primary origin in the brain. Myxomata have been repeatedly seen, so have osteomata. Concretions which may occasionally form small psammomata are not rare in the ventricular ependyma.

Cysticerci coming out from the pia mater often in great numbers take up their seat in the cortex. Echinococcus-vesicles and dermoid cysts are very rare.

The clumps of grey substance which are sometimes found within the medullary centre must not be looked upon as tumours; in minute structure they resemble the neighbouring grey masses or the cortex. They are termed heterotopes, and are most common in the cerebellum. They are always to be traced to abnormalities in development.

SECTION VII.—THE CAVITIES OF THE CENTRAL NERVOUS SYSTEM.

THE whole of the central nervous system is enveloped with a three-fold covering of fibrous membrane.

The outermost coat, or dura mater, D (fig. 181), lies within the skull-case, close to the bone, but in the spinal canal [it is not so intimately united to the surrounding bones], although standing at some distance away from the spinal cord. The innermost layer or pia mater, P, clings close to the nerve-mass. The middle coat, or arachnoidea, A, lies everywhere close inside the dura mater, touching it in many places; but only united to it by scanty threads of connective-tissue. To the pia mater, from which in many places it is separated by a considerable space, it is tied by a great quantity of connecting plates and filaments (subarachnoid tissue), especially around the brain. So abundant are the connections that pia and arachnoid are often mistaken for a single membrane.

Two spaces are confined between these three membranes; the subdural space, sd (or arachnoidal sac), between dura and arachnoidea; the subarachnoid space, sa, between arachnoidea and pia mater.

Owing to the close proximity of the arachnoid and the dura, the SUBDURAL SPACE is very narrow, and contains but little fluid. *Schwalbe's* investigations seem to have proved that it is a lymph space. Colouring masses injected into the subdural spaces, enter the lymphatic vessels and glands of the neck, and the lumbar lymphatic glands, as well as the subdural spaces around the nerve-roots. From the lymph spaces around the nerve-roots the injection-mass travels on into the lymphatic space in the olfactory membrane, in the labyrinth of the ear, and in the bulbus oculi (perichoroidal space). All these channels do not seem to exist in Man; at any rate it cannot be proved that the lymphatics of the neck are in direct communication with the subdural space. There is no communication between the subdural and subarachnoid spaces (*Merkel*).

Owing to the peculiar configuration of the brain the SUBARACHNOID SPACE is divided into a considerable number of larger and smaller spaces, which communicate with one another, and, through the foramen of Magendie and the aperturæ laterales ventriculi quarti, with the ventricles of the brain.

Merkel, as well as *Mierzejewsky*, maintains that a slit-like communication exists between the subarachnoid space and the inferior horn of the lateral ventricle (*cf.* p. 60).

The cerebro-spinal fluid (liquor cerebro-spinalis) circulates within the subarachnoid spaces and the ventricles. It finds an outlet in the lymph-paths surrounding the peripheral nerves (particularly the optic and auditory) and beneath the olfactory membrane (*A. Key & Retzius, Fischer*). The subarachnoid spaces are also, by means of the arachnoid plexuses (p. 384), in connection with the venous sinuses of the dura mater.

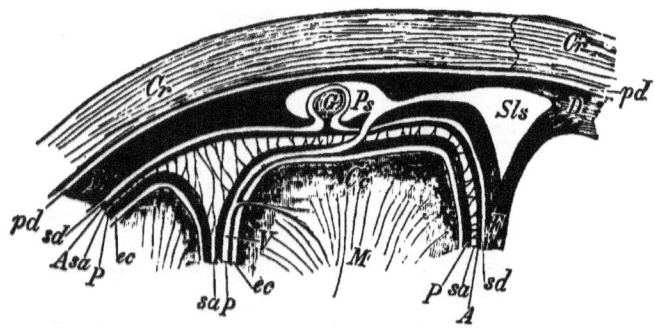

Fig. 181.—Diagram of the membranes of the brain.—*Cr*, Cranial bones; *pd*, peridural space; *D*, dura mater; *F*, falx cerebri; *sd*, subdural space; *A*, arachnoid; *sa*, subarachnoidal space; *P*, pia mater; *ec*, epicerebral space; *Cc*, cortex cerebri; *M*, white matter of the brain; *Sls*, sinus longitudinalis superior; *Ps*, parasinoidal space; *G*, glandula Pacchioni; *V*, veins of the pia mater.

A. DURA MATER (Meninx Fibrosa, μῆνιγξ παχεῖα).

We distinguish a dura mater cerebralis and a dura mater spinalis. The former of these lies, with the exception of certain prolongations to be mentioned immediately, entirely within the skull-case in close apposition with its inner table; the latter is divisible into two layers, the outer and thinner being the periosteum of the vertebræ, the inner being the dura mater spinalis, in the strict sense of the word. Between the two laminæ, which unite with one another, and with the dura cerebralis at the foramen magnum, little is interposed but plexuses of veins and loose fatty tissue (epidural tissue).

PROLONGATIONS OF DURA AND ARACHNOID.

The dura mater cerebralis is a firm white fibrous membrane, which, within the skull-case, gives off several reduplications, viz., the falx cerebri (processus falciformis major), tentorium cerebelli, and the inconsiderable falx cerebelli (processus falciformis minor).

To allow of the interposition of the venous sinuses and venous lacunæ about to be described, the dura mater is split into two layers, the one visceral the other parietal; the same arrangement obtains, too, wherever nerve-structures such as the trunks of the third, fourth, and sixth cranial nerves or the ganglion Gasseri of the trigeminal (in the cavum Meckelii) are embedded in the substance of the membrane.

It would lead us too far were we to examine in detail the anatomical disposition of the dura mater. We must point out, however, that on either side the middle line near the sinus longitudinalis superior, Sls (fig. 181), peculiar hollow spaces are met with in the substance of the dura, Ps (parasinoidal spaces), into which the veins of the brain, V, embouch before reaching the sinus.

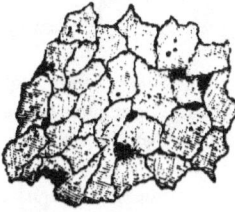
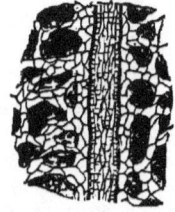
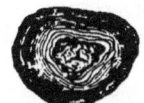

Fig. 182.—Epithelium on the inner surface of the dura mater of the guinea-pig. Silver-impregnation. Magn. 400.

Fig. 183.—Dura mater of a new-born puppy. Silver-impregnation. Magn. 200.

Fig. 184.—A corpus arenaceum from the dura mater. Magn. 300.

From twenty to twenty-three triangular plates of connective-tissue fix themselves by their points to the inner surface of the dura spinalis, their broad bases resting upon the pia mater along the whole lateral border of the spinal cord (ligamentum denticulatum).

Short isolated threads of connective-tissue unite the dura spinalis with the arachnoidea.

The spinal as well as the cerebral dura forms fibrous sheaths around the issuing spinal nerves. The dural sheath of the optic nerve is connected at one end with the periosteum of the orbit, and at the other with the sclerotic. At the caudal end of the spinal cord the dura mater forms a sheath around the filum terminale, and fuses at last with the periosteum of the sacrum.

A tesselated epithelium can be shown to cover the surfaces of both the parietal and visceral dura mater (fig. 182). To see the epithelium it is best to take the dura of a young animal just killed. The membrane is spread out in a watch-glass or shallow dish, and acted upon for several minutes by a 0·2 to 0·5 per cent. solution of nitrate of silver. . The preparation is then thoroughly rinsed in distilled water and preserved in glycerin or, after dehydration, in dammar varnish.

After exposure to the light, the boundaries of the epithelial cells become evident. Small dark areas appear between them which must probably be regarded as stigmata. The substance of the dura is composed of coarse connective-tissue, with only a few elastic fibres. Numbers of Waldeyer's plasma cells (connective-tissue cells with abundant coarsely granular protoplasm) are said to be found in the dura mater in addition to the usual connective-tissue cells.

Without further preparation one can see with the naked eye that the large blood-vessels of the dura lie nearer to the parietal layer of the dura than to its visceral layer. If a very thin dura which has been obtained from a suitable animal is prepared as recommended above, it is seen that the relation to it of its vessels is peculiar (fig. 183).

The arteries first attract attention, the boundaries of their epithelial cells, as well as the cement substance between the muscle fibres, being distinctly shown. In addition, outlines of epithelial cells are seen on either side of the artery; these cells line a straight space which adjoins the wall of the artery and sends out numerous irregular branches on either side. The side spaces are connected with one another into a characteristic network. Not rarely the branches cross the artery.

The meaning of these spaces is still in dispute. They are injected with more or less ease from the blood-vessels. Sometimes blood-corpuscles are seen in them; but, despite this, they are not to be regarded as genuine veins, but as peculiar appendages to the system of blood-vessels, for no blood normally circulates through them. If they were filled with blood, the dura of a living animal would appear of a dark violet colour, so close is the network. It may be accepted that they communicate with the subdural space by means of the stigmata on the visceral surface of the dura; and that on the other side they open into the real blood-vascular system.

Langer says that fine arteries of the dura mater pass over into veins of much greater calibre by means of funnels resting by their bases upon the veins and receiving the arteries into their apices.

The nerves of the dura are not numerous, nevertheless it possesses a proper network of non-medullated fibres in addition to the nerves which merely traverse it and the nerves intended for its blood-vessels.

The question as to the sensitiveness of the dura must now be answered unconditionally in the affirmative.

The following are the most important **pathological changes** in the dura mater :—

Not rarely, in old people especially, concentrically laminated glancing concretions, corpora arenacea (fig. 184), are found in the dura. They hardly exceed 80 μ in diameter and are surrounded by a many-layered envelope of connective-tissue. They consist of phosphate and carbonate of lime, and when in large numbers are easily detected on touching the membrane. Their favourite situation is on the dura of the basis cerebri, and especially on the clivus. Concretions are also scattered about in many tumours of this membrane. When present in excessive quantity they constitute psammomata. The latter have usually an abundant coarse connective-tissue scaffolding which supports large numbers of round mulberry-shaped or elongated corpora arenacea.

OSSIFICATION of the dura occurs under otherwise normal conditions; and in many animals it is in places always converted into bone. As instances of this may be quoted, the falx of the dolphin and to a less extent of the seal, probably also of the ornithorhynchus; the tentorium of certain carnivores, especially the cat and the bear, and to a smaller degree of such ungulates as the horse, pachyderms, &c. Bony neoplasms are more common in the dura of people suffering from cerebral disease, especially epilepsy, than in healthy brains; in women ossification is less common than in men. Ossification occurs most commonly in the falx cerebri or its immediate neighbourhood, where the bony plate may reach a diameter of 8 cm. It is more common on the left side of the falx than on the right.

Ossification of the spinal dura is extremely rare.

Warty thickening of the dura, which may involve the pia mater and the cortex, is a frequent result of syphilis.

Amongst the real TUMOURS OF THE DURA, fibromata, fibro-sarcomata, and endothelial overgrowths must be placed in the first rank. The pure or mixed fibromata of the dura have a tendency to take on a rounded alveolar form (tumor fibro-plasticus). Primary tuberculosis of the dura does not occur.

Inflammation of the dura mater is termed PACHYMENINGITIS.

Simple purulent pachymeningitis, in which the dura is found to be occupied by pus-corpuscles, is rare; such inflammation is usually traumatic or due to extension from other localities.

A chronic process, in which a false membrane (or new membrane) containing blood-pigment is deposited on the inner side of the dura, is more common. It is termed pachymeningitis interna hæmorrhagica (pigmentosa). The deposition on the inner side may amount to no more than the formation of a delicate, usually rust-coloured, spotted membrane, or may (owing to the deposition of many such membranes, one above the other) constitute a thick crust almost half a centimeter through.

The processes by which this deposition is effected are of two kinds resulting in membranes of different structure.

It may happen that a small hæmorrhage is effused into the subdural space from the vessels of the dura mater, and that the fibrin of the effused blood, which usually is small in quantity, coagulating on its visceral side, the blood is encapsuled in a sac with the dura mater for its outer wall, and the coagulated fibrin for its inner wall, hæmatoma duræ matris. By-and-bye the contents of the sac are absorbed, and a layer of fibrin coloured with blood-pigment is left as a false membrane (fig. 185).

The second and more important variety of pachymeningitis interna runs quite a different course. At first a very delicate layer is formed on the inner side of the dura, as the result of a condition of irritation accompanied probably with emigration of leucocytes out of the dura. The lymphoid cells then begin to exhibit their formative activity, constructing a thin connective-tissue membrane, in which vessels of wide calibre, but thin walls, make their appearance. They do not at first contain blood (fig. 186). There is thus developed not a pseudomembrane but an organised neo-membrane devoid at first of blood-corpuscles, and not stained with blood-pigment. The vessels of the neo-membrane form communications with the vessels of the dura, and so obtain a supply of blood. The connecting vessels also are very delicate; they easily rupture, and thus occur hæmorrhages between the dura and the neo-membrane. In turn the effused blood is absorbed, but some of its colouring-matter remains behind as the contents of what were once lymphoid cells. The pigment-containing cells are very common in the vicinity of the blood-vessels. Such neo-membranes may be deposited layer upon layer (so too, may, perhaps, the pseudomembranes of the other form of pachymeningitis), and thus give rise to the already-mentioned thick covering. Concretions are frequently met with in these neo-membranes.

Large encapsuled hæmorrhages which are only partially absorbed constitute permanent hæmatomata.

The anatomical condition resulting from pachymeningitis interna is most often found in chronic cerebral disease, especially dementia paralytica and alcoholism. The dura mater spinalis is normally very thick on the ventral side of the cervical cord. An abnormal overgrowth of the dura also occurs in this situation, which may attain such dimensions as to cause pressure on the cord (*Joffroy*), pachymeningitis cervicalis hypertrophica.

B. ARACHNOIDEA (Meninx Serosa, Visceral Layer of the Arachnoid).

The arachnoidea does not follow the irregularities in the contour of the brain; on the contrary it is adherent to the dura mater, both cerebral and spinal, being separated from the pia by a considerable space bridged across by connecting threads or platelets (subarachnoid tissue).

Thus the arachnoid does not dip into the fissures, but retires in many places a considerable distance from the brain, leaving large spaces, the subarachnoid sinuses (cisternæ subarachnoidales). Of these, two deserve especial mention—

(1) SINUS SUBARACHNOIDALIS POSTERIOR (cisterna magna cerebello-medullaris) between the back of the cerebellum and the medulla oblongata. Here the arachnoid spreads like a veil from the vermis superior

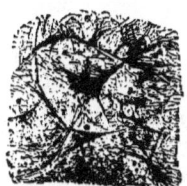
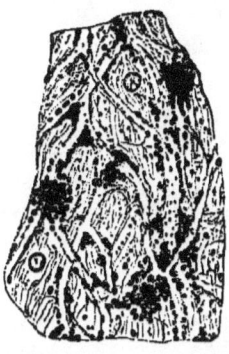

Fig. 185.—Pseudomembrane formed on the dura mater after a small hæmorrhage. *Magn.* 40.

Fig. 186.—Neo-membrane of the dura mater resulting from pachymeningitis hæmorrhagica. *Magn.* 40.

and the posterior part of the dorsal cerebellar surface over the vallecula to the medulla oblongata below the calamus scriptorius.

(2) SINUS SUBARACHNOIDALIS BASALIS is shaped like a star with five rays. The body of the star is formed by the arachnoid sweeping forwards from the anterior border of the pons over the corpora candicantia, the infundibulum, and the optic chiasm. The rays of the star are thus formed—the subarachnoid sinus extends around the crura cerebri on either side; in front two paired lateral diverticula extend into the Sylvian fossa, while the fifth ray is directed forwards and upwards into the space which commences in front of the chiasm, and is prolonged in the median fissure of the brain above the corpus callosum.

STRUCTURE OF ARACHNOID.

All the nerves which leave the skull-case receive a sheath from the arachnoid.

The arachnoid membrane of the cord retains its individuality throughout; it takes part in the formation of the sheath of the filum terminale as well as of the nerves. Numerous threads of sub-arachnoid tissue connect it with the pia spinalis especially on its dorsal surface.

Certain peculiar excrescences of the arachnoid (which are most numerous in the neighbourhood of the great longitudinal fissure, but occur also on the side of the cerebellum, sometimes at the apex of the temporal lobe, and rarely in other places) are termed arachnoid plexuses [glandulæ Pacchioni]. They will be described in connection with the histology of the membrane.

The arachnoidea consists of connective-tissue fibres (which are not, as a rule, united into definite bundles), and of the nuclei belonging to them. It contains neither vessels nor nerves. On either side it is covered with an exceedingly delicate tesselated epithelium. The threads, after which the web-like membrane is named, spread out from the arachnoid like roots. They always consist of a central core or connective-tissue bundle invested with an epithelial sheath. For the rest, its structure, despite the closest examination by *Key* and *Retzius*, is hardly understood. When acted on by acetic acid, the threads exhibit circular constrictions formerly attributed to elastic fibres. Besides this, many bundles are surrounded with a homogeneous or faintly striated sheath, as to the meaning of which it is impossible at present to draw any conclusions.

Fig. 187.—Trabeculæ of the arachnoid membrane after treatment with acetic acid. *Magn.* 200.

To examine the structure of the arachnoid membrane it is well to take it from situations in which it is widely separated from the pia, as around the cauda equina or in the posterior subarachnoid sinus.

Subarachnoid tissue is best obtained from the sinus basalis. After treating the fibres for some time with acetic acid the preparation may be washed and mounted in glycerin.

The ARACHNOID PLEXUSES (Pacchionian granulations, corpuscles, or glands) are knobby, cauliflower-shaped, pedunculated excrescences of the arachnoid with which they agree in minute structure; they consist of loose connective-tissue like that of the arachnoid covered with epithelium. Scattered arenaceous corpuscles may be found in these plexuses as everywhere else over the arachnoid.

They grow out into the subdural space, but do not stop there; for they force their way into the substance of the dura where it offers least resistance or presents preformed cavities. Thus they press themselves into the sinuses as well as into the venous spaces, *Ps* (fig. 181), on either side of the superior longitudinal sinus.

The parasinoidal spaces (lacunæ laterales) are found at the spots where the great veins which come up from the surface of the hemispheres enter the dura. Through them the blood is discharged into the superior longitudinal sinus. Like the sinuses they are lined inside with epithelium. If the dura mater is stript off from the parasinoidal spaces, or even from the sinus, the floor of the spaces is seen to be lifted up by the Pacchionian granulations. The Pacchionian body does not make its way through the wall of the space, but pushes it in front of itself, and so comes to be invested by its epithelium (*G*) When the body grows farther towards the bony wall (in which it usually produces a depression) it pushes in front of it the outer wall of the sinus also, in which case the epithelium lining the sinus usually disappears under the pressure.

Pigments injected into the subarachnoid space find their way into the meshes of the arachnoidal plexuses, and through its complicated epithelial coat into the parasinoidal spaces and the veins. In purulent and bloody effusions into the subarachnoid space pus-corpuscles and blood-corpuscles are found in the Pacchionian bodies.

In children these bodies may be quite absent, and they are never much developed before the tenth year. They are found in many animals, especially the larger ones, but they are certainly not so well developed as in Man.

Of the **pathological changes** which affect the arachnoid only the following will be mentioned:—

Small plates of bone without any essentially pathological significance are sometimes met with in the arachnoid. They are most common in the arachnoidea spinalis, especially over the dorsal surface of the lumbar and lower dorsal cord. These little plates, as thin as paper, frequently occur in cases of chronic disease of the spinal cord and in dementia paralytica.

The arachnoid may become extensively thickened and opaque in consequence of chronic irritation, especially over the convexity of the great brain and near the middle line. This condition is most often found in lunatics and drunkards, but even in simple atrophy (especially senile atrophy) opaque patches are observed.

Quite different to the above are certain small disseminated white patches of thickened arachnoidea which may be scattered over the whole convexity after chronic brain-disease, in idiots more especially.

C. PIA MATER (Meninx Vasculosa, Tunica Propria).

The pia mater cerebralis, which closely invests the surface of the brain, not only sinks down into all the fissures of the cerebrum and cerebellum, but also enters by means of special slits into the interior of the ventricles, where it forms the telæ choroideæ. In a similar way the pia mater spinalis adheres to the spinal cord sending the triangular folds already mentioned, the ligamentum denticulatum, to the dura. The connections between pia and arachnoidea have been already described. The filum terminale, as well as all nerves coming out of the cerebro-spinal axis, receive sheaths from the pia mater.

The pia mater cerebralis consists of two layers. The outer layer is a somewhat delicate connective-tissue membrane rich in nuclei. In this membrane arteries and veins as well as capillaries, the latter relatively the least abundant, spread out before giving off their lateral branches, which descend vertically into the brain-substance. The vessels of the pia are surrounded by lymph-spaces, and carry sheaths from the pia with them into the substance of the brain. The epithelium which covers the arachnoid is continued on to the pia.

Fig. 188.—Pia mater in meningitis tuberculosa. *Magn.* 15.

The inner layer of the pia cerebralis is an exceedingly delicate membrane, a basal membrane, which can only be recognised easily in certain places; as, for instance, where it covers the cerebellum. Its appearance in this situation has been already described. It sends off processes into the interior of the brain-substance, which take part in the formation of the connective-tissue scaffolding.

In the spinal cord, too, the pia consists of two layers; but both are coarser than those which invest the brain. The external layer, composed for the most part of longitudinal fibres, supports the blood-vessels which are disposed less closely than in the pia cerebralis. The inner non-vascular layer is composed of strong circular connective fibres.

The whole of the pia enters into the sulcus longitudinalis ventralis, but only the inner layer enters the dorsal sulcus.

Beneath the pia, on the surface of the cerebrum and cerebellum, *Fleischl* describes a layer, double for the most part, of very small epithelial cells which he calls cuticulum cerebri et cerebelli.

Numerous branched pigment-cells are often found in the pia, especially in old persons. They are most numerous on the ventral side of the medulla which often appears, in consequence, of a smoky colour to the naked eye. These pigment-cells may be met with throughout the whole length of the spinal cord in the external layer of the pia mater. On the other side, on the base of the brain the pigment-cells can be followed forwards as far as the olfactory bulb and the fossa Sylvii. There is no relation between the colour of the skin and hair and the pigment in the pia mater.

Amongst **pathological changes** in the pia, HYPERÆMIA and HÆMORRHAGE may be mentioned.

The same alterations in the contents of the LYMPHATIC SHEATHS of its blood-vessels are met with, as in the case of the sheaths of the intra-cerebral vessels (p. 146).

Secondary PURULENT MENINGITIS (lepto-meningitis purulenta) is often observed. Different kinds of bacteria must be taken into consideration in studying the ætiology of this disease. The occurrence of pus is not limited, however, to the substance of the pia; it is found in the subarachnoid space and also in the lymphatic sheaths around the cerebral vessels.

TUBERCULAR MENINGITIS is characterised chiefly by the appearance of little elevations on the pia, varying from the smallest dimensions up to the size of a millet seed. They consist of tubercle-cells, and are deposited by preference around the vessels on the base of the brain and in the Sylvian fossa (fig. 188).

Many TUMOURS of the brain and spine originate in the pia mater.

We have already described the topographical disposition of the **telæ choroideæ** and the **plexus choroidei** of the ventricles of the cerebrum and cerebellum.

It will suffice to recall here the fact that these structures are reduplications involuted from the pia mater. The telæ choroideæ correspond exactly in structure with the pia mater.

The histology of the plexus choroidei demands closer examination. In them the pia mater seems to be reduced to a nearly structureless membrane, in the substance of which no blood-vessels run. On the other hand, peculiar vessels, capillaries of wide calibre, are pushed forwards in the reduplication of the membrane so that they are enclosed by it on all sides. The numerous convolutions of these capillaries during their long course give to the plexus its characteristic knotted appearance. The plexus is covered on its ventricular surface by a single layer of epithelial cells of a peculiar kind (fig. 189). Although of various shapes these cells incline to a cubical form. Their edges and corners are drawn out into processes, by means of which they dovetail with one another.

Fig. 189.—Epithelium of the choroid plexus. *Magn.* 200.

A round nucleus lies in a coarsely granular protoplasm, but almost every cell contains in addition to the nucleus a highly refractive, bright, yellow or yellowish-brown body. As this body is coloured dark on the addition of osmic acid it may be looked upon as probably formed of some substance related to fat. Sometimes it assumes a rod-like or ring-like form. The structure of the choroidal plexus is in a high degree suggestive of an evoluted gland. [It recalls to mind the glomerular plexus of a renal tubule.] We may look upon it as a distinctly glandular organ concerned with the secretion of the liquor cerebro-spinalis. We are the more entitled to take this view, because the cerebro-spinal fluid cannot by any means be regarded as a simple serous exudation; its chemical constitution shows that it is a specific fluid. It contains but few formed elements.

Among the accessory bodies found in the choroidal plexus may be mentioned fat granules, brown pigment, and especially chalky concretions of smaller or larger size. No pathological significance can be attributed to these substances. Tumours, such as lipomas, may affect the choroidal plexus. Cysts are hardly ever absent in old people; they may, however, be found at birth. Their favourite location is the glomus of the plexus lateralis. *Schnopflagen's* view that they are formed by the dropsical separation from one another of the two layers of the pia which constitute the tela and plexus is most probably correct.

In the horse, concretions of inorganic material (carbonate or phosphate of lime) or of cholesterin are constantly met with (*Faivre*); they may be very numerous and even as large as hen's eggs.

D. THE GREAT VESSELS OF THE BRAIN.

Within the skull-case arteries and veins do not run in company as they do in other organs. Leaving out of consideration the great venous sinuses of the dura mater, we may say that all the larger arteries are found on the base of the brain, while the larger veins are directed towards its convexity.

The method in which the vessels within the brain divide is well known, thanks to the investigations of *Heubner* and *Duret;* here, however, the subject can only be treated in outline.

The brain is supplied with blood by two arteries on each side, the internal carotid and the vertebral.

The internal carotid artery, Ci (fig. 190), advances along the side of the tuber olfactorium, and divides, after it has given off the ophthalmic artery which courses forwards, into its two chief branches, the anterior and middle cerebral arteries.

CIRCLE OF WILLIS.

Arteria cerebri anterior, *Ca* (arteria corporis callosi), winds at first towards the middle line, slips over the optic nerves, and then bends round into the great longitudinal fissure where it can be followed a long way backwards on the upper surface of the corpus callosum.

At the spot where the anterior cerebral arteries of the two sides assume a sagittal direction, they are placed so close together that a very short

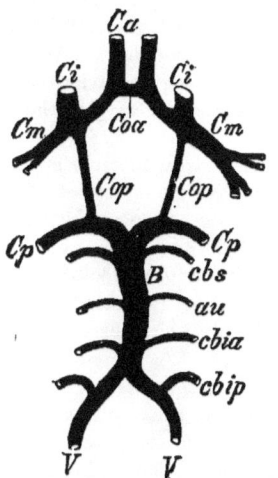 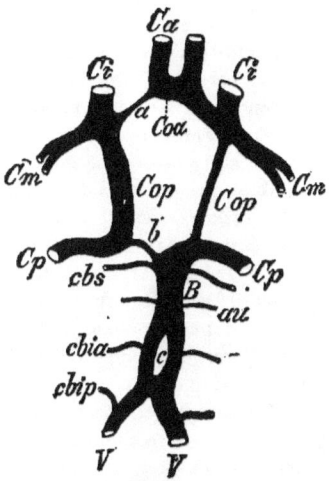

Fig. 190.—Arteries at the base of the brain; circle of Willis.— *Ci*, a. carotis interna; *Ca*, a. cerebri anterior; *Cm*, a. cerebri media; *Cp*, a. cerebri posterior; *Coa*, a. communicans anterior; *Cop*, a. communicans posterior; *V*, a. vertebralis; *B*, a. basilaris; *cbs*, a. cerebelli superior; *cbia*, a. cerebelli inferior anterior; *cbip*, a. cerebelli inferior posterior; *au*, a. auditiva.

Fig. 191.—Anomalies of the circle of Willis. *Lettering as in fig.* 190. —*a*, a. cerebri anterior; *b*, a. cerebri posterior, reduced to the dimensions of communicating arteries; *c*, island in the vertebral artery.

connecting branch, *Coa* (arteria communicans anterior), suffices for their anastomosis. At the place where the artery slings itself round the corpus callosum it gives off a fine branch for the dura mater, which courses backwards, along the lower edge of the falx (*Langer*).

The arteria cerebri media, *Cm* (arteria fossæ Sylvii *seu* transversa cerebri), must be looked upon as the direct continuation of the internal carotid. Hence it is that emboli which have come up the latter vessel are more likely to continue their course into the middle than into the anterior cerebral artery. It turns sidewards so as to enter the fossa Sylvii, where it soon divides into from three to five branches.

Each of the two vertebral arteries (V) gives off an arteria cerebelli inferior posterior ($cbip$), and then unites with its fellow at the back of the pons into a single trunk, the basilar artery (B). The basilar artery passes forwards in the median line with, usually, a slight convexity to the left, and, as a rule, gives off at right angles three small arteries from either side; arteria cerebelli inferior anterior ($cbia$), arteria auditiva (au), and arteria cerebelli superior (cbs). At the proximal border of the pons the basilar artery again divides into two branches directed straight outwards, arteriæ cerebri posteriores, Cp (arteriæ profundæ cerebri).

Immediately before the internal carotid passes over into the middle cerebral or just from the commencement of the latter, comes off a usually rather narrow vessel, arteria communicans posterior, Cop, which passes backwards towards the posterior cerebral which it joins at a distance of scarcely as much as 1 mm. from the point where the arteria basilaris bifurcates. In this way there is formed on the base of the brain a hexagonal or heptagonal ring, circulus arteriosus Willisii (hexagon, polygon of Willis).

After giving off the posterior communicating, the internal carotid, or as it is here called the middle cerebral, gives off backwards a second fine branch, arteria choroidea. This artery courses along the tractus opticus, by which it is conducted into the plexus choroideus of the descending horn of the lateral ventricle.

From these large vessels the arteries for the brain-substance come off in a two-fold way. So long as the chief arteries lie on the base of the brain they give off fine branches for the brain-substance, which, as they do not anastomose with one another, belong to the class of vessels termed "end arteries" (Heubner's basilar area). Over the whole of the rest of the surface of the brain the larger arteries gradually break up dichotomously (Heubner's cortical area). Neighbouring vascular territories are in connection with one another by means of numerous anastomoses in the pia mater. Anastomoses between the vessels of the cortical areas of the two hemispheres are very uncommon.

The direction of the fissures on the surface of the brain is almost entirely unconnected with the course of the vessels. The central parts of the brain, including the central ganglia, are exclusively supplied from the basal system, the cortex is nourished from the vessels of the cortical system.

The whole surface of each hemisphere may be divided up into three areas corresponding to the three chief arteries of the great brain—

(1) Area of the arteria cerebri anterior, which includes (on the con-

vex surface of the brain) the greater part of the upper and middle frontal convolutions, the mesial surface as far backwards as the cuneus, as well as the mesial portion of the orbital surface.

(2) Area of the arteria cerebri media—both central convolutions, the whole of the rest of the convex surface of the parietal lobe, the upper temporal convolution, the island of Reil, and the lateral part of the orbital surface; on the median surface it supplies at the utmost a small branch in the neighbourhood of the uncus.

(3) Area of the arteria cerebri posterior—the whole occipital lobe and the greater part of the temporal lobe. The superficial **veins** form a freely anastomosing network in the pia mater, and open into the several sinuses of the dura. The largest anastomosis lies horizontally above the temporal lobe, vena magna anastomotica temporalis. It is fairly constant.

The veins of the various internal parts of the brain collect into the vena cerebri interna communis (vena magna Galeni). It is formed chiefly by the confluence of the two venæ cerebri internæ, which run in the tela choroidea media in the roof of the third ventricle.

The vena cerebri interna communis comes out through the great transverse fissure, and discharges its contents into the sinus perpendicularis.

The brain more than any other organ stands in need of a sufficient supply of blood. The united cross-section of the four arteries which supply it, however, bears a by no means constant relation to the size of the brain, on the contrary it varies within wide limits (*Löwenfeld*).

[Not only does the brain require a large supply of blood, but the amount supplied to the organ as a whole (and still more the amount distributed to its several parts) must be capable of rapid and extensive fluctuations. Nervous tissues, in a more conspicuous degree than others, are in immediate relation to lymphatic spaces. Every nerve-cell lies in a little bath of lymph,* while the whole cerebro-spinal axis is suspended in a sea of the same fluid, from which it extracts its nutrients, and into which it discharges its waste products; whilst the balancing of the spinal cord in the centre of the vertebral canal, by means of the ligamentum denticulatum, and the support of the brain upon the sinus subarachnoidalis basalis, protects the cerebro-spinal axis from sudden jars. The enclosure of the brain within the skull places the circulation under certain difficulties, but the abundant lymph serves to carry off any excess of pressure due to vascular turgescence. The venous sinuses, unlike veins in other parts of the

* The cerebro-spinal fluid is functionally equivalent to lymph.

body, are incapable of distension. The communications between the intracranial sinuses and the veins outside the skull, established by the emissary veins of Santorini (which traverse the frontal, parietal, mastoid, posterior condyloid, and other foramina), serve to keep down the pressure in the sinuses. Doubtless the rearrangement of the central grey matter, which occurs where the medulla oblongata replaces the spinal cord, is due to the need for distributing to the general lymph the pressure produced by local turgescence of the very important centres in this part of the axis, instead of allowing an active centre injuriously to compress its neighbours. The membranous roofs of the fourth and third ventricle confine the lymph less closely than would solid nerve-tissue. It has been suggested that the veins, which fill up the spaces between the great arteries that enter the base of the skull and the margins of the holes through which they pass, may serve as a self-regulating apparatus for the supply of blood to the brain, the veins when distended, pressing upon, and so diminishing the calibre of the arteries; but the difference in pressure between the arterial and venous blood, in favour of the former, seems to make this impossible.

The recent experiments of *Corin* show that the pressure in the several arteries which supply the brain is very uniform. Even when three out of the four great affluents of the circle of Willis were tied, the pressure in the circle was unaffected].

In most animals the part played by the carotid and vertebral arteries respectively in supplying the brain with blood is different to what it is in Man.

As compared with the carotids the vertebrals in most rodents and some other animals are very much the more strongly developed. On the other hand, in ruminants (as well as in the pig and, probably, in the leopard) the vertebral arteries do not directly reach the brain. In these animals both carotids form a rete mirabile on the base of the skull outside the dura mater; not till it has passed this plexus does the carotid reach the base of the brain and unite with the basilar to form the circulus Willisii. The basilar artery is continued backwards on the ventral surface of the spinal cord as the arteria spinalis anterior. The two vertebral arteries remain all the time outside the dura mater, and only anastomose with the rete mirabile.

The above described typical arrangement of the great arteries on the base of the brain is subject to frequent variations of greater or less physiological importance.

We will mention the commoner **variations in the circle of Willis.** There may be several anterior communicating arteries; or, on the other hand, this artery may be altogether wanting, the anterior cerebrals growing together after an independent course of some distance.

Sometimes both anterior cerebral arteries are almost entirely derived from the same carotid artery (fig. 191). In this case there is, as a rule, a small branch of communication with the other side, connecting the anterior cerebral artery which comes from the opposite carotid with the carotid of its own side (*a*). In the same way it may happen that the posterior cerebral artery does not come off from the basilar, but from the carotid of the same side; being only connected with the anterior end of the basilar artery by an inconspicuous anastomosis (*b*). In this case the posterior communicating artery must be very strongly developed. The posterior communicating artery may on the contrary be wanting on one side.

Very often the origin of the basilar artery from the junction of the two vertebrals is indicated by the presence of a longitudinal septum in its interior; indeed the artery may even be doubled again for a certain distance forming an "island," *c*. Very often the two vertebral arteries are not equally strong, usually it is the right which is the thinner.

The vessels of the brain differ in no respects from those of other organs in their **minute structure;** the veins of the brain are devoid of valves.

Only the most important **diseases of the large vessels** of the brain will be mentioned.

EMBOLISM is frequent. In three-fourths of the cases it is the arteria fossæ Sylvii which is affected. Left and right equally often.

Autochthonous THROMBOSIS of the arteries of the brain must be distinguished from embolism.

Thrombosis of the sinuses is not rare. ANEURYSMAL ENLARGEMENT of the basal vessels is relatively uncommon. According to tables drawn up by *Lebert*, out of 86 cases of aneurysms of the vessels of the brain the basilar artery was affected 31 times, the middle cerebral 21. The remaining cases were divided amongst the other arteries. The left side is more frequently affected than the right.

ATHEROMATOUS DEGENERATION is almost always met with in the vessels of the brains of old people. Sometimes it is difficult to distinguish an atheromatous degeneration from SYPHILITIC DISEASE. In the latter, disease is due to the growth of granulations or infiltrations, originating, probably, in the capillaries which supply the muscular coat of the vessel (vasa vasorum). The infiltration is spread out in the intima, and especially between the epithelium and the membrana fenestrata. Often in such cases of syphilitic disease a many-layered fenestrated membrane is seen, indicated in cross-section by a number of clear, highly refracting, waving lines. *Heubner* thinks that this is new formation; but it is possible that it is due to splitting of the fenestrated membrane by the intercalation of granulations (*Rumpf*).

Syphilitic disease is distinguished from atheromatous disease by its greater tendency to active overgrowth and extension (even leading to thrombosis of the artery); while atheroma soon ends in a retrogressive process leading to calcification and fatty degeneration.

APPENDIX A.

ROTATION OF THE GREAT BRAIN.

In the spring of 1885 the *translator* propounded, as part of a general scheme of the structure of the central nervous system, the theory, that the mammalian brain during its growth rotates upon itself.* The rotation produces a loop or kink, giving to the whole brain somewhat the form of a ram's horn, and bringing the part which was at first in front, on to the under side of the back. At its first formation the cerebral hemisphere of the mammalian brain is directed, like that of the reptile, straight forwards; the foramen of Monro opens into the back of its ventricle, the anterior end of the hemisphere bears the olfactory bulb. In the adult condition the foramen of Monro opens into the front of the ventricle, the olfactory apparatus is attached through the pyriform lobe (gyrus uncinatus) to the inner side of the temporo-sphenoidal lobe.

This view as to the alteration in position of the great brain was the almost necessary deduction from certain conclusions as to the connections of the olfactory tract with the anterior end of the optic thalamus, *viâ* the fimbria, fornix, descending pillar of the fornix, corpus mammillare, and bundle of Vicq d'Azyr.

There are many reasons for believing that the olfactory nerve is connected in this way with the anterior end of the optic thalamus, and the circuitous route just sketched out is by no means difficult to follow. Its path is easily broken off from a hardened brain, of which it seems to form an organically distinct part, as shown in the accompanying sketch; and apart from the question of the arrangement in adult anatomy, we have, as will be shown presently, a certain amount of embryological evidence in favour of this extensive displacement of the olfactory tract. The formulation of such a path for the fibres of the olfactory tract almost necessarily presupposes a rotation of the brain; but it is not this connection only which the theory renders

* *The Plan of the Central Nervous System.* Cambridge: Deighton Bell & Co., 1885.

intelligible. Many other features in the plan of formation of the great brain are accounted for by this supposition of its rotation.

Among EXTERNAL APPEARANCES, the most suggestive is the progressive closing in of the fossa of Sylvius. The fissure of Sylvius, as has been shown (p. 92), is not, for the greater part of its extent, in any way comparable with the other fissures of the brain, since, instead of being formed like them, as a narrow slit-like depression on the smooth rounded surface, it makes its appearance before any other

Fig. 192.—Fornix, pyriform lobe and olfactory bulb, broken off from the hardened brain of the Ox.

fissure, if we except the so-called rhinal fissure, as a dimpling of the mid part of the outer surface which gradually deepens into an extensive shallow fossa. The appearance of this fossa is due, apparently, to the traction exercised upon a part which will subsequently become the temporo-sphenoidal lobe by its attachment through the olfactory tract and bulb, to the cribriform plate of the ethmoid. As the great brain swells, a pitting of the outer surface is the necessary result of the attachment of its olfactory part to the skull. The fossa of Sylvius is subsequently converted into the fissure of Sylvius, a Y-shaped fissure, with a common portion dividing into an anterior and a posterior limb, by the outgrowth in three swellings of the surface of the brain which borders it; the operculum coming down between the limbs of the Y; the frontal and temporo-sphenoidal lobes closing up beneath the operculum. The want of homology between the fissure of Sylvius and other fissures is shown, by its enclosing a considerable cortical area, the island of Reil, which originally formed the floor of the fossa. The fan-like arrangement of the fissures of the island of Reil indicates plainly enough the direction in which traction was exerted.

When viewing the outer surface of any mammalian brain, it is easy to convince oneself that is has undergone a rotation, and that the position of its parts has been altered from the form maintained amongst fishes, amphibia, and reptiles (for special reasons it is better to exclude from the comparison the bird's brain) in such a way that the temporo-sphenoidal lobe of the mammal's cerebrum corresponds to the anterior end of the brain of lower vertebrates.

Following closely upon the external form, the delimitation of the cortex into areas associated with peripheral nerves, presents itself as a test by which to try the theory. Recently, chiefly as the result of ablation experiments, the function of almost every part of the cortex has been determined, and although we are as yet far from possessing an accurate industrial map, we can nevertheless speak with assurance as to the general distribution of work among its several territories. Undoubtedly the anterior end of the brain, particularly the Rolandic area, is in connection through the mediation of the grey matter of the spinal cord with the body nerves. The localities in which sensations are received through the 8th and 5th nerves, lie somewhere in the neighbourhood of the fissure of Sylvius; the 2nd nerve has its area of cortical distribution in the occipital lobe—the 1st nerve on the inner side of the temporo-sphenoidal lobe. Better even than direct experiment, the observation of brains of animals remarkable either for preponderance or for deficiency of particular senses, will help us to map out the cortex into functional areas (see p. 278, figs. 149, 150, 151). Whatever method be adopted for obtaining the necessary data, any sketch of the brain in which functional distribution is indicated, shows clearly enough that the order from before backwards in which the areas connected with the several nerves are situate, is in its largest features the reverse of that which obtains among the nerves themselves. In considering the bearing upon the question of the territorial allocation of the brain surface, it is important to have in one's mind a clear picture of the process of histogenetic differentiation of the nervous system. The cells from which nerve processes (fibres) extend into the cortex lie in the central grey tube. As yet we have no means of telling how they are guided to the particular areas in the cortex in which they are distributed. Is the connection between these areas and the primary centres in the cord determined from the first involution of the neuro-epithelial tube, or is such a connection established in a way which, for want of knowledge of the laws by which it is governed, may for the time be called "by chance"? When the processes from a certain group of cells in the central grey tube take possession of a particular area of the cortex, are they guided to this area merely by its situation in relation to the skull-case or head, or owing to its primitive relation to the axis of the brain? Does the differentiation of the motor cells in the cortex precede the reception in this tissue of the sensory cell-processes, or does it depend upon their reception and the need thus established for a descending limb of the reflex arc? Many similar problems present themselves to warn us that we must not lay too much stress upon the allocation of the cortex to specific functions until we are far better equipped than at present with genetic data. It is sufficient for our purpose to consider only the larger groups of cortical centres. The arrangement of their sub-centres, as determined, for instance, in the marginal gyrus by Horsley and Schäfer, may for the present be neglected.

The body nerves have their cortical areas in front, and are followed by the nerves of the head; of these latter the optic nerve is indirectly connected with the back part of the upper surface, while the olfactory has its area in what we take to be the reflected portion. It seems as if the rotation over from before backwards were accompanied by a certain displacement from above downwards and outwards.

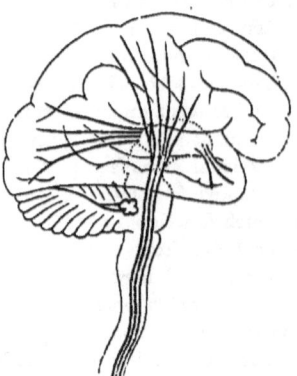

Fig. 193.—Diagram showing the tracts of fibres which connect the central grey tube with the cortex. The front of the thalamus is connected with the temporal lobe, the back of the thalamus with the occipital lobe, and the spinal cord with the Rolandic area.

Passing now to the INTERNAL STRUCTURE of the hemisphere, so far as the course of fibres within its white surface is sufficiently well known to help us, we find further evidence of rotation. Leaving out of consideration all doubtful tracts, we have the classical observations of *Gratiolet*, which have been frequently confirmed, to show that the anterior end of the optic thalamus is connected through its inferior peduncle with the temporal lobe, the posterior end of the thalamus through the "optic radiations" with the occipital lobe. Numberless recent observations have placed the connection of the grey matter of the spinal cord with the frontal and parietal lobes by means of the pyramidal tracts which occupy the anterior part of the internal capsule, beyond doubt. These three sets of fibres (1) of the internal capsule, (2) optic radiations, and (3) inferior peduncle, obviously cross one another, as shown in the accompanying diagram (fig. 193), plainly indicating that the cortex of the cerebrum and the grey matter of the basal part of the system extend in contrary directions. The force with which the crossing of these three sets of fibres appeals to the *translator*, will become apparent if his view as to the fundamental plan of construction of the central nervous system is borne in mind. The system arises as an involuted tube of epiblast. Throughout the spinal region the inner wall of the tube becomes grey matter; the outer wall constitutes the white columns of fibres. In the cephalic region a second layer of grey matter, the cortex, is added on the outer side of the tube; all the grey matter of the system grows from one or other of these two grey layers. In the spinal cord, which only contains the tube of grey matter which borders the central canal, are situate all the primary centres of the body nerves; indeed, the grey matter is entirely used up in forming nerve-cells and plexus for the origin of motor, and the reception of sensory, fibres. Travelling up into the brain, no break in the continuity of the grey matter is to be observed. It lies in the floor of the fourth ventricle, around the aqueduct, and in the sides and floor of the third ventricle. No one doubts the continuity of plan as far as the anterior end of the iter; but hitherto it has not been recognised that the grey matter bordering the third ventricle is in functional continuity with the grey matter of the cord. The optic thalami are treated of as "basal ganglia." The olfactory and optic nerves are deprived of primary centres such as are accorded to all other sensory nerves. The modifications in structure of the central system, due to the situation peripherally in the olfactory bulb and retina of some of the grey matter, have been already pointed out (pp. 127 and 270). If the optic thalamus is part of the anterior end of the central grey tube, the attachment of the extreme anterior end of the thalamus with the temporo-spenoidal lobe, of the back of the thalamus with the occipital lobe, and of the grey matter of the spinal cord with the front of the hemisphere, supplies the strongest possible evidence of rotation.

Passing now to the consideration of one of the chief inter-cerebral tracts, we find in the arrangement of the fibres of the anterior commissure evidence of rotation of a most convincing kind. Fig. 194, which is a copy of the picture on page 490 of Schwalbe's *Neurologie*, accurately represents the peculiar twisting which the anterior commissure has undergone. It looks as if the middle of the commissure had been fixed, while its ends were twisted in such a way that the fibres which cross the median line on its under side, pass round in front to reach

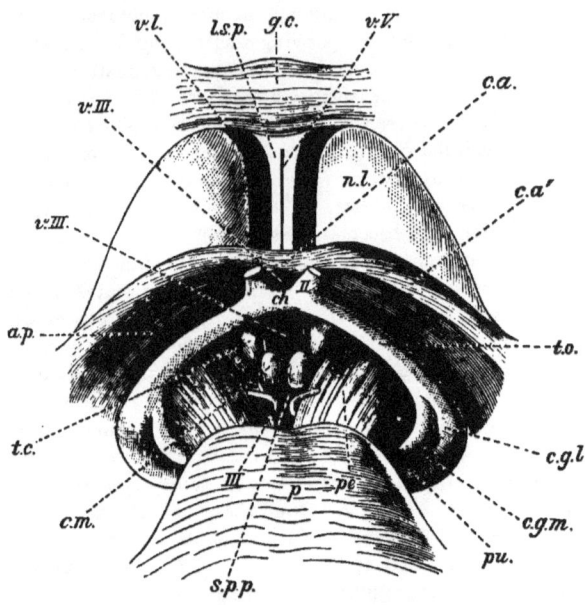

Fig. 194.—Diagrammatic view of the under side of the brain, after *Schwalbe*. In front of the optic tracts the superficial substance of the base of the brain has been dissected away so as to expose the ansa pedoncularis, anterior commissure, and basal aspect of the nuclei lenticulares.—*l.s.p.*, Septum pellucidum; *v.l.*, anterior horn of lateral ventricle; *n.l.*, nucleus lenticularis; *c.a.*, anterior commissure; *c.a'.*, twisted portion of ditto; *v.III.*, third ventricle; *g.c.*, genu corporis callosi; *c.g.m.*, corpus geniculatum mediale; *c.g.l.*, corpus geniculatum laterale; *c.m.*, corpus mammillare; *pu.*, pulvinar; *v.V.*, fifth ventricle; *II*, optic nerve; *ch*, chiasma nervorum opticorum; *t.o.*, tractus opticus; *t.c.*, tuberculum cinereum; *III*, oculomotor nerve; *s.p.p.*, substantia perforata posterior; *pe*, pedunculus cerebri; *p*, pons.

the upper side at its extremities. Just such a torsion as would be produced by the rotation of the hemisphere round an axis situate just below and in front of the foramen of Monro.

In addition to the demonstration which the twisting of the anterior commissure affords of the fact of rotation, peculiar interest attaches to it, since it enables us to fix approximately the time at which rotation takes place. It must occur after this commissure is formed. There can be no doubt that the anterior commissure appears very early in the mammalian brain, for although we have no exact observations as to the time of its appearance in the individual, this conclusion may be drawn from its great prominence in the lower forms of vertebrates. It is *par excellence* the commissure of birds and reptiles. The existence of a rudimentary corpus callosum in birds was pointed out by *Meckel*, but the view of *Stieda*, that no such structure exists, is endorsed by recent observers such as *Rabl-Rückhard* and *Bellonci*. On the other hand, Stieda recognises the existence of a corpus callosum in reptiles, but denies it in amphibia, where *Leuret*, *Blattmann*, and *Reissner* had found it. In fishes, the commissures appear to be fused, so that no distinction between corpus callosum and anterior commissure is possible. Quite recently *Osborn* has shown that both commissures may be found in reptiles, birds, and amphibia, although in the two former classes the corpus callosum, owing to the rudimentary condition of the cortex in these animals, is very ill-developed. The cerebral hemisphere of the bird, and to a less extent of the reptile, is homologous, not with the entire hemisphere of the mammal, but, as pointed out by *Cuvier*, with little more than the corpus striatum.

The MINUTE STRUCTURE of the brain also affords some evidence of rotation, for the margin of the cortex mantle is for a varying distance, where it skirts the under side of the crus cerebri, invested in a sheath of grey matter quite different in structure from the cortex itself. This sheath of small-celled matter, the fascia dentata, which forms part of the cornu Ammonis, receives the thin edge of the cortex mantle. It commences beneath the uncus at the apex of the temporal lobe, and extends along the course of the posterior pillar of the fornix, reaching in some animals (particularly rodents) almost to the point where the fornix dips down in the anterior pillar. It has been already (p. 367) suggested that the fascia dentata is a continuation backwards of the nuclear (glomerular) layer of the olfactory bulb. In straight-brained animals the hemisphere terminates in the olfactory bulb. In twisted brains the olfactory bulb is attached to the apex of the pyriform lobe (the gyrus uncinatus of human anatomy). The continuation of its nuclear formation, as the fascia dentata, obliquely beneath, behind, and above the crus, simply shows the direction in which rotation of the great brain has taken place.

ONTOGENY.—Evidence obtained from direct observation of the brain in successive stages of its growth must obviously, when the subject is thoroughly worked out, be the first to be submitted in proof of the rotation of the hemisphere. As yet, however, our knowledge of this subject is very incomplete owing in part to the absence of research, in part to the special difficulties which depend upon the nature of the tissue of which the brain is composed. Unlike other tissues of the body, the nervous tissue is peculiarly plastic. In the course of its growth it may receive marks of torsion and other displacements, but while elsewhere such marks remain permanently, in the brain they are obliterated by further growth. The manner in which its mass is added to also has an important bearing upon its form. It must be remembered that only the cellular elements enter into its original constitution; the fibres are but processes of the cells. Of the primitive epiblastic cells involuted to form the nervous tube, some are favoured above the rest and become the functional

Fig. 195.—Early fœtal brain.

Fig. 196.—Brain of fœtal sheep in which the rhinal fissure is developed and separates the pyriform lobe from the rest of the cerebrum. Its distinctness is exaggerated in the wood-cut.

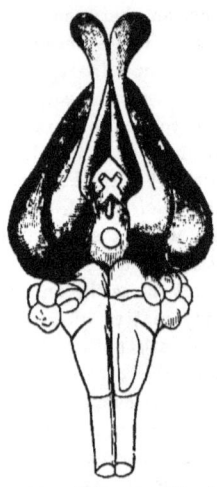

Fig. 197.—Diagrammatic view of the under surface of the brain of a young rabbit.

nervous elements, while to the others only subsidiary, supporting, nutritive and insulating functions are assigned; the latter become the neurogleia and myelin cells; just as in the case of each little group of epithelial cells involuted into a follicle of the ovary, one only becomes an ovum, the others minister to its wants. It may be, therefore, that the line along which a nerve-fibre is to grow is laid down, before the fibre itself appears, by the chain of accessory cells which have taken up their position along its route; or, on the other hand, it is open to us to suppose that the route chosen by a fibre depends upon convenience at the time of its appearance, and that the marshalling of the myelin-cells in files depends upon some stimulus which the growing fibre affords. If, owing to a displacement of the cortex, it can reach its destination by a shorter route than the tract occupied by neighbouring fibres, which being fully developed at the time of the rotation of the brain, shared in the displacement of the cortex, it is open to it to leave the original tract, and thus the primitive arrangement of the constituent parts of the brain may be masked. Despite, however, the obliteration of the indications of rotation which these two causes are likely to produce, we still can see in the changes of form which the great brain undergoes appearances suggestive of rotation. This at least is the conclusion likely to be drawn from the examination of three such pictures as are here inserted.

In fig 195, copied from Löwe,* a very early stage in the growth of the brain is shown; the hemisphere is directed forwards and bears the olfactory bulb at its anterior end. Fig. 196 represents the appearance presented by almost any mammalian brain after the formation of the rhinal fissure. The olfactory bulb is now attached to the extremity of the pyriform lobe, and the suggestion of a folding over of the whole hemisphere is very strong. The pitting of the side of the brain due to the tension on its posterior end produced by the attachment of the olfactory bulb to the cribriform plate, is beginning to appear as the fossa of Sylvius. Fig. 197, again after

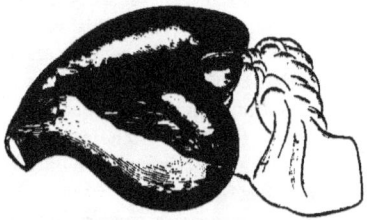

Fig. 198.—Brain of rabbit. The upper part of the cerebrum has been cut away so as to expose the hippocampus.

Löwe, is inserted for the purpose of showing the continuity of the olfactory bulb, pyriform lobe, and gyrus uncinatus in a rabbit's brain. Fig. 198 shows the hippocampus exposed from above by cutting away the roof of the lateral ventricle. It is curious to notice how the backward growth of the hemisphere increases the size of the

* Löwe, *Entwickelungsgeschichte des Nervensystems*, Pl. I., figs. 1 and 2.

loop. In the brain of a sheep, for instance, the fimbria (posterior pillar of the fornix) is transverse at its first appearance. In marsupial brains it never becomes more than very slightly oblique. The higher the animal in the mammalian scale, the more nearly does its fornix-system assume a longitudinal direction.

Of the earliest stages of the growth of the brain we have very few observations, but *Marshall's* discovery that the olfactory nerve arises, like all other sensory nerves, from the neural ridge, and that it is then pushed down in front by the outgrowth of the cerebral hemispheres, to be again caught up in the cortex mantle, opens up great possibilities with regard to the relation of the olfactory apparatus to the hemisphere, and prepares us, although the research has gone no further, to find an arrangement in the first nerve quite unlike that of the others.

PHYLOGENY.—The reference already made to the marsupial brain has drawn attention to the fact, that in lower mammals the rotation has gone less far than it has in the higher members of the class. Whether or not it is a phase in the development of any sub-mammalian animals, we are not prepared to state. Certain it is that, if we descend as far as the reptiles, rotation is out of the question. As already pointed out, the reptilian brain, like the brain of the early mammalian embryo, is directed forwards, its ventricle communicating at the back with the third ventricle through the foramen of Monro, and ending in the ventricle of the olfactory bulb in front. The great gap between the bird's and the mammal's brains, although bridged over by the brain of the monotreme, renders it difficult to point out the first animal in which the brain has adopted a looped disposition. The peculiar deficiency of cortex in the bird's brain renders rotation unnecessary, since the reason for rotation is not far to seek. If we compare such an extreme type as the crocodile with any mammal, we notice a contrast in the whole architecture of the head. The reptilian head is directed forwards; is elongated in the direction of the axis of the body. The mammalian head approaches a globular form. With the greatly increased size of the cerebrum as we ascend in the vertebrate scale, economy of space for its reception becomes necessary. Doubtless, for other reasons also, a shortening of the head is desirable, and the larger brain can only be packed inside the shorter skull by throwing its hemispheres over into loops.

GLOSSARIAL INDEX.

German names with the English equivalents used in this translation.

Latin and English names for parts of the central nervous system with common synonyms and their German and French equivalents.*

A

                                                                                 PAGE

Abducenskern, abducent nucleus.
Abducent nerve, ... 226, 292
  ,, nucleus, ... 226, 292
Aberrant bundle of the lateral column, ... 265
Aberrirendes Seitenstrangbündel, aberrant bundle of the lateral column.
Abscess in the brain, ... 373
Accessorius Willisii, ... 171, 312
Acervulus cerebri, ... 370
Acusticushauptkern, auditory nucleus, chief.
Acusticuskern accessorischer, auditory nucleus, accessory.
  ,,   **grosszelliger**, auditory nucleus, large-celled.
Acusticuswurzel, aufsteigende, auditory root, ascending.
  ,,   **laterale**, auditory root, lateral.
  ,,   **mediale**, auditory root, mesial.
Adergeflechte des Grosshirns, choroid plexus.
  ,,   ,, **Kleinhirns**, choroid plexus of cerebellum.
Adergeflechtsfurche, choroidal fissure, transverse fissure.
Aditus ad aquæductum, ... 79
Adventitia of blood-vessels, ... 137
Adventitial lymph space, ... 138
Aeussere Kapsel, external capsule.
Aeusserer Kern der Keilstränge, nucleus funiculi cuneati.
After-brain, medulla oblongata, **Nachhirn**, . . . 36, 42
Ala cinerea, ... 58
  ,, lobuli centralis, ... 52
  ,, pontis, ... 59, 219
Alum-hæmatoxylin, ... 11
Alveus, **Muldenblatt**, *feuillet de la conque*, . . . 365
Ammonia-carmine, ... 10

* No attempt has been made to render this Glossary complete. French terms are not given when their form of expression is very similar to English.

GLOSSARIAL INDEX.

PAGE

Ammenshorn, cornu Ammonis.
Amygdala cerebelli, **Mandel**, tonsil, ... 52
,, *seu* nucleus amygdaleus, ... 52
Amyloid bodies, ... 156, 207, 273
Annectant convolutions, ... 97
Anomalies of the circle of Willis, ... 393
,, in convolution, ... 102
Ansa intergenicularis, ... 65
,, lentiformis, ... 248, 341
,, nuclei lenticularis, **Linsenkernschlinge**, *anse du noyau lenticulaire*, 341
,, peduncularis, *seu* substantia innominata, **Hirnschenkelschlinge**,
    l'anse pédonculaire (Gratiolet), ... 248, 339, 343
,, Rolandica, ... 345
Anterior column of spinal cord, ... 41
    ,, horn of lateral ventricle, ... 80
    ,, ,, spinal cord, ... 171
    ,, spinal nerve-roots, ... 21
    ,, or direct pyramidal tract, ... 193, 253
    ,, of the two hinder vesicles, ... 36
Apertura inferior ventriculi quarti, *seu* foramen Magendii, . . 60
    ,, lateralis ,, ,, ... 60
Apex cornu posterioris, ... 175
Apoplexy, cerebral, ... 375
    ,, spinal, ... 203
Aquæductus Sylvii, ... 62, 238
Arachnoidea, ... 383
Arachnoid plexus, ... 385
Arbor vitæ, ... 53
Arrhinencephalia, ... 104
Arme der Vierhügel, brachia corporum quadrigeminorum.
Arnold's arcuate fibres, ... 349
Arteries (minute structure), ... 135
    ,, at the base of the brain, ... 389
    ,, of the cerebellum, ... 333
    ,, of the cerebrum, ... 388
    ,, of the spinal cord, ... 202
Association fibres, ... 163
    ,, systems, ... 168, 344
Asymmetry of the cerebral convolutions, ... 102
Atheroma, ... 144
Atheromatosis of the large blood-vessels, ... 393
Atrophy of the cerebellum, ... 334
    ,, of the great brain, ... 103
    ,, of nerve-cells, ... 134
    ,, senile, of the cerebrum, ... 372
Auditory nerve, ... 299
    ,, nucleus accessory, ... 302
    ,, ,, chief, ... 300
    ,, ,, large-celled, ... 301
    ,, root, ascending, ... 224, 300

GLOSSARIAL INDEX.   407

	PAGE
Auditory root, lateral,	224, 300
,, ,, mesial,	224, 300
Axis-cylinder, **Primitivband, Axenfaser**,	. 109
,, ,, process,	. 125
,, ,, sheath *or* rind,	111, 113
,, ,, staining of,	. 16
,, fibre,	. 109

B

Bahnen, nervöse, nerve tracts.
Baillarger's stripes, . .   350, 352, 356, 357
Balken, corpus callosum.
Balkenwindung, gyrus corporis callosi.
Bandelette accessoire de l'olive supérieure, . . . 246
Bandelette externe, ... 194
Bandkern oder **Vormauer,** claustrum.
Basal membrane of the cerebellum, . . 328
Bechterew's nuclei, ... 305
Beinerv, nervus accessorius Willisii.
Belegungskörper, nerve-cells.
Berg, monticulus.
Bindearme, brachia conjunctiva.
Bismarck-brown, staining, ... 12
Blood-pigment in the adventitia, ... 140
Blood-vessels (minute structure), ... 135
,,   of the base, ... 389
,,   of the cerebellum, ... 333
,,   of the great brain, ... 388
,,   of the spinal cord, ... 201
Bodencommissur graue, grey commissure forming the floor of the third ventricle.
,,   **weisse,** white commissure in floor of the third ventricle.
Bogenbündel des Grosshirns, fasciculus arcuatus.
,,   **der Medulla oblongata,** fibræ arcuatæ.
Bogenfurche, arcuate fissure (comprising sulcus corporis callosi et fissura hippocampi).
Brachia conjunctiva, *seu* conjunctoria cerebelli ad cerebrum, **Bindearme**, *pédoncules cérébelleux superieurs*, processus a cerebello ad testes, 55, 237, 317
Brachia corporum quadrigeminorum, **Vierhügelarme**, *bras des tubercules quadrigumeaux, bras conjonctifs* (of Charcot), . . 62, 242, 282
Brachia cerebelli conjunctoria, **Arme der Vierhügel,** *bras conjonctifs* (Charcot), corp. quadrig., ... 55
,,   ,, inferiora, ... 44
,,   ,, media, ... 46
,,   ,, superiora, ... 55
Brain-stem, **Stamm,** *souche ou tronc des hémisphères,* . . 35
Broca's convolution, ... 93
Brücke, pons.
Brückenarm, crus pontis.

Brückenbahn, frontale, frontal pontine tract.
Brückenfasern, fibres of the pons.
Brückenkerne, nuclei pontis.
Bulbar paralysis, progressive, ... 315
Bulbus cornu posterioris, ... 81
  ,, nervi olfactorii, ... 76, 271
Burdach's column, ... 41
  ,, nucleus, ... 215, 256

C

Calamus scriptorius, ... 43, 216, 219
Calcar avis, *seu* pes hippocampi minor, **Vogelklaue,** *Ergot de Morand,* . 81, 85
Calcareous degeneration of blood-vessels, ... 142
  ,, ,, nerve-cells, ... 133
Canalis centralis, ... 186
Capillary vessels, ... 139
Capsula corporis dentati, **Vliess des Kleinhirns,** 316
  ,, externa, ... 68
  ,, extrema, ... 68
  ,, interna, **innere Kapsel,** *capsule interne,* internal capsule, geminum centrum semicirculare (*Vieussens*), . . . 67, 70, 254
Carrefour sensitif, ... 255
Caruncula mammillaris, ... 267
Cauda equina, ... 39
Caudex, brain-stem, ... 36
Cavum Meckelii, ... 379
Cella media of the lateral ventricle, ... 80
Celloidin method, ... 8, 9
Central canal, ... 186
Central grey tube, the grey matter bordering the central canal or central grey formation of the cerebro-spinal axis, comprising with the addition of the optic thalamus, &c., Meynert's **Centrales Höhlengrau,** *substance grise des cavités centrales de l'isthme, substance grise centrale, substance grise du canal encéphalo-médullaire,* 169
Centrales Höhlengrau (*Meynert*), central grey matter, part of central grey tube (*Hill*).
Centrallappchen, lobulus centralis.
Centralspalte, fissure of Rolando.
Centralwindung, hintere, ascending parietal convolution.
  ,, **vordere,** ascending frontal convolution.
Centrum semiovale Vieussenii, ... 68, 344
Cerebellar peduncles, ... 55
  ,, tract, direct, ... 197, 321
Cerebellum, **Kleinhirn,** *cervelet,* epencephalon (*embryologically*), . 47, 315
  ,, cortex of, ... 323
  ,, granule layer of the, **Körner Schichte, rostbraune Schichte,** 323
  ,, histology of, ... 323
  ,, pathological changes in, ... 333

GLOSSARIAL INDEX. 409

	PAGE
Cerebellum posterior commissure,	321
,, vessels of,	333
Cerebral vesicle middle,	36
,, ,, primary,	36
Cerebro-spinal fluid,	376, 388
Cerebrum, **Grosshirn**, *hémisphères cérébraux*,	65, 335
Cervical Anschwellung, cervical enlargement.	
Cervical enlargement, intumescentia cervicalis	38, 174
Cervix cornu posterioris,	211
Chiasma nervi acustici,	318
,, ,, optici,	64, 247, 277, 279
Choroid plexus, **Adergeflechte**, *plexus choroides*,	81
Chromophilous and chromophobic cells,	127
Cingulum, **Zwinge**,	349
Circulus Willisii,	389
,, ,, varieties of,	393
Cisternæ subarachnoidales	383
Clarke's vesicular column,	181, 185, 262
Claustrum, **Vormauer**, *avant-mur*,	68
Clava,	46, 213
Coagulation products in the blood-vessels,	147
Colatorium, *seu* hypophysis,	370
Collateral tracts,	161
Colliculus subpinealis,	61
Colloid degeneration of blood-vessels,	145
,, ,, nerve-cells,	133
,, extravasation,	147
Columna fornicis,	73
,, vesicularis of Clarke,	181, 185, 262
Combined systemic diseases of the cord,	207
Commissura alba med. spin.,	175, 180, 187, 199
,, anterior cerebri,	74, 247, 274
,, arcuata posterior N. optici,	280
,, baseos alba,	347
,, ,, grisea,	76, 337
,, grisea med. spin.,	171, 187
,, inferior N. optici,	280
,, media,	64, 245, 337
,, medullæ spinalis,	171
,, mollis,	64, 245, 337
,, posterior,	65, 245, 285
Commissural fibres of the cerebrum,	346
Commissure,	166
,, of the nuclei of the roof,	320
Communication of lateral ventricle with surface,	83
Comparative method,	24
Conarium, *or* pineal gland,	65
Conductor sonorus, **Klangstab**, *baguette d'harmonie*,	58, 305
Connective-tissue,	150
Conus medullaris, *seu* terminalis, **Markkegel, Endzapfen**,	41, 177

GLOSSARIAL INDEX.

	PAGE
Conus terminalis,	41, 177
Convolutio trigemini,	293
Convolution, arcuate of Arnold,	34
Cornu Ammonis,	81
,, anterius,	171
,, posterius,	171
Corona Reili, *seu* radiata, **Stabkranz**, *couronne rayonnante*,	256, 342
Corpora bigemina,	37
,, quadrigemina, **Vierhügel**, *tubercules quadrijumeaux*,	56, 237
Corpus callosum, **Balken**, *corps calleux*,	34, 83
,, candicans, *tubercules mammillaires ou pisiformes*, corpus mammillare,	63, 74, 244, 345
,, ciliare,	53, 316, 318
,, dentatum,	53, 316
,, fimbriatum, *seu* fimbria, *bandelette de l'hippocampe*,	74, 81, 366
,, geniculatum, laterale, *seu* externum, **äusserer Kniehöcker**, *corps genouillé externe*,	62, 242, 281
,, ,, mediale, *seu* internum, **innerer Kniehöcker**, *corps genouillé interne*, internal or mesial geniculate body,	62, 242, 282, 307
,, mammillare, *seu* candicans, **Markkügelchen**, *tubercule mammillaire*,	63, 74, 244, 345
,, quadrigeminum anterius,	61, 237
,, ,, post.,	61, 237
,, restiforme, **Kleinhirnstiel, Strickkörper, unterer Kleinhirnarm**, *corps restiforme*, restiform body,	45, 219, 261–317
,, rhomboideum,	53, 316
,, striatum,	65, 247
,, ,, relation to cortex,	36
,, subthalamicum, *seu* nucleus amygdaliformis, *seu* discus lentiformis, **Forelscher Körper, Luyscher Körper**, *bandelette accessoire de l'olive supérieure*,	242, 246, 343
,, trapezoides, **Trapezkörper**, *corps trapézoïde*,	225, 226, 228, 306
,, ,, cerebelli,	53, 319
Corpuscula arenacea, **Sandkörper**,	379
,, Pacchioni,	385
Cortex cerebri, **Mantel**, structure of,	350
Cortical centre of hearing,	307
Crus pontis, crus cerebelli ad pontem, **Brückenarm**, *pédoncule cérébelleux moyen*, middle peduncle of cerebellum,	44, 54, 231, 317
Crura cerebri,	60
,, fornicis, *piliers postérieurs du trigone*,	73
Crus ,,	83
Crusta pedunculi, *seu* crusta, **Hirnschenkelfuss**,	63
Csokor's carmine,	12
Culmen, **Gipfel**,	52
Cuneus, **Zwickel**, *le coin*,	101
Cuticulum cerebri,	386
Cylinder, axis—axis-cylinder,	107
Cystic degeneration of the cortex,	374

D

Dachkern, *noyau du toit,* nucleus of roof of Stilling.
Declive, ... 52
Decussatio, **Kreuzung,** 166
  ,,   Lemnisci, ... 214, 258
  ,,   Pyramidum, ... 211, 251
Defibering, ... 4
Degeneration of blood-vessels, ... 141, 145
  ,,   method, ... 22-23
  ,,   of nerve-fibres, ... 20, 118
  ,,   of nerve-cells, ... 131
  ,,   secondary, ... 19
  ,,   ascending in spinal cord, . . . 198, 205
  ,,   descending in spinal cord, . . . 196, 205
  ,,   Wallerian, ... 118
Deiters' nucleus, ... 225, 303
  ,,   cells, development, ... 32
Dementia paralytica, ... 373
Depigmentation of nerve-cells, ... 133
Development, ... 28
  ,,   of great brain, ... 35
  ,,   of nerve-fibres, ... 117
  ,,   of spinal nerves, ... 31
Dilatation of the adventitia, ... 146
Direct *or* lateral cerebellar tract, . . . 197, 262
Dorsal nucleus of *Stilling,* columna vesicularis of *Clarke,* . . 175
Dura mater, meninx fibrosa, **Faserhaut,** *dure mère,* . . 378

E

Ecker's nomenclature, ... 84
Einkerbungen von Lantermann, Lantermann's cones.
Einschnürungen von Ranvier, nodes of Ranvier.
Embolism, ... 147
  ,,   in the cerebrum, ... 375
Embolus, ... 147
Eminentia collateralis Meckelii, ... 81
  ,,   olivaris, ... 42
  ,,   teres, ... 58
Encephalitis, ... 375
Endothelium of arteries, ... 136
Epiblast, ... 29
Eosin cells of cerebellum, ... 223
Ependyma, ... 148
Epiphysis, *seu* conarium, *seu* glandula pinealis, **Zirbeldrüse,** *glande pinéale,* 369
Epithelia, ... 148
Erhlich's methyl-blue method, ... 25
Erlitzky's fluid, ... 6
État criblé, ... 14

GLOSSARIAL INDEX.

	PAGE
Étranglements annullaires,	110
Exner's perosmic acid method,	12

F

	PAGE
Facial nucleus,	225, 297
,, knee,	298
Faisceau en écharpe,	252
Falx cerebelli,	379
,, cerebri, **Hirnsichel** oder **Sichel**, la grande faux du cerveau,	84
Fascia dentata,	81, 82, 366
Fasciculus arcuatus, seu longitudinalis superior, **Bogenbündel, oberes Längsbündel**, faisceau arqué,	349
,, seu funiculus cuneatus,	43, 173
,, longitudinalis inferior, **unteres Längsbündel**,	81, 349
,, ,, posterior,	222, 264, 341
,, ,, superior, seu arcuatus,	349
,, obliquus pontis, ruban fibreux oblique,	55
,, retroflexus or Meynert's bundle,	244, 337
,, uncinatus, **Hakenbündel**, faisceau cunéiforme,	349
Fasciola, seu tæniola cinerea,	46
Faserhaut, dura mater.	
Fat and pigment degeneration of ganglion-cells,	132
,, granule-cells,	156
,, ,, ,, in the spinal cord,	207
,, in adventitia,	139
Fatty degeneration of the adventitia,	146
,, ,, muscularis,	141
Fibræ arciformes, seu arcuatæ, **Bogenfasern**, fibres arciformes,	43
,, ,, arcuatæ Arnoldi, seu fibræ propriæ,	349
,, ,, externæ,	216
,, ,, ,, posteriores,	261
,, ,, internæ,	216
,, heterodesmoticæ,	160
,, homodesmoticæ,	160
Fibres of the pons,	225
Fibril-sheath,	111
Fibro-plastic bodies,	153
Fillet or lemniscus,	55, 257
,, cortical,	258
,, lateral,	258
,, mesial,	258
Filum terminale,	41
Fimbria, seu corpus fimbriatum, seu tænia hippocampi, corps bordant, bandelette de l'hippocampe,	74, 81, 366
Fissura calcarina,	89
,, centralis,	87
,, choroidea, seu transversa, **Querspalte des grossen Gehirns, Randspalte, Adergeflechtsfurche**, scissure transverse,	89
,, cornu Ammonis,	81

GLOSSARIAL INDEX. 413

		PAGE
Fissura hippocampi,		98
,, lateralis,		85
,, longitudinalis cerebri,		65
,, ,, medullæ oblongatæ anterior,		43
,, ,, ,, posterior,		43
,, ,, med. spin. ant.,		38, 41
,, ,, ,, post.,		38, 41
,, occipitalis horizontalis,		87
,, ,, perpendicularis,		87
,, parietalis,		95
,, parieto-occipitalis,		87
,, paroccipitalis,		95
,, subiculi interna,		81
,, Sylvii,		85
,, transversa,		87
Fissures of cerebellum,		49
,, principal or total (fissuræ, scissuræ, primary fissures),		85
,, typical secondary (sulci secundarii),		85
,, atypical tertiary (sulci tertiarii),		85
,, and convolutions on the surface of the great brain,		84
,, of great brain,		84
Fixing media,		6
Flechsig's method,		18
Flocculus, **Flocke**, *Lobule du pneumogastrique*,		51
Flocke, Flocculus.		
Floor of fourth ventricle, **Rautengrube**, *sinus rhomboïdal*,		55
,, third ventricle, grey commissure,	64, 75, 333,	337
,, ,, white commissure,		347
Fol's fluid,		6
Folium cacuminis, **Wipfelblatt**,		52
Fontanal decussation,		242
Fontainenartige Haubenkreuzung, Fontanal decussation of the tegment, Meynert's decussation.		
Foramen cæcum posterius,		43
,, Magendi,		60
,, Monroi,		33, 83
Forceps anterior,		347
,, posterior,		80, 347
Fore-brain, **Vorderhirn**, prosencephalon,		36
Forel's nucleus, corpus subthalamicum,		246
,, ventral tegmental decussation,		242, 265
Formatio reticularis,		159
,, ,, alba,		216
,, ,, grisea,		216
Fornix, **Gewölbe**, *voûte à trois piliers, voûte à quatre piliers, trigone cérébrale, trigone*,		73, 247, 345
Fossa interpeduncularis,		61
,, rhomboidalis,		55
,, Sylvii,		85
Fovea anterior,		58

	PAGE
Frenulum lingulæ,	52
,, veli medullaris, *le frein de la valvule de Vieussens*,	59
Freud's method,	16
Friedmann's staining of cortex,	356
Fromann's lines,	110
Frontale Brückenbahn, frontal pontine tract.	
Frontal pole,	86
,, pontine-tract,	253, 345
Funiculus cuneatus, **Keilstrang**, *cordon cunéiforme*, Burdach's column,	41, 173
,, gracilis, **Zarter Strang**, *cordon grêle*, Goll's column,	41, 173
,, respiratorius, **Respirationsbündel** *of Krause*, **Solitärbündel** *of Stilling*, ascending root of vagus, ascending root of the lateral mixed system,	174, 309, 313
,, siliquæ, **Hülsenstränge**,	43
,, teres,	53
Fuss des Hirnschenkels, pes pedunculi cerebri *seu* crusta.	

G

Ganglienzellen, always translated nerve-cells.	
Ganglion geniculatum laterale (ext.),	62, 242, 281
,, ,, mediale (int.),	62, 242, 282, 307
,, habenulæ,	65, 245, 337
,, interpedunculare,	337
Gefässhaut, pia mater.	
Gennari's bands,	352, 360
Genu capsulæ internæ,	70, 254
,, corporis callosi, **Knie des Balkens**,	72
,, nervi facialis,	298
Geschwanzter kern, oder Striefenhügel, nucleus caudatus.	
Gesichtsnerv, optic nerve.	
Gewölbe, fornix.	
Giacomini's method for dry-brains,	27
Gipfel, culmen.	
Gitterschichte des Thalamus, stratum reticulatum thalami.	
Glandula Pacchioni,	385
,, pinealis, *seu* conarium, *seu* epiphysis, **Zirbeldrüse**, *glande pinéale*,	61, 65
,, pituitaria, *seu* hypophysis,	370
Globus pallidus,	68
Glomeruli olfactorii,	269
Glomus,	78
Glossopharyngeal flock of Roller,	312
,, nucleus,	308
,, root ascending,	218, 312
Glossopharyngeusherd, glossopharyngeal flock.	
Gold, impregnation with,	12
Golgi, mercury staining,	17
Goll's column, funiculus gracilis, **Zarterstrang**, *cordon grêle*,	41
,, nucleus,	215, 256
Gowers' bundle *or* tract,	198, 263, 265

GLOSSARIAL INDEX. 415

	PAGE
Granular degeneration of vessels,	144
Granulation of ependyma,	156
Granules or nuclei, **Körner**, *myélocites*,	127
Grenacher's carmine,	12

Grenzstreif, stria cornea.

Ground-bundle of anterior column,	198, 263
,, posterior column,	194, 199

Grosshirn, cerebrum.

Gudden's commissure,	280
,, method,	23
,, microtome,	7

Gürtelschichte des Thalamus, stratum zonale thalami.

Gyri breves insulæ,	101
,, frontales,	93
,, occipitales,	97
,, operti,	98
,, recti,	101
,, temporales,	97
,, ,, transversi,	98
,, transitivi,	85
Gyrus angularis,	96
,, ascendens frontalis, **hintere Centralwindung**,	93
,, ,, parietalis, **vordere Centralwindung**,	95
,, centralis anterior,	93
,, ,, posterior,	95
,, cinguli, *seu* corporis callosi, *part of* g. fornicatus, **Zwinge**,	98, 358, 360
,, corporis callosi, **Balkenwindung**, *circonvolution de l'ourlet, pli du corps calleux*, gyrus fornicatus,	98
,, fornicatus, *seu* corporis callosi, *circonvolution de l'ourlet*,	82, 83, 360
,, fusiformis,	98
,, hippocampi, subiculum cornu Ammonis,	81
,, inframarginalis,	97
,, lingualis,	98
,, occipitalis descendens,	101
,, occipito-temporalis lateralis,	98
,, ,, medialis,	98
,, parietalis,	95
,, paroccipitalis,	97
,, postcentralis,	95
,, præcentralis,	92
,, supramarginalis,	96
,, uncinatus, **Hakenwindung**, *circonvolution en crochet*,	99

H

Habenula, *seu* pedunculus conarii,	05, 245, 337
Hæmatoidin,	140
Hæmatoma of the dura mater,	382
Hæmatomyelia,	203

GLOSSARIAL INDEX.

	PAGE
Hæmatoxylin staining,	11
,, method of Weigert,	13
,, ,, ,, applied to Purkinje's cells,	327
Hæmorrhagia cerebri,	374
,, medullæ spinalis,	203
Hardening processes,	5, 27

Hakenbündel, fasciculus uncinatus.
Hakenwindung, gyrus uncinatus.
Hals des Hinterhornes, cervix cornu posterioris.
Halsanschwellung, cervical enlargement.
Haube des Hirnschenkels, tegment.
Haubenbahn, centrale, central tegmental tract.
Haubenfeld, tegment.
Haubenkern, nucleus tegmenti.
Haubenkreuzung, tegmental decussation.

Heterodesmotic fibres,	160
Heterotopia in the cerebellum,	332
,, ,, cerebrum,	376
,, ,, spinal cord,	332
Hexagon or pentagon or circle of Willis,	389
Hilum corporis dentati,	316
,, fasciæ dentatæ,	367
,, olivæ inferioris,	261
Hind-brain, **Hinterhirn**, *cerveau postérieur*, metencephalon (*in part*),	36, 46

Hintere Rückenmarkswurzeln, posterior spinal roots.
Hinteres äusseres Feld, postero-external tract.
,, **Langsbündel**, posterior longitudinal bundle.
Hinterhauptslappen, occipital lobes.
Hinterhirn, hind-brain.
Hinterhorn, cornu posterius.
,, **(seitenventrikel)**, posterior horn (lateral ventricle).
Hintersäule, posterior column of grey matter.
Hinterstrang, posterior white column.
Hinterstrangbahnen, connections of posterior columns of cord.
Hinterstrangsgrundbündel, ground-bundle of posterior column.
Hinterstrangskerne, nuclei funiculorum gracilis et cuneati.

Hippocampus major, **Seepferdefuss**,	81
,, minor,	81

Hirnanhang, hypophysis.
Hirnschenkel, pedunculus cerebri.
Hirnschenkelfuss, pes pedunculi cerebri *seu* crusta.
Hirnschenkelschlinge, ansa peduncularis.
Hörcentrum, corticales, cortical centre of hearing.
Hörnerv, auditory nerve.

Homodesmotic fibres,	160

Horngerüste, neurokeratin network.
Hornstreif, tænia semicircularis, stria cornea.
Hülsenstränge, funiculi siliquæ.

Hyaline degeneration of nerve-cells,	133
Hydromyelia,	206

GLOSSARIAL INDEX. 417

	PAGE
Hypertrophy of the axis-cylinder,	121
,, of brain,	103
,, of nerve-cells,	133
,, of vessel walls,	143
Hypoglossal nucleus,	219, 314
Hypophysis, glandula pituitaria, colatorium, **Hirnanhang**, *hypophyse*, *corps pituitaire*, pituitary body *or* gland,	64

I

Incisura, *seu* fissura pallii longitudinalis, great longitudinal fissure,	65
,, ,, ,, transversa, rima transversa cerebri, transverse fissure of Bichat,*	54
,, cerebelli marsupialis,	47
,, ,, semilunaris	47
Indusium griseum,	82
Inferior descending horn,	81
Inflammatory swelling of connective-tissue corpuscles,	153
Infundibulum, **Trichter**,	64
Innere Kapsel, internal capsule.	
Ink-staining,	26
Inoccipitia,	104
Insel, island of Reil.	
Inselpol, most projecting part of the island of Reil.	
Inselschwelle, limen insulæ.	
Interolivary tract, **Olivenzwischenschichte**,	257
Interpeduncular space,	61
Intumescentia cervicalis, **Halsanschwellung**, *renflement cervical*, cervical enlargement,	38
,, lumbalis, **Lendenanschwellung**, *renflement lombaire*, lumbar enlargement,	38
Iron staining,	26

K

Kapsel, äussere, external capsule.
,, **innere**, internal capsule.
Karyomitosis in inflammation,	134
,, in hydrophobia,	144

Keilstrang, fasciculus cuneatus.
Kern des Keilstranges, nucleus funiculi cuneati.
,, **Vorderstranggrundbündels**, nucleus funiculi anterioris.
,, **zarten Strangs**, nucleus funiculi gracilis.
Kernblatt, lamina medullaris involuta.
Klangstab, conductor sonorus.

* In the text (p. 84) the term Incisura pallii is applied to the great transverse fissure. When this term is used without a qualifying adjective, it is better to restrict it to the great longitudinal fissure or cleft between the two cerebral hemispheres, and to term the space between the cerebrum and cerebellum incisura pallii transversa, *seu* fissura marsupii cerebri. The terms longitudinal fissure and transverse fissure are undesirable since they suggest an homology with the true fissures of the brain.

Klappdeckel, operculum.
Klappenwulst, tuber valvulæ.
Kleinhirn, cerebellum.
Kleinhirnarme, cerebellar peduncles.
Kleinhirnseitenstrangbahn, direct or lateral cerebellar tract
Kleinhirnstiel, corpus restiforme.
Knie des Balkens, genu corporis callosi.
„ **der inneren Kapsel**, genu capsulæ internæ.
Kniehöcker, äusserer, corpus geniculatum externum.
„ **innerer**, corpus geniculatum mediale.
Knötchen, nodulus.
Körnerschichte des Kleinhirns, nuclear layer of cerebellum.
Kreuzung, decussation.
Kugelkern, nucleus globosus.

L

Lacunæ venosæ laterales,	385
Lamina cribrosa,	75
„ fossæ Sylvii,	68
„ medullaris involuta, **Kernblatt**, nuclear layer of cornu Ammonis,	365
„ „ nuclei lentiformis,	67, 340
„ „ thalami optici,	66, 338
„ terminalis, **Schlussplatte**,	77

Längsbündel, hinteres, posterior longitudinal bundle.
„ **oberes**, fasciculus arcuatus *seu* longitudinalis superior.
„ **unteres**, fasciculus longitudinalis inferior.

Lantermann's cones,	112
Laqueus, *seu* lemniscus, **Schleife**, *ruban de Reil*, fillet,	55, 258
Lateral *or* crossed pyramidal tract,	193, 249
„ column of spinal cord,	41
„ horn of spinal cord,	171
„ sclerosis,	204
„ ventricle,	84

Lebensbaum, arbor vitæ.

Leber's corpuscles, ... 157
Lemniscus, *seu* laqueus, **Schleife, Schleifenschicht**, *ruban de Reil, faisceaux triangulaires lateraux de l'isthme (Cruveilhier)*, fillet, 55, 221, 258

Lendenanschwellung, lumbar enlargement.

Ligamentum denticulatum,	370, 386
„ tectum, striæ longitudinales laterales,	361
Ligula (velum medullare inferius), *seu* ala pontis *seu* ponticulus, **hintere Marksegel**,	59
Limen insulæ, **Inselschwelle**,	101
Limiting membrane of blood-vessels,	137
Lineola albida Gennari,	352
Lingula, **Züngelchen**,	55

Linsenkern, nucleus lenticularis.
Linsenkernschlinge, ansa lenticularis.

Liquor cerebro-spinalis, . . . 378, 391

	PAGE
Lobe, olfactory,	271
Lobes of cerebellum,	50
,, cerebrum,	84
Lobulus centralis,	46
,, cuneiformis,	50
,, gracilis,	50
,, lunatus,	50
,, paracentralis,	101, 359
,, parietalis,	96
,, quadrangularis,	50
,, quadratus, Quader, Vorzwickel, *l'avant-coin*, præcuneus,	101
,, semilunaris,	50
,, triangularis,	101
,, vagi, *seu* flocculus, Flocke, *lobule du pneumogastrique*,	51
Lobus falciformis,	99
,, frontalis, Stirnlappen,	92
,, limbicus,	99
,, occipitalis, Hinterhauptlappen,	96
,, olfactorius, Riechlappen,	273
,, parietalis, Scheitellappen,	95
,, pyriformis,	276
,, temporalis, Schläfenlappen,	97
Localisation in the cortex cerebri,	104
Locus cæruleus,	59, 233, 294, 295
Longitudinal fissure, great, Mantelspalte, incisura, *seu* fissura pallii,	65
Luys' nucleus,	246
Lymphatic cysts,	146
Lymph spaces,	133
Lymph vessels,	138
Lyra Davidis,	74

M

Mandel des Kleinhirns, tonsil.
Mandelkern, nucleus amygdaleus.
Mantel, cortex.
Mantelspalte, great longitudinal fissure.
Markkegel, conus medullaris.
Markknopf, medulla oblongata.
Markkügelchen, corpora mammillaria.
Marklager, sagittales, radial fibres of centrum semiovale, "optic radiations" of Gratiolet.
Marklose Nervenfasern, non-medullated nerve-fibres.
Markmantel des Rückenmarkes, white matter of cord.
Markscheide, medullary sheath.
Markscheidenentwickelung, myelination.
Marksegel, hinteres, velum medullare posterius.
    ,,   **vorderes,** velum medullare anterius.
Mastzellen, "plasma-cells" (*see footnote*, p. 129).

	PAGE
Medulla oblongata, **Markknopf** *oder* **verlängertes Mark,** *moelle allongée ou bulbe,*	42
,, spinalis,	38
Medullary sheath, **Markscheide,**	111
Melanin,	141
Membrana fenestrata,	136
,, limitans,	112
Meningitis spinalis,	207
Meninx fibrosa, *seu* dura mater,	378
,, vasculosa, *seu* pia mater,	386
Methods,	3
Methyl blue-injection,	25
Meynert's bundle, fasciculus retroflexus,	244, 337
,, commissure,	283
,, scheme,	168
,, tegmental decussation, fontanal decussation of the tegment,	242
Micromyelia,	205
Microtomes,	7
Mid-brain, **Mittelhirn,** *vésicule cérébrale moyenne,* mesencephalon,	36, 62
Miliary aneurysm,	145
,, sclerosis,	375
Mittelhirn, mid-brain.	
Mixed lateral zone,	198, 264
Molecular layer of cerebellum,	326
Mons, *seu* Monticulus, **Berg,**	52
Morphology,	26
Motor nerve-roots,	165
Muldenblatt, alveus.	
Müller's fluid,	6
Myelination,	18, 19
Myelitis acuta,	204
,, annularis,	207
,, centralis,	205
,, transversa,	205
Myélocytes,	323

N

Nachhirn, after-brain.
Nebenhorn, hinteres laterales, nucleus funiculi cuneati.
  ,,   ,, **mediales,** nucleus funiculi gracilis.
Nebenkern, gezackter, *Meynert's* name for nuclei globosus et emboliformis,
Nebenolive, vordere, nucleus pyramidalis.
  ,,   **obere (äussere),** nucleus olivaris accessorius externus, *seu* superior.

Neoplastic elements in lymph spaces,	147
Nerve-cells,	121
,, ,, relation to axis-cylinder,	20
,, fibres,	107
,, olfactory,	33

GLOSSARIAL INDEX.

	PAGE
Nerve roots, development,	30
,, ,, physiology,	41
,, tracts,	160
Nervi olfactorii, **Riechnerven**,	273
Nervus abducens,	40, 220, 292
,, accessorius Willisii, *seu* recurrens, **Beinerv**,	46
,, acusticus, **Hörnerv**,	46, 221, 300
,, cochlearis,	209
,, facialis, **Gesichtsnerv**,	225, 297
,, glossopharyngeus, **Zungenschlundkopfnerv**,	46
,, hypoglossus, **Zungenfleischnerv**,	46, 213, 314
,, intermedius Wrisbergi,	299
,, Lancisii,	35, 82
,, oculomotorius, **gemeinsamer Augenmuskelnerv**,	61, 241, 287
,, opticus, **Sehnerv**,	33, 279
,, patheticus,	290
,, pneumogastricus, *seu* Vagus, **herumschweifender Nerv**,	312
,, recurrens,	312
,, trigeminus,	54, 231, 292
,, trochlearis, **Rollmuskelnerv**,	63, 235, 290
,, vagus,	46, 220, 312
,, vestibularis,	299
Neurilemma,	112
Neuroblasts,	30
Neurokeratin network,	114
Neurogleia,	153
,, cells, development,	32
Nodes of Ranvier,	32
Nodulus, **Knötchen**, nodule,	52
Non-medullated fibres,	115
Nuclear division of nerve-cells,	134
,, stains,	11
Nucleus ambiguus,	216
,, amygdaleus,	81
,, amygdaliformis, *seu* amygdaleus, **Mandelkern**,	68, 246
,, angularis,	303
,, anterior thalami, *centre antérieur de la couche optique*,	66
,, arcuatus,	217, 261
,, caudatus, *seu* corpus striatum (in limited sense), **geschwanzter Kern, Schweifkern** *oder* **Schwanzkern, Streifenhügel,** *Noyau caudé*, intraventricular nucleus of corpus striatum, 37, 65, 340, 343	
,, centralis inferior,	223, 263
,, ,, superior,	237, 265
,, dentatus	53
,, denticulatus, *seu* dentatus, *noyau dentelé, corps rhomboïdal du cervelet*, dentate nucleus *or* cerebellar olive,	53
,, emboliformis, *seu* embolus, **Pfropf**,	53, 235, 316
,, fastigii,	53
,, fimbriatus,	53, 212
,, funiculi anterioris,	214

	PAGE
Nucleus funiculi cuneati,	262, 256, 213
,, ,, gracilis,	212, 256, 260
,, ,, lateralis,	213
,, ,, teretis,	219, 301
,, globosus, **Kugelkern**,	53, 235, 316
,, lateralis thalami optici,	66
,, lemnisci lateralis,	233, 259
,, ,, medialis,	301
,, lenticularis,	36
,, lenticulatus cerebelli,	53
,, lentiformis, *seu* lenticularis, **Linsenkern**,	67, 340, 341, 343
,, medius thalami optici,	66, 216
,, of roof,	54
,, olivaris,	216
,, ,, accessorius anterior, *seu* pyramidalis, **vordere Nebenolive**, -	216
,, ,, ,, externus, *seu* superior, **äussere Nebenolive**,	219
,, pontis,	225, 253
,, reticularis tegmenti,	229, 254
,, ruber tegmenti, **rother Kern**,	241
,, superior thalami,	64
,, tæniæformis (*Arnold*), claustrum,	68
,, tecti, **Dachkern**, *noyau du toit*, nucleus of roof of Stilling,	54, 316
,, tegmenti, **rother Kern**, *noyau de la calotte*, Stilling's red nucleus,	241
,, trapezoides, **Trapezkern**, .	306

O

Oberwurm, vermis superior.	
Obex, **Riegel**, *le verrou*,	59, 219
Obliteration of small vessels,	143
Occipitalbündel, senkrechtes, optic radiations.	
Occipital lobe, **Hinterhauptlappen**,	96
,, pole,	96
Oculomotor nucleus,	241
Olfactory bulb,	76, 268
,, lobe,	275
,, nerve,	76, 267, 268
,, tract,	272
Olive, inferior,	227, 306
,, superior,	42, 216
,, ,, of Luys,	241
Olivenzwischenschichte, interolivary tract.	
Operculum, **Klappdeckel**, *opercule de la fosse de Sylvius*,	96
Ophthalmoplegia externa nuclearis,	292
Optic ganglion basal of Wagner,	283
,, nerve,	279
,, radiations,	338

GLOSSARIAL INDEX. 423

Optic radiations of Gratiolet, **sagitalles Marklager** (in narrower sense),
    senkrechtes Occipitalbündel, ... 338
,, thalamus, **Sehhugel**, *couche optique*, ... 64, 245, 336

P

Pacchionian bodies *or* granulations *or* glands, ... 385
Pachymeningitis, ... 385
    ,, cervicalis hypertrophica 207
Palladium and gold impregnation, ... 12
Pal's medullary stain, ... 14
Parietal eye, ... 369
  ,, lobe, ... 96
Paracentral lobule, ... 101, 359
Parallel convolution, ... 97
  ,, fissure, ... 97
Paralysis labio-glossopharyngeal, ... 315
Paralytic idiocy, ... 373
Parasinoidal spaces, ... 385
Par quintum, ... 292
Peduncle of the corona radiata, ... 345
Pedunculus bulbi olfactorii, ... 272
Pedunculus cerebelli medius, ... 46
    ,, cerebri, *seu* crus, **Hirnschenkel**, *pédoncule*, . . 63
    ,, conarii, ... 65, 337, 367
    ,, corp. mammillaris, ... 241, 345
    ,, cunei, **Zwickelstiel**, ... 90
    ,, flocculi, ... 41
    ,, septi pellucidi, ... 72, 77
    ,, substantiæ nigræ Soemmeringi, . . . 253
Pericellular lymph spaces, ... 138
Periencephalitis chronica, ... 373
Peripheral grey tube, grey matter formed on the surface of the cerebro-
    spinal axis outside the white tube, ... 169
Perivascular lymph spaces, ... 138
Pes hippocampi major, ... 81
  ,, ,, minor, ... 81
,, pedunculi, *seu* crusta, **Hirnschenkelfuss**, *pied du pédoncule*, . 63, 238, 252
Pfropf, nucleus emboliformis.
Physiological method, ... 24
Pia mater, **Gefässhaut**, *pie mère*, ... 386
Picrocarmine, ... 11
Pigment in the adventitia, ... 146
  ,, in nerve-cells, ... 122
Pillars of fornix, **Säulen des Fornix**, *piliers antérieurs du trigone*, . 74
Pineal body or gland, **Zirbel**, ... 34, 65, 369
  ,, gland, ... 34, 65, 369
Piniform decussation, ... 214

GLOSSARIAL INDEX.

	PAGE
Plexus choroideus cerebelli,	59
,, ,, cerebri,	77
,, ,, lateralis,	89
Pole, frontal,	86
,, occipital,	96
,, temporal,	97
Polioencephalitis inferior,	315
,, superior,	292
Poliomyelitis anterior acuta,	204
,, ,, chronica,	204
Polster, pulvinar.	
Polygon, heptagon or circle of Willis,	389
Polygyry,	103
Pons Varolii, **Brücke**, *pont de Varole, protubérance annulaire*,	46, 223, 335
Ponticulus,	55, 59, 219
Porencephalia,	104
Posterior columns of cord, cerebral connections,	41
,, longitudinal bundle, **hinteres Längsbündel, Acusticusstrang** (*Meynert*), **oberer weisser Saum der reticulären Substanz**(*Henle*), **oberes Längsbündel** (*Stieda*), *fibres spinales des régions postérieures* (*Luys*)	222, 264, 341
,, ,, or dorsal fissure,	43
,, spinal roots,	173, 189
,, of the two hinder vesicles,	36
,, vesicular columns of cord,	174
,, white columns of cord,	34, 173
Præcuneus, **Vorzwickel**,	101
Primitivband, axis-cylinder.	
Primitive fibrillæ,	109, 111
Processus cervicalis medius	174
,, mammillaris,	212
,, reticularis,	175
Progressive bulbar paralysis,	315
Projection systems of *Meynert*,	168
Proliferation of nuclei,	146
Propons, *seu* ponticulus,	55, 59, 219
Protoplasmic processes,	125
Psalterium,	74
Pseudo-hypertrophy of vessels,	144
Pulvinar, **Polster**,	63
Purkinje's cells,	324
Putamen,	68
Pyramid,	43, 214
,, posterior,	46, 214
Pyramidal decussation,	211, 251
,, tract,	193, 249
Pyramidenseitenstrang, lateral pyramidal tract.	
Pyramidenvorderstrang, anterior or direct pyramidal tract.	
Pyramis cerebelli,	52

GLOSSARIAL INDEX. 425

Q

Quader, lobulus quadratus.
Querspalte des grossen Gehirns, transverse fissure.
Quintusstränge (also called by Meynert fasciculi marginales aquæducti), quintus columns.
Quintus tracts or columns, . . 242

R

Radiatio corporis callosi, ... 70
Radices ascendens et descendens fornicis, . . . 74
Randbogen, äusserer, external arcuate convolution.
Randspalte, fissura choroidea.
Randzone, border-zone of Lissauer.
Ranvier's nodes, *étranglements annulaires*, . . 112
Raphe, ... 214, 215
,, of the corpus callosum, ... 70
Rautengrube, floor of fourth ventricle.
Recessus chiasmatis, ... 79
,, infrapinealis, ... 65, 242
,, infundibuli, ... 64
,, lateralis ventriculi quarti, ... 50
Red nucleus, ... 241
Regeneration of nerve-fibres, ... 117
Regio subthalamica, ... 68, 246
Reichert's sledge microtome, ... 7
Remak's fibres, ... 115
Respirationsbündel (*Krause*), **Solitärbündel** (*Stilling*), ascending root of vagus.
Restiform nucleus, . . 213
Riechkolben, olfactory bulb.
Riechlappen, olfactory lobe.
Riechnerv, olfactory nerve.
Riechwurzeln, olfactory roots.
Riegel, Obex.
Rima transversa cerebri, transverse fissure of Bichat, . 7
Rindenschleife, cortical fillet.
Root-zone, **Wurzelzone**, *zone radiculaire*, . . . 193
Rostrum corporis callosi, **Schnabel des Balkens**, . . 77
Rother Kern, red nucleus of tegment.
Roy's microtome, ... 8
Ruban de Reil, fillet, ... 55, 258
,, *fibreux oblique*, ... 55
Rückenmark, spinal cord.

S

Safranin staining of Adamkiewicz, . . . 17
Sagittales Marklager, optic radiations and other vertical fibres.
Sandkörper, corpuscula arcuacea.

GLOSSARIAL INDEX.

                                                                           PAGE

Secondary anterior cerebral vesicle, ... 36
Scheidewand, durchsichtige, septum pellucidum.
Scheitellappen, parietal lobe.
Schenkel des Gewölbes, pillar of the fornix.
Schläfenlappen, temporal lobe.
Schläfenpol, temporal pole.
Schleife, fillet.
Schleifenbündel zum Fuss, tract from the fillet to the crusta.
Schleifenkern, nucleus of the fillet.
Schleifenkreuzung, decussation of the fillet.
Schleifenschicht, fillet.
Schlussplatte, lamina terminalis.
Schnabel des Balkens, rostrum corporis callosi.
Schnürringe von Ranvier, nodes of Ranvier.
Schwalbe's method for dry-brains, ... 27
Schwanzkern, nucleus caudatus.
Schweifkern, nucleus caudatus.
Scissura limbica, ... 276
Secondary degeneration, ... 21
Seepferdefuss, grosser, hippocampus major.
        ,,   **kleiner,** hippocampus minor.
Sehcentrum, cortical visual centre.
Sehhügel, optic thalamus.
Sehnerv, optic nerve.
Sehstrahlungen, optic radiations.
Seitenhorn, lateral horn of spinal cord.
Seitenstrang, lateral column of spinal cord.
Seitenstrangkern, nucleus funiculi lateralis.
Seitenstrangzone, gemischte, mixed lateral zone.
Seitenventrikel, lateral ventricle.
Senile atrophy of cerebrum, ... 372
Sensory ganglia, development, ... 28
Septum paramedianum dorsale, ... 170, 173
Septum pellucidum, **durchsichtige Scheidewand,** *cloison transparente,* 72, 362
Series of sections, *Weigert's* method, ... 14
Sheath of Schwann, ... 22
Sichel, falx major.
Sillon collateral postérieur, see sulcus lateralis dorsalis, . . . 43
Sinus longitudinalis superior, ... 379
Sinus rhomboidalis, ... 58
   ,, subarachnoidalis, ... 383
Sclerosis of the brain, ... 156
   ,, of the cornu Ammonis, ... 375
   ,, of nerve-cells, ... 133
   ,, disseminated, ... 206
Softening of the brain, ... 376
Solitäres (Stilling) oder Respirations (Krause) Bündel, fasciculus solitarius.
Spatium suprachoroideum, ... 78
Spinal cord, ... 38

GLOSSARIAL INDEX.

	PAGE
Spinal ganglia,	164
,, ,, development,	28
,, ,, nerve-roots,	21

Spindelwindung, fusiform convolution.
Spinnenzellen, spider-cells.

Splenium corporis callosi, *bourrelet du corps calleux*,	82
Spongioblasts,	29

Stabkranz, corona radiata.
Stamm, brain-stem.
Stiel des Kleinhirns, pedunculus cerebelli.

Stimulation,	24

Stirnlappen, frontal lobe.
Stirnpol, frontal pole.

Stratum cellularum pyramidalium,	365
,, granulosum fasciæ dentatæ,	367
,, intermedium, **Zwischenschicht**,	63, 246
,, lacunosum *seu* reticulare *seu* medullare medium,	365
,, marginale,	367
,, moleculare,	367
,, nigrum,	63
,, oriens,	366
,, radiatum,	361
,, reticulare cornu Ammonis,	367
,, ,, thalami,	245, 338
,, zonale,	338
,, ,, olivæ inferioris, **Vliess der unteren Olive**,	261
,, ,, thalami,	66, 338

Streifenhügel.

Stria alba tuberis,	346
,, cornea, *seu* terminalis, tænia cornea, **Grenzstreif, Hornstreif**, *la lame cornée*,	64
,, longitudinalis corporis callosi,	72, 82, 361
,, medullaris thalami optici, *seu* stria pinealis, **oberes Markblatt des Conariums**, *pédoncule antérieur de la glande pinéale*, the basal or attached portion of the tænia thalami,	65, 245
,, terminalis,	64
Striæ acusticæ, *barbes du calamus scriptorius*,	58, 221, 305
,, longitudinales mediales,	72, 82, 361
,, medullares,	35, 58, 221, 305
,, obtectæ,	35

Strickkörper, corpus restiforme.
Stützgewebe, supporting-tissue.

Subarachnoid sinus,	383
,, space,	383
,, tissue,	384
Subdural space,	377
Subiculum cornu Ammonis,	81, 363
Sublimate colouring of Golgi,	16, 17
Substantia ferruginea,	59, 233, 293
,, ,, cerebelli,	54

428 GLOSSARIAL INDEX.

	PAGE
Substantia gelatinosa,	172, 183, 185
,, ,, centralis,	171
,, Rolandi,	184, 217
,, innominata, *seu* ansa peduncularis,	339
,, nigra Soemmeringi,	63, 238, 344
,, perforata anterior, **vordere Siebsubstanz**, *substance perforée antérieure*,	75, 274
,, ,, posterior,	61
,, reticularis alba,	216
,, ,, Arnoldi,	81
,, ,, grisea,	213
,, ,, spongiosa,	171
Sulcus arcuatus, **Bogenfurche**, *sillon arqué*,	84, 93
,, calloso-marginalis,	101
,, centralis,	87
,, choroideus,	64, 78
,, corporis callosi,	89, 98, 360
,, corporum quadrigeminorum longitudinalis,	61
,, ,, ,, transversus,	61
,, cruciatus,	92
,, flocculi,	49
,, frontalis,	92
,, fronto-marginalis,	65
,, interbrachialis,	62, 238
,, intermedius posterior,	41
,, interparietalis,	95
,, lateralis dorsalis medullæ spinalis,	39, 43
,, ,, mesencephali,	60
,, ,, ventralis medullæ spinalis,	38
,, longitudinalis inferior cerebelli,	49
,, ,, superior ,,	47
,, magnus horizontalis,	49
,, medianus sinus rhomboidalis,	43
,, nervi oculomotorii,	61
,, occipitalis lateralis,	96
,, ,, transversus,	96
,, olivæ internus,	43
,, orbitalis,	94
,, paracentralis,	101
,, paramedianus dorsalis,	43
,, parapyramidalis,	43
,, parietalis,	95
,, postcentralis,	95
,, postolivaris,	43
,, postrolandicus,	95
,, præcentralis,	93
,, rectus,	95
,, Rolandi,	87
,, subparietalis,	101
,, substantiæ perforatæ posticæ,	62, 75

GLOSSARIAL INDEX. 429

	PAGE
Sulcus triradiatus,	94
Sutura corporis callosi,	72
Sylvian fossa,	97
Syphilis of brain-vessels,	394
Syringo-myelia,	205

T

Tabes dorsalis, ... 204
Tænia, *seu* stria cornea, ... 64
  ,, hippocampi, *seu* fimbria, . . . 65, 67, 77, 337
  ,, pontis, ... 55
  ,, thalami optici, stria pinealis, **oberes Markblatt des Conariums**,
    pédoncule antérieur de la glande pinéale, . . . 65, 245
  ,, ventriculi quarti, ... 59
  ,, ,, tertii, ... 65
Tœniola cinerea, ... 46, 221
Tangential cortex fibres, ... 356
Tapetum Reili, ... 81, 347
Tegmental decussation, ... 214, 258
  ,, ,, of Forel, ... 242, 265
  ,, ,, of Meynert, ... 242, 265
  ,, tract central, ... 228, 265
Tegmentum, **Haubenfeld**, *Calotte, Coiffe* (Gratiolet), *étage supérieur du pédoncule,* 62
Tela choroidea inferior ventriculi quarti, ... 50, 84
  ,, cerebelli, ... 59
  ,, superior, ... 78
Temporal or temporo-sphenoidal lobe, **Schläfenlappen**, . . 97
  ,, pole, ... 97
Tentorium cerebelli, **Kleinhirnzelt**, ... 379
Terrain désert of Broca, ... 99
Thalamencephalon, ... 63
Thalamus opticus, **Sehhügel**, *Couche optique*, . . 63, 245, 336
  ,, anterior peduncle of, **vorderer Thalamusstiel**, *racine antérieure,* 338
  ,, inferior peduncle, ... 248, 338
  ,, internal peduncle, ... 248, 338
  ,, posterior peduncle, **Gratioletsche Sehstrahlungen, hinterer Thalamusstiel, sagittales Marklager**, *faisceaux optiques de Gratiolet*, optic radiations, . . . 338
Thalamusstiel, hinterer, oder Sehstrahlungen, optic radiations of Gratiolet, &c.
  ,, **innerer** (*Wernicke*), internal peduncle of the thalamus.
  ,, **unterer**, inferior peduncle of the thalamus.
  ,, **vorderer**, anterior peduncle of the thalamus.
Tonsil, **Mandel**, amygdala, ... 52
Trabecula cinerea, ... 64
Tractus intermedio-lateralis, ... 171

GLOSSARIAL INDEX.

	PAGE
Tractus olfactorius,	76, 268, 272
,, opticus, *seu* **Sehstreif**, *bandelette optique*,	61, 64, 244, 279
,, peduncularis transversus,	62, 283

Trapezkern, nucleus trapezoides.
Trapezkörper, corpus trapezoides.
Transverse fissure of Bichat or incisura marsupii cerebri, *grande fente cérébrale*, 84
Trichter, infundibulum.

Trigeminus, motor nucleus,	231, 295
,, sensory nucleus,	231, 295
,, ascending root,	231, 293
,, descending root,	212, 295
Trigonum habenulæ,	65
,, hypoglossi,	58
,, intercrurale,	61
,, subpineale,	61, 65
,, vagi,	58
Trochlear nucleus,	237, 289
Tuber cinereum,	64
,, olfactorium,	273
,, valvulæ,	52
Tuberculum acusticum,	58, 221, 302
,, anterius thalami, corpus album subrotundum, *tubercule supérieur et antérieur de la couche optique*,	66
,, cinereum Rolandi,	43, 211
,, cuneatum,	46, 215
,, fasciæ dentatæ,	82
Türck's column,	253
'Tween-brain, **Zwischenhirn**, *cerveau intermédiaire, ventricule des couches optiques*, thalamencephalon,	63

U

Uebergangswindungen, gyri annectantes.
Uncus, **Hakenwindung**, *crochet ou circonvolution en crochet*, . 99
Unterwurm, vermis inferior.
Uvula, **Zäpfchen**, ... 52

V

Vacuoles in nerve-cells,	133
Vagus nucleus,	312
Vago-glossopharyngeal, chief nucleus,	220
,, motor nucleus,	216
Vallecula,	49
Valvula semilunaris Tarini, *seu* medullare posterius, **hinteres Marksegel**,	55
,, Vieussenii, *seu* lamina tectoria anterior, **vorderes Marksegel, Klappe**, *valvule de Vieussens*,	61
Varicose axis-cylinders,	115
,, degeneration of the axis-cylinder,	121

GLOSSARIAL INDEX. 431

	PAGE
Varieties of the circle of Willis,	393
,, of convolution,	102
Vasocorona medullæ spinalis,	202
Vela medullaria,	59
Velum confine,	30
,, interpositum,	34, 83, 84
,, medullare anterius, *seu* valvula Vieussenii,	55
,, ,, posterius Tarini, valvula semilunaris, **hinteres Marksegel**, *voile sur le quatrième ventricule*,	53
,, triangulare,	78
Ventriculus bulbi olfactorii,	268
,, conarii,	64
,, corporis callosi,	360
,, lateralis, **Seitenventrikel**, *ventricule latéral*,	79
,, quartus,	55
,, quintus,	72
,, septi pellucidi,	72
,, tertius,	78
,, tricornus,	79
,, Vergæ,	74
Verga's ventricle,	74
Verlängertes Mark, medulla oblongata.	
Vermis,	52
,, cerebelli inferior, **Unterwurm**,	49
,, ,, superior, **Oberwurm**,	48
Vicq d'Azyr's bands, lineola albida Gennari,	351
,, bundle,	74, 247, 339
Vierhügel, corpora quadrigemina.	
Vierhügelarme, brachia corporum quadrigeminorum.	
Virchow-Robin's lymph space,	137
Visual centre, cortical,	284
Vliess des Kleinhirns, capsula corporis dentati.	
,, **der unteren Olive**, stratum zonale olivæ.	
Vorderhirn, fore-brain.	
Vorderhorn des Rückenmarks, cornu anterius.	
,, ,, **Seitenventrikels**, anterior horn of lateral ventricle.	
Vordersäule, anterior columns of grey matter of spinal cord.	
Vorderstrang, anterior white column.	
Vorderstranggrundbündel, ground-bundle of the anterior column.	
Vormauer, claustrum.	
Vorzwickel, præcuneus.	
Voute à trois piliers, fornix,	73, 247, 345

W

Waller's degeneration,	23
Weigert's hæmatoxylin staining,	12, 13, 18
Weisser Kern der Haube, brachia conjunctiva in mid-brain.	
Wernekink's commissure,	318

                                                                            PAGE
Westphal's trochlear nucleus, ... 291
Wipfelblatt, folium cacuminis.
Wulst des Balkens, gyrus corporis callosi.
Wurm, vermis.
Wurmpyramide, pyramis vermis.
Wurzelzone, root-zone.

Z

Zäpfchen, uvula.
Zarter Strang, funiculus gracilis, Goll's column.
Zelt oder Kleinhirnzelt, tentorium cerebelli.
Zirbel, glandula pinealis.
Zirbelauge, pineal eye.
Zirbeldrüse, pineal gland.
Zirbelstiel, pedunculus conarii.
Zona incerta, ... 247
Zone, radicular, ... 19
Zuckerkandl, callosal convolution of, ... 83
Züngelchen, lingula.
Zungenfleischnerv, hypoglossal nerve.
Zungenschlundkopfnerv, glossopharyngeal nerve.
Zwickel, cuneus, *le coin*.
Zwickelstiel, pedunculus cunei.
Zwinge, gyrus cinguli.
Zwischenhirn, 'tween-brain.
Zwischenmarkscheide, membrane at node of Ranvier (*Obersteiner*) at Lantermann's segments (*Kuhnt*).
Zwischenschichte, stratum intermedium.

THE END.

BELL AND BAIN, PRINTERS, GLASGOW.

www.ingramcontent.com/pod-product-compliance
Lightning Source LLC
Chambersburg PA
CBHW022139300426
44115CB00006B/255